Agriculture in the Middle Ages

University of Pennsylvania Press
MIDDLE AGES SERIES
Edited by

Ruth Mazo Karras,
Temple University

Edward Peters,
University of Pennsylvania

A listing of the available books in the series
appears at the back of this volume

Agriculture in the Middle Ages

Technology, Practice, and
Representation

Edited by
DEL SWEENEY

University of Pennsylvania Press
Philadelphia

Library of Congress Cataloging-in-Publication Data
Agriculture in the Middle Ages : technology, practice, and representation / edited by
Del Sweeney.
    p.   cm. — (Middle Age series)
    Papers from two conferences, the first held at Pennsylvania State University and the
second held at the University of British Columbia.
    Includes bibliographical references (p.   ) and index.
    ISBN 0-8122-3282-8 (alk. paper). — ISBN 0-8122-1511-7 (pbk. : alk. paper)
    1. Agriculture—Europe— History—Congresses. 2. Europe— History—476–1492—
Congresses.   3. Europe—Economic conditions—To 1492— Congresses. 4. Europe—
Social conditions—To 1492—Congresses. 5. Middle Ages— History—476–1492—
Congresses. 6. Civilization, Medieval—Congresses. I. Sweeney, Del. II. Series.
S452.A47 1995
306.3′49′0940902—dc20
                                        95-24416
                                          CIP

Contents

Acknowledgments

The essays in this volume originated as papers presented at two conferences, "People of the Plough: Land and Labor in Medieval Europe," at the Pennsylvania State University, and "The Image of the *Rusticus*," at the University of British Columbia. For their suggestions about the manuscript I am grateful to Janós Bak, who organized the *Rusticus* conference, to Bruce Campbell, Robert Worth Frank, Jr., Jeanne Krochalis, Vickie L. Ziegler, and to László Bartosiewicz and Alice M. Choyke of the Archaeological Institute of the Hungarian Academy of Sciences. Kathryn Jones helped to prepare the index. It is a pleasure to acknowledge my great debt to Brian Tierney and R. H. Hilton, who first sent me off along medieval pathways. The Office of the Senior Vice President for Research and Graduate Studies, Pennsylvania State University generously provided a subvention for publication. I am grateful to Jerome E. Singerman and Alison A. Anderson of the University of Pennsylvania Press for guiding me through the publication process. Finally, I wish to thank James Ross Sweeney for his advice and encouragement at every stage.

D. S.

Figures

Del Sweeney

I. Introduction

No sphere of activity was more important or more central to life in medieval Europe than farming. Throughout the Middle Ages the overwhelming majority of the population lived off the land. The demographic expansion of the eleventh century and the rise of urban centers would have been impossible without an increasingly productive agricultural base. The great contribution of the past fifty years of scholarship on the economic and social history of northern Europe in the Middle Ages has been to demonstrate the diversity of responses to broadly similar challenges.[1] A notable feature of recent investigations has been the focus on regional studies. Medieval Europe was characterized by important differences in climate, soils, topography, and culture; historical patterns of settlement and agricultural production varied considerably. The pattern that typified the English Midlands, perhaps the most widely known model in the English-speaking world, cannot be used to characterize the French Midi or eastern Saxony. Within these larger geographical areas, considerable internal variation existed—for example, Normandy versus Burgundy, or northern England versus the southeast. Significant regional differences also are observed in the relationship between arable, pasture, and forest, each of which sustained the human and animal population in different ways. Where the documentation permits, studies have been made even of individual villages. Yet we are, perhaps, on the verge of new syntheses. Medieval historians are again asking broad questions. How do agricultural populations adapt to different ecological conditions? In what circumstances is new technology adapted or rejected? What were the realities of medieval peasant life? What did peasants themselves believe about the world in which they lived?

The limitations of medieval sources for investigating economic history are well known. As Carlo Cipolla has written:

> We would like to know the size of the population, the patterns of consumption, the level of production of, let us say, the province of Reims in France at

the beginning of this millennium. The documents of the time give us instead detailed information of the miracles performed by St. Gibrian in the area.[2]

Nevertheless, for selected regions and particular points in time, surviving documents and other types of evidence provide detailed information about the organization of agrarian production and about social institutions and relationships. Earlier generations of scholars staked out the terrain and established the first lines of investigation by devoting critical attention to a series of land inventories, the best known of which are the ninth-century Carolingian surveys called polyptychs, and the late eleventh-century English survey known as the Domesday Book. Such documents describe, according to the purpose for which the survey was made, the land held by lords and their dependents; the status of those dependents and the dues and labor services owed for the use of the land; the division of land into arable, vineyards, meadow, pasture, and woodland; and other kinds of property such as mills, fisheries, and beehives. Later in the Middle Ages, governments, seeking to increase their financial resources in the face of the erosion of traditional types of revenue and increased expenses resulting from frequent wars, undertook detailed surveys about the number, property, and wealth of their subjects. The mid-fourteenth-century Becerro of Castile, the English poll tax surveys of 1377–81, and the Tuscan Catasto of 1427–30 are examples of such surveys. Other types of documentary sources supply a more dynamic perspective, particularly the account rolls of the great English ecclesiastical estates that describe agrarian production on the demesne lands; court records dealing with the enforcement of seignorial rights and communal regulations as well as the resolution of disputes between tenants; and a vast number of charters conveying land and other property.

Without in any way diminishing the importance of these written texts, modern investigators have supplemented the picture drawn from the documentary record with insights derived from anthropology, archaeology, art history, literary history, and geographical studies. The study of modern peasant societies has contributed not only to a more complex picture of communal and family structures but also to a more subtle understanding of popular beliefs and reactions to changes in the natural and cultural environment. Archaeological investigations of medieval settlements have recovered tools, household possessions, animal bones, the structure of peasant houses, and field patterns. These have made possible a much more accurate rendering of the details of peasant life as well as a deeper understanding of the patterns of settlement and resettlement. Art historians and literary historians have penetrated conventions and, by asking questions about intent and audience, have delineated more clearly the complex web of attitudes, aspirations, and fears evidenced by the ideological representation of peasants in artistic and literary works. Finally, climatic and ecological studies are contributing to a better base for understanding long-term changes in settlement and farming patterns.

Each generation of historians has asked different questions about the medieval village. In the late nineteenth and early twentieth century, interest centered on the legal status of the rural population and their obligations, drawing distinctions among the many terms used to describe free and unfree persons. The legal and social status of individual peasants varied not only by region but also by lordship and village. Moreover, there was no uniform correlation between status and the type of tenure and obligations. Interestingly, questions about legal and social status are being examined again, particularly the chronology of the decline of slavery and the rise of seignorial institutions.[3] Karl Brunner, in Chapter 2, sees the Carolingian period as "a last impulse of continuity" from Roman times. Recent studies have suggested that true slavery persisted through the Carolingian period and that the large ecclesiastical estates, which showed the earliest trend toward the settlement of slaves on holdings, may have been more the exception than the rule.[4] A consensus appears to be developing that the characteristic seignorial institutions of the eleventh and twelfth centuries developed much more slowly than had been previously thought and emerge in their full-blown form in western Europe only in the eleventh century.[5]

Beginning in the late 1930s, but significantly in the 1960s and 1970s, students of agrarian history turned their attention to the organization of agricultural production on large estates. Such studies were based primarily on the seignorial accounts of ecclesiastical estates, the best examples coming from thirteenth-century England. Studies of the estates of Ramsay Abbey and the bishoprics of Winchester and Worcester, for example, produced data on cropping patterns, grain yields, and market prices. A recent study of the Peterborough Abbey estates provides evidence about large-scale pastoral farming.[6] A persistent question is the degree of entrepreneurialism among the great ecclesiastical and lay lords—their willingness to take advantage of new technological and organizational possibilities, their interest in capital investment and maximization of profits, and their responsiveness to changing economic conditions.

The adaptation and diffusion of new technology, a critical subject in today's developing countries, is of considerable interest for the period of the Middle Ages. The advances in the design of the plough, particularly the heavier wheeled plough fitted with a moldboard, capable of cutting more deeply into the heavy clay soils found in much of northern Europe; the growing use of horses in addition to oxen for drawing the plough and for hauling; the spread of water mills used not only for grinding grain but also in the textile industry; the use of marl for fertilizer in England and parts of France; and the more intensive cropping patterns that produced higher yields—all undoubtedly contributed to agrarian growth in western Europe, which is evident from at least the eleventh century.[7] The adoption of such innovations points to a medieval willingness to embrace new techniques of demonstrable utility. Yet complex political, social, and cultural factors could inhibit the diffusion of such technology, as Andrew Watson demonstrates in Chapter 4. One of these factors was the high degree of communal regulation of farming activities in medieval villages. The chronology of the development of common fields and whether there was a conversion from a two-field system to a three-field system, or, more properly, from a two-course to a three-course rotation or cropping pattern, remain vexed questions.[8] Scholars are more cautious in postulating a widespread change from one pattern to another, and it is clear from Bruce Campbell's discussion in Chapter 5 that many different types of cropping patterns existed simultaneously.

Communal rights over pasture and woodland became more important as the arable was extended in favorable areas at the expense of pasture. Particularly in the thirteenth and fourteenth centuries, seignorial and village authorities more strictly regulated grazing rights on the arable and the common lands. The dominant interpretation of the "crisis" of the early fourteenth century has been a Mathusian one. Increasing population led to extension of the arable to marginal lands that produced lower yields and to lands previously used for pasture, reducing the number of animals that could be maintained and thus the available amount of fertilizer. This sequence of events led to the cycle of famine and disease that occurred even before the Black Death in 1348–50 killed a third of the population.[9] The interpretation has been challenged but still looms large in the thinking of economic historians and medievalists in general. Bruce Campbell's essay challenges this interpretation, while Robert Frank, in Chapter 11, evaluates the effect of these crises on the author of *Piers Plowman*.

Whereas most of the evidence about the organization of medieval agrarian and pastoral production pertains to large estates, the most common unit of cultivation was the small peasant holding. Small-scale production for the subsistence of the household is the defining characteristic of peasant societies. But the circumstances of medieval peasants varied considerably by place and time. Some families were able to accumulate large holdings through inheritance, marriage, and purchases, and in turn often sublet some of this property to others. Other villagers—cottagers or landless peasants—depended on a variety of means to scrape out a living, including leasing land from wealthier villagers, working as farm laborers or household servants, and becoming part-time retailers and artisans. The trends in peasant production in the Middle Ages are much more difficult to trace because of the small amount of documentation. Nevertheless studies have been made of peasant land transfers, and evidence has been collected about peasant participation in local markets.[10]

In the historical reconstruction of peasant life, the interplay of family, household, and community is assuming increasing importance. It is not coincidental that among the most influential medieval historians of this century have been those who have examined rural life to discover the mechanisms of more general social change. Strongly influenced by sociologists and anthropologists, the foremost practitioners have been the historians known as *Annalistes*, the name taken from the seminal French journal *Annales: Économies, Sociétés, Civilisations*.[11] Drawing on the works of such historians as Marc Bloch and Georges Duby, many scholars, particularly in the United States, have directed their attention to the family, the household, and the institutions of the village. Traditional documents such as seignorial court records, charters, and custumals, surviving from the late thirteenth century in England and from the fourteenth century elsewhere in Europe, are being employed in new ways to investigate demographic patterns, define gender roles, reconstruct family structures, and describe norms of behavior.[12] In Chapter 7, Ludolf Kuchenbuch explores these questions for the less well-known region of Eastphalia.

Another aspect of peasant studies focuses on the structure of peasant beliefs both about society and about the natural world. This is a very difficult subject for study because the peasants of the Middle Ages left few documents. Descriptions of their character, activities, and beliefs were composed for an upper-class audience. The beliefs of the peasants themselves must often be reconstructed from those who wrote to combat those beliefs, as we see in Paul Dutton's contribution on peasant beliefs in the ninth century (Chapter 6).[13] Michael Camille (Chapter 12) and Jane

Welch Williams (Chapter 13) show that artistic representations of the peasants frequently reflect the dominant medieval ideology of the three orders of society—*oratores, bellatores, laboratores*—those who prayed, those who fought, and those who toiled. The appearance of this tripartite scheme has been traced to the eleventh century, but attitudes toward the *laboratores* and toward manual labor changed substantially over the course of the Middle Ages.[14] The prevailing view in the early Middle Ages was that manual labor was a punishment for Adam's sin; but as Jacques Le Goff has noted, the Bible offered "an ideological arsenal containing weapons . . . in favor both of labor and nonlabor."[15] The rule of Saint Benedict, which prescribed manual work as an integral part of service to God, was a constant reference point in the later monastic debates about the relative merits of spiritual and manual labor.[16] The term "laboratores" evolved, by the tenth century, to denote especially those who cleared the land and whose work involved acquisition and profit.[17] Yet the portrayal of these workers in the art and literature of the thirteenth and fourteenth centuries is overwhelmingly derisory. The rapid changes taking place in the economic and social structures of Europe clearly aroused unease if not fear among the ruling classes. As Herman Braet notes in Chapter 9, such portrayals are an acknowledgment that "change is on the way."

The essays in the present volume deal with two major themes: first, the necessity of understanding the cultural framework within which changes in agricultural technology and economic organization occur; and, second, how changes in the social fabric influence attitudes toward rural work and the peasantry. The essays span the entire Middle Ages, from the end of the Roman Empire through the fifteenth century. The geographical emphasis is on western and central Europe, but the essays also touch on eastern Europe and the Mediterranean.

A notable feature of these essays is the variety of methodological approaches employed. The contributors use documentary sources such as chronicles, tax records, land registers, and seignorial accounts to analyze patterns of agricultural production; archaeological analysis of bones and tools to discuss agricultural technology, diet, and village structure; anthropological approaches to illuminate contemporary reactions to violent natural events and disasters; and literary and artistic analyses of poems, plays, *fabliaux*, manuscript illuminations, and stained-glass windows to understand medieval upper-class attitudes toward the peasantry and toward social change.

A common theme is the need for great care in the use and interpretation of the sources for reconstructing the conditions of peasant life. The authors stress the importance of using all of the available sources, with the understanding not only that descriptions and interpretations differ between sources but that even the same source may have contradictory elements. It is obvious that the peasants were responsible for only a small proportion of the evidence that relates to them, primarily the archaeological remains. As Michael Camille comments, "The man who held the spade did not hold the pen." Documentary sources pertaining to economic production were drawn up by the lord's agents for the purposes of record-keeping and the collection of rents and other dues owed by tenants. Literary descriptions of the peasants are often derogatory and generally have a cautionary purpose. Artistic images are idealized for didactic ends. It is only by taking these sources in their totality, in attempting to understand the intent, the audience, and the contemporary context, that we can hope to reconstruct, with some degree of assurance, the material culture of the medieval world.

The book is divided into four sections. The first part, "Agricultural Continuity and Diffusion," deals with agricultural technology and the organization of production. The second section, "Rural Society," examines the various sources for reconstructing the material culture and mentality of the peasantry. The final two sections, "Literary Representations" and "Artistic Representations," analyze the depictions of peasants in literary and artistic works and what these convey not only about the peasantry but also about the mentality of the upper-class audiences for these works.

Agricultural Continuity and Diffusion

Part I focuses on the political, social, and cultural context within which technical innovations are adopted, modified, or rejected and efforts to increase investment and specialization are encouraged or hindered. In the two hundred years since the publication of Edward Gibbon's *Decline and Fall of the Roman Empire*, historians have been examining the question of disruption and continuity after the disintegration of Roman imperial organization in western Europe. In Chapter 2, Karl Brunner critically reviews earlier interpretations that saw a radical decline in the level of agricultural knowledge and technology with the collapse of Roman government in the West. Brunner emphasizes "the creative assimilation of technology" that developed in response to the changed circumstances of the early Middle

Ages and sees the rural middle classes as the critical element in this process. Analyzing a hoard of tools discovered in Germany, now dated to the fifth century, Brunner notes the existence of transitional forms of tools that anticipate later medieval models, but that were better adapted to small-scale production. While acknowledging that the primary means of transmitting agricultural technology was undoubtedly from person to person, Brunner also stresses the transmission of agricultural knowledge through books. The classical Latin works of agriculture, which had been an accepted part of a Roman aristocratic education, continued to be read, at least in compendia, especially in monastic houses, which had the resources for more specialized agricultural production.

The question of change and continuity in the rural economy is surveyed also by Sándor Bökönyi, in Chapter 3, whose subject is animal husbandry practices. On the basis of extensive archaeological evidence from a broad range of sites in central and eastern Europe, Bökönyi discusses the retreat from Roman selective breeding practices in the early Middle Ages and the consequent decrease in the size of domestic animals, especially cattle. He sees a reversal of these trends beginning only in the fourteenth century: the reintroduction of selective breeding practices; an increase in the size and productivity of draft animals and those raised for meat; and the development of new breeds and the introduction of new domestic species. In a careful examination of the types of meat consumed in various regions of Europe, Bökönyi concludes that the environment and the lifestyle of a people have a greater influence than ethnic group on the types of animals raised.

In Chapter 4, Andrew Watson returns to the question of the transmission of technology across political and cultural borders. Watson asks why Islamic agricultural techniques such as new crops, more intensive rotations, and irrigation of semi-arid land, which seem clearly to have been more advanced, were adopted so little and so late by the Christian West. Watson attributes the lack of diffusion to differences in the demographic, economic, legal, and social environments of the two societies. He points especially to lower population densities in Europe, the relative weakness of market forces, the strong influence of the community on agricultural practices, and decreasing personal contact as Muslims retreated in advance of the Christian reconquest. The medieval case accords with what we have learned in modern times about the attempts to transfer agricultural technologies to developing countries. Watson comments in this connection: "What seems to be a superior technology may, as it attempts to move into a new context, encounter economic, social, cultural, and even environmental obstacles that make it inappropriate and, to all intents and purposes, inferior."

In the final chapter in this section, Bruce Campbell addresses the question of whether the overexploitation of agricultural resources was the source of the early fourteenth-century economic crisis in England and much of western Europe. Campbell marshals a vast array of evidence from English seignorial accounts to delineate regional patterns of arable cultivation and livestock production ranging from relatively simple one-crop systems to very complex and intensive patterns of production. He notes that specialization and market production could occur even in difficult environmental conditions, if there was good access to markets and if institutional constraints were not burdensome. Campbell views the breakdown in the fourteenth century as a result not so much of intrinsic limitations as of "agriculture's inability to maintain established levels of biological reproduction in the face of excessive external demands." While acknowledging the challenges of population growth and natural disasters in the early fourteenth century, he believes that it was war and the taxation levied to support it that tipped a precarious balance in the direction of decline.

Rural Society

The three authors represented in Part II, using different approaches and different types of evidence, contribute to the reconstruction of the material culture and the beliefs of the peasantry in the Middle Ages. In Chapter 6, Paul Dutton analyzes the popular belief in "storm-makers" from a ninth-century Carolingian text written by Agobard, bishop of Lyons and a member of the reform group in the time of Louis the Pious. By scrutinizing contemporary sources and comparing them with similar belief systems in other societies, Dutton attempts to reconstruct what the people in the region of Lyons in the ninth century actually believed. Similar studies have been done for later periods, for example by Carlo Ginzburg in *Night Battles*, but this is an unusually early example.[18] Trust in those persons who can make and control rain and storms is common in societies where the weather is unpredictable and where storms can do significant damage to crops. Such ideas survived in Europe at least as late as the sixteenth century. Dutton, however, also places these beliefs in a historical context and notes the concern of the Carolingians about the association between paganism and resistance to central authority. He suggests that the particular belief in aerial sailors who arrived in cloud ships and stole the grain cut down by the storm may have been a way of hiding grain from the tax collectors and that the "protection money" or "insurance" that farmers claimed they had paid to the storm-makers may have been offered as an excuse not to pay tithes.

Ludolf Kuchenbuch's essay (Chapter 7) examines social structure and family relationships in peasant communities in Eastphalia, north of the Harz mountains. Although the evidence for this region is sparse, Kuchenbuch teases out, from such sources as lists of tenants and custumals, a structure of family relationships in which status depends on the principles of masculinity, seniority, and marital status. He also discusses marriage, pointing up the importance of companionship and reciprocity as well as the husband's authority. The sex-linked orientation of activities and objects, implicit in the Eastphalian sources, we know from other sources to be an ideological view rather than the "reality" of the situation. Kuchenbuch elucidates, as well, the forms of organization in the village in the fourteenth century, a period in which seignorial rights were weakening. The sources show that the villagers were active in maintaining rights over the use of common lands and also in ensuring that their spiritual welfare was provided for.

The theme of Gerhard Jaritz's essay (Chapter 8) is the need for a critical approach to the interpretation of sources, particularly the literary and artistic sources, that have been used to reconstruct the material culture of the peasantry. By means of several cautionary examples, he stresses the importance of taking into account all the available sources, acknowledging that these are sometimes contradictory. He draws attention, too, to the necessity of examining not only the purpose for which these sources were composed but also the context in which they would have been interpreted in the Middle Ages. Things often were written down only because they were "unusual, special, new, deviant, or subject to criticism." The commonplace was taken for granted. Although the portrayals of peasants are grossly exaggerated in literary and artistic works, Jaritz reminds us that exaggeration depends on some core of reality. It is these recognizable features that give us clues about what might have been "real" in descriptions of peasant clothing, diet, and the tools used in agriculture. For example, Jaritz cites the depiction in a mid-fourteenth-century illustration of the use of a scythe for mowing grain. It might be assumed that this is an early example of use of the scythe for such a purpose, but, in fact, medieval observers would have known that this represented deliberate destruction.

Literary Representations

Medieval vernacular literature has long been used by scholars as a rich source of descriptions of rural society. In Chapter 9, Herman Braet compiles a portrait of the peasant from the *fabliaux*, short tales written in the thirteenth and fourteenth centuries. In examining the evidence about attitudes toward the peasantry that can be derived from the *fabliaux*, Braet notes that the emphasis is on types rather than individuals. Like Jaritz, however, he finds that what makes evidence from such a source reliable is the frequency and the "gratuitous" nature of the evidence. In the *fabliaux*, the picture of the peasantry is overwhelmingly negative. The peasants are physically unattractive—ugly, dirty, animal-like in their features. These portrayals suggest the extent to which the lower classes were despised. Even more, this revolting outward appearance is seen to mirror their inner corruption. A similar view had developed early in the Middle Ages,[19] and its reemergence in the late Middle Ages is perhaps a reaction to another period of significant social change. In the thirteenth and fourteenth centuries, peasant resistance to seignorial demands was becoming more overt. Some enterprising peasants were able to accumulate significant landholdings and became important village figures. Braet suggests that the *fabliaux* and other literary genres illustrate the writers'—and readers'—fears about the social and economic aspirations of the peasants and the threat these posed to the established orders. A particular concern was the peasant who married above his station. Such marriages reflect real pressure on noble families to improve their financial circumstances. But these liaisons, the authors of the *fabliaux* tell us, are never successful; such a misalliance is compared by one author to a pear grafted on a turnip.

Looking at another genre, Jane Dozer-Rabedeau (Chapter 10) analyzes the role of rural characters in early French miracle plays. She notes the development in these plays, written for an urban audience, of a secular comic element manifested in a new setting—the tavern. Peasants and rural characters enjoy themselves by eating, gambling, thievery, and playing tricks on others. The peasants, who often serve the purpose of comic relief, inhabit the mundane world. Chance controls their fate, and their activities, games, and songs "repeat recognizable forms of entertainment probably known to the audience." Dozer-Rabedeau suggests that such obvious enjoyment may reflect a more optimistic attitude stemming from improving economic conditions and changing theological views.

A much more somber view of peasant life is seen in Robert Worth Frank's essay (Chapter 11), which explores the devastating episodes of hunger and starvation that began in western Europe in the early fourteenth century. He argues that Langland's description of Hunger in *Piers Plowman* reflects the real suffering experienced in this period and suggests that Langland himself may have encountered near-starvation. Frank distinguishes between the "hungry gap," the period in the summer between the depletion of stores and the new grain harvest; "bad years," when crops failed; and, finally, famine conditions brought on by a succession of bad years. By analogy with the Potato Famine in Ireland and twentieth-century famines in Asia and Africa, he suggests that even more terrifying than the widespread deaths from starvation is the loss of societal constraints on behavior. In fourteenth-century England contemporary writers describe the flight of the hungry to the cities, the increase in crime, and even reports of cannibalism. Conditions of scarcity and famine led Langland not only to wrestle with the moral question of the deserving and undeserving poor—the "wastours"—but even, Frank thinks, obliquely to question the contributions of the upper classes to society. It was the person who attempted to forestall hunger by working the land—the ploughman—that Langland chose as his hero.

Artistic Representations

Medieval images of peasants and their agricultural tasks are discussed in Part IV. Adam as digger rather than ploughman is the subject of Michael Camille's essay (Chapter 12). The peasant who possessed a plough and a team of oxen (the *laboreur*) was comparatively well-to-do; the landless peasant or cottager (*manouvrier*) usually owned only hand tools, most significantly the spade. That medieval artists placed Biblical figures in a medieval context is well known. Camille shows that these images also reflect the current technology. In English examples of Adam working, the spade predominates as the tool of ordinary cultivation. In Sicily, however, Adam is shown with a heavier mattock or an axe, used for working drier soils and clearing trees. Camille's primary interest is in the ideological views inherent in these images. For example, the archetypal thirteenth-century images of Adam and Eve laboring—Adam digging and Eve spinning—show the accepted roles for men and women, Adam's province being the outside world of the fields, Eve's role that of caretaker of the house and children. Recent scholarship has demonstrated that women's work encompassed a wide range of activities, not only field work but commercialized activities such as dairying and brewing. The purpose of such images, however, was to reinforce the idea of traditional and "proper" roles. In analyzing images of peasants such as those in the Canterbury Cathedral windows, Camille suggests that their purpose was to "[remind] the medieval masses of their duties in the divinely ordained social order." Yet there are strands here, too, of the more positive associations labor acquired in the eleventh and twelfth centuries, in the images of Adam being taught to dig by an angel and even, in one instance, by God himself.

In Chapter 13, Jane Welch Williams examines the depiction of peasants in the bottom portion of cathedral windows, in the space normally reserved for a portrayal of the donor of the window. She concludes that at Chartres and Tours peasants were not the actual donors of the windows, but rather that the images of peasants in the windows signify property being donated to the church or possibly owned by the church. At Le Mans, however, the donation of a window by vineyard workers is documented. This window shows peasants not merely working but also enjoying the fruits of their labor. Williams suggests that these scenes may represent peasants who had contracts of *complant*, also known as *méplant*, which on increasingly favorable terms would make the lessees ultimately the proprietors of the vineyards. She also sets the Le Mans window in its political context, interpreting the imagery of the various ranks of society as a demonstration of support for the bishops of Le Mans in their struggles with the count of Anjou. Williams concludes, however, that "it was precisely in areas where ecclesiastical lords sought to hold back the trend of peasant enfranchisement, at Chartres and Tours, that peasants were first given artistic stature in place of legal status" although this artistic prominence was not long-lasting.

In Chapter 14, Bridget Henisch finds that the idealization of the peasant does persist for a long time in the calendar tradition. The favorable portrayals of the peasants in the calendar illustrations are markedly different from the caricature, criticism, and contempt found in the *fabliaux*, for example. Henisch comments on the sense of "calm, quiet, purposefulness" in these pictures: the sun is always shining, tools are appropriate, everyone seems in good health. Henisch notes the surprising omission of a religious framework in these pictures and even the lack of a communal context until late in the tradition. The constraints of miniature painting result in a "doll's house" world in which everything is depicted on a small scale. Here, too, we find the overriding ideological viewpoint: the illustrations of peasants are meant to depict "honest laborers," those who work contentedly at their assigned tasks and accept their place in society. In the most luxurious of these manuscripts, Henisch sees a "beautiful veil of illusion" that contrasted with the generally harsh realities of peasant life.

In the Afterword, Alfred Crosby, a self-confessed "world historian," relates these essays to "the big picture." He views Europe in the Middle Ages as "the wild west" of contemporaneous civilization, and asks how this comparatively backward region came to world prominence. This, of course, is the question historians of the late Middle Ages and the early modern period have been asking in this century: What developments in medieval Europe laid the groundwork for the economic expansion of the late fifteenth and sixteenth centuries and the development of a highly commercialized society? Crosby thinks that a critical factor may have been the number of large animals available for food, power, and manure, giving Europe more "muscle power" than other societies. He notes also the more positive attitudes toward labor, and the growing markets in bulk commodities such as foodstuffs. Finally, he acknowledges an important theme implicit in this volume: that the peasantry was "the cultural center of gravity" in the Middle Ages.

Taken as a whole, the essays in this volume, although deriving from different fields of interest and different sources, encompass several common themes. They reemphasize the importance of change in the Middle Ages and of the reaction and adaptation to change. Often change came in the form of the violence and chaos of war, the disruption caused by natural disasters, the strains on social institutions caused by famine and disease. Nevertheless, in the course of the Middle Ages, at varying rates and despite setbacks, we see the adaptation of technical improvements in agriculture and animal production, a significant development of production for small and large markets, and the emergence of a variety of patterns of agrarian organization. The essays further reinforce the importance of regional analysis for medieval economic and social studies. Institutions, economic trends, and the reactions to political events varied by time and place in Europe, and it is only by the patient examination of all available sources pertaining to particular areas that a more accurate and detailed picture of the forms and techniques of production, social relationships, and the effects of the institutional framework can be assessed. In different ways, too, the authors demonstrate the necessity of taking an interdisciplinary approach to reconstructing the material culture and mentalities of the peasantry from sources that exist for quite different purposes. The growing economic importance of the *laboratores* and their perceived threat to the established order can be seen underlying, on the one hand, the reassuring images of peasants doing their accustomed tasks in an orderly way and, on the other hand, the cautionary tales that warn the upper classes against allowing peasants to move up in the social scale. The cycle swings between admiration and contempt, between models of endurance and examples of behavior to be avoided. The essays in this volume make an important contribution to their particular fields of inquiry but also give us a fuller and more nuanced view of the Middle Ages. They show us how to think more carefully about how we use sources and the context in which we interpret them.

Notes

1. For recent surveys of the medieval peasantry see Werner Rösener, *Peasants in the Middle Ages*, trans. Alexander Stützer (Urbana: University of Illinois Press, 1992), and Robert Fossier, *Peasant Life in the Medieval West*, trans. Juliet Vale (Oxford: Basil Blackwell, 1988). Very valuable still is Georges Duby, *Rural Economy and Country Life in the Medieval West*, trans. Cynthia Postan (Columbia: University of South Carolina Press, 1968).

2. Carlo M. Cipolla, *Before the Industrial Revolution: European Society and Economy, 1000–1700*, 2nd ed. (New York: W. W. Norton, 1980), p. xiv; also see Cipolla, *Between Two Cultures: An Introduction to Economic History*, trans. Christopher Woodall (New York: W. W. Norton, 1991).

3. A discussion of the problem and the previous literature may be found in Pierre Bonnassie, *From Slavery to Feudalism in South-Western Europe*, trans. Jean Birrell (Cambridge: Cambridge University Press, 1991), pp. 1–59.

4. Note Guy Bois's caution, "Let us not succumb to 'polyptychomania'": *The Transformation of the Year One Thousand: The Village of Lournand from Antiquity to Feudalism*, trans. Jean Birrell (Manchester: Manchester University Press, 1992). See also Adriaan Verhulst, "La genèse du régime domanial classique en France au haut moyen âge," in *Agricoltura e mondo rurale in Occidente nell'alto Medioevo*, Centro italiano di studi sull'alto Medioevo, Settimane di Studio 13 (Spoleto: Presso la sede del Centro, 1966): 135–60.

5. A recent study argues that serfdom in Catalonia developed fully only in the thirteenth century: Paul Freedman, *The Origins of Peasant Servitude in Medieval Catalonia* (Cambridge: Cambridge University Press, 1991).

16   Del Sweeney

6. J. Z. Titow, *Winchester Yields: A Study in Medieval Agricultural Produc-tivity* (Cambridge: Cambridge University Press, 1972); Christopher Dyer, *Lords and Peasants in a Changing Society: The Estates of the Bishopric of Worcester, 680–1540* (Cambridge: Cambridge University Press, 1980); J. Ambrose Raftis, *The Estates of Ramsey Abbey* (Toronto: Pontifical Institute of Mediaeval Studies, 1957); Kathleen Biddick, *The Other Economy: Pastoral Husbandry on a Medieval Estate* (Berkeley: University of California Press, 1989).

7. There is a large literature on this subject. See, among others, Lynn White, Jr., *Medieval Technology and Social Change* (New York: Oxford University Press, 1962); Rösener, *Peasants in the Middle Ages*, pp. 107–21; John Langdon, "Agricultural Equipment," in *The Countryside of Medieval England*, ed. Grenville Astill and Annie Grant (Oxford: Basil Blackwell, 1988); Langdon, *Horses, Oxen and Techno-logical Innovation: The Use of Draught Animals in English Farming from 1066 to 1500* (Cambridge: Cambridge University Press, 1986). For a critical view, see Adriaan Verhulst, "The 'Agricultural Revolution' of the Middle Ages Reconsidered," in *Law, Custom and the Social Fabric in Medieval Europe: Essays in Honor of Bryce Lyon*, ed. S. Bachrach and D. Nicholas (Kalamazoo, Mich.: Medieval Institute Publications, 1990), pp. 17–28; reprinted in Verhulst, *Rural and Urban Aspects of Early Medieval Northwest Europe* (Aldershot, Hants.: Variorum, 1992).

8. See H. S. A. Fox, "The Alleged Transformation from Two-Field to Three-Field Systems in Medieval England," *Economic History Review*, 2nd ser. 39 (1986): 526–48; and Robert A. Dodgshon, *The Origin of British Field Systems* (London: Academic Press, 1980), pp. 1–28.

9. M. M. Postan, "Medieval Agrarian Society in Its Prime, Pt. 7: England," in *The Cambridge Economic History of Europe*, 1: *The Agrarian Life of the Middle Ages*, 2nd ed., ed. Postan (Cambridge: Cambridge University Press, 1966), pp. 549–632; J. Z. Titow, *English Rural Society, 1200–1350* (London: Allen & Unwin, 1969); Duby, *Rural Economy and Country Life*, pp. 293–311.

10. For the development of market relationships within villages and between villages and towns, see R. H. Britnell, *The Commercialisation of English Society, 1000–1500* (Cambridge: Cambridge University Press, 1993). On peasant land markets see Bruce M. S. Campbell, "Population Pressure, Inheritance and the Land Market in a Fourteenth-Century Peasant Community," in *Land, Kinship, and Life-Cycle*, ed. Richard M. Smith (Cambridge: Cambridge University Press, 1984); P. D. A. Harvey, ed., *The Peasant Land Market in Medieval England* (Oxford: Clarendon, 1984), esp. pp. 1–28; and Teofilo F. Ruiz, *Crisis and Continuity: Land and Town in Late Medieval Castile* (Philadelphia: University of Pennsylvania Press, 1994), pp. 140–74.

11. For a recent work of this type, see Bois, *Transformation of the Year One Thousand*; an example for the later Middle Ages is Otto Brunner, Land *and Lord-ship: Structures of Governance in Medieval Austria*, trans. Howard Kaminsky and James Van Horn Melton (Philadelphia: University of .Pennsylvania Press, 1992). An intimate portrait of a late medieval village in the Pyrenees is the subject of Emmanuel Le Roy Ladurie, *Montaillou: The Promised Land of Error*, trans. Barbara Bray (New York: George Braziller, 1978).

12. See, for example, David Herlihy, *Medieval Households* (Cambridge, Mass.:

Harvard University Press, 1985); David Herlihy and Christiane Klapisch-Zuber, *Tuscans and Their Families: A Study of the Florentine Catasto of 1427* (New Haven, Conn.: Yale University Press, 1985); Judith M. Bennett, *Women in the Medieval English Countryside: Gender and Household in Brigstock Before the Plague* (New York: Oxford University Press, 1987); Barbara A. Hanawalt, *The Ties That Bound: Peasant Families in Medieval England* (New York: Oxford University Press, 1986).

13. See also Aron J. Gurevich, *Medieval Popular Culture: Problems of Belief and Perception* (Cambridge: Cambridge University Press, 1988).

14. Georges Duby, *The Three Orders: Feudal Society Imagined*, trans. Arthur Goldhammer (Chicago: University of Chicago Press, 1980).

15. Jacques Le Goff, *Time, Work, and Culture in the Middle Ages*, trans. Arthur Goldhammer (Chicago: University of Chicago Press, 1980), pp. 71–86.

16. See George Ovitt, *The Restoration of Perfection: Labor and Technology in Medieval Culture* (New Brunswick, N.J.: Rutgers University Press, 1987).

17. Le Goff, "Labor, Techniques, and Craftsmen," in Le Goff, *Time, Work, and Culture in the Middle Ages*, p. 86.

18. Carlo Ginzburg, *The Night Battles: Witchcraft and Agrarian Cults in the Sixteenth and Seventeenth Centuries*, trans. John and Anne Tedeschi (Baltimore: Johns Hopkins University Press, 1983).

19. Le Goff, "Labor, Techniques, and Craftsmen," pp. 87–97.

PART I

Agricultural Development and Diffusion

Karl Brunner

2. Continuity and Discontinuity of Roman Agricultural Knowledge in the Early Middle Ages

Almost exactly one hundred years ago, in 1897, a large hoard of iron tools was found in a safe retreat inside the rampart ditch of a Roman fort in Osterburken, situated on the Roman *limes* between Würzburg and Heilbronn. It was assumed that a Roman blacksmith had hidden it in the third century in connection with raids by Alamannic tribes. In a recent reinvestigation of these finds it has been found that the tools are at least a hundred years later than had been assumed, perhaps fifth century, and have to be looked upon as part of the inventory of an Alamannic estate.[1] The implications of this discovery are not to be underestimated, as they will change part of our picture of the transition from Roman antiquity to the early Middle Ages.

Lynn White's book *The Transformation of the Roman World* generated a change in our overall view of history. Its title signaled the final overthrow of the "catastrophe theories" as promoted, for example, in Edward Gibbon's classic *Decline and Fall of the Roman Empire*.[2] Since then, two important theses have been advanced. The first is that the initial stage of the transition from antiquity to the Middle Ages is to be sought in various changes that took place within the Roman world. The second thesis is that the so-called "barbarians" made use of classical institutions in the course of their invasions and their subsequent ethnogenesis, the process in which a group of people becomes a tribe.[3] The first thesis has been covered from many points of view, but has focused primarily on law, politics, and intellectual life. The differences between antiquity and the Middle Ages in everyday life seemed to be too pronounced and obvious to researchers to be called into question. But this is true only for certain aspects of civilization: those connected with the lifestyles of a relatively small leading group.

In any period, agrarian and social historians need to have a clear idea about the process of creative assimilation of technology if their subject is not to be relegated to an existence without history. In every age we encounter people who are prone to see only heroic deeds in history. This is the reason why peasants, the people of the plough, have been called people without history.[4]

The answer to the question of continuity is different for Rome itself, for the Mediterranean, for the Roman provinces, for their border zones, and for the regions outside. In certain regions we find not only a more or less fragmentary continuation of Roman styles of life, as in the countries with a Roman tradition, but also a merging with other cultures, such as Celtic, Germanic, or Slavic. This is true of a broad margin along the Roman *limes*, inside and outside the Roman Empire. We find a type of "inner border" as well in places within the Roman Empire where adverse features of the environment generated a need for special technical and organizational solutions in times of crisis. This is especially true for the Alpine region and parts of North Africa.[5]

In the monastery of Lucullanum, in Italy, the abbot Eugippius wrote, in 511, a biography of Saint Severin, which is a valuable source of information concerning the last years of the Roman presence north of the Alps. In this *Vita* we are told that the Romans withdrew from these regions under Severin's leadership. According to this account there should have been no continuity in respect to political institutions or persons, nor any form of technological transfer.[6]

However, the opposite is the case. The Christian church and the secular powers in Raetia, Noricum, and Pannonia refer to late classical models. In many cases remarkably strong traces of the Roman population are to be detected in the names of places and people. Nevertheless Eugippius has not deluded himself or us. The Romans who departed were members of an upper class and members of the government. They were not "the people," in any case not "the people of the plough." In the search for continuity our best chances lie in a rural middle class: the people who could pass on technical skills were rooted in their homeland and had little chance of making a new life inside the borders of the shrinking Empire. We are talking not only about peasants, breeders of livestock, beekeepers, hunters, and cooks but also about members of the lower nobility.[7]

Sources for these topics are scarce. Economic records are fragmentary, and their survival throughout Europe varies in quantity. The surviving sources are written in Latin, but should offer a glimpse into a world where very few persons knew any Latin, not to say good Latin.[8] With official regulations such as the famous *Capitulare de villis* of Charlemagne the problem is to what extent these were descriptions of an ideal state, never, or at least never fully, implemented in reality.[9] We find some help in the lives of the saints. As in a medieval miniature, the background scenes are the most valuable since the saints performed their miracles in surroundings that had to be as realistic as possible.

Nevertheless, we need the aid of the allied disciplines, especially archaeology, philology, and art history. Historians of literature have already done a respectable job. Their achievement, however, would have been even greater if they had sometimes diverted their attention from heroes and bards and concentrated more on the various aspects of everyday life. For example, there is probably no song about summer in medieval literature. Even if songs are called "Summer Song," they are, like most of the others, songs about springtime.[10] During the summer the nobles, like everybody else, were concerned with practical affairs—for example, the harvest—and not with heroism. Knightly attitudes resumed in the autumn.

Different views exist about the survival of technical aids and methods of production in the early Middle Ages. For a long time the prevailing view was that a new start was made on a lower cultural level. A number of observations strengthened this idea. As a matter of fact, we know of a large number of destroyed forts, towns, and villages. It seemed that a pronounced gap in archaeological finds could be observed toward the beginning of the fifth century.

This has implications for our perspective of historical development. For example, the economic development of the Carolingian age has been called a "renaissance," a rebirth, meaning a reclaiming of classical knowledge that had been lost. A sharp cultural distinction was seen between central Europe—as the zone of barbarian states—and western Europe as the zone of Roman continuity. Henri Pirenne's thesis could be valid only for those parts of Europe with a Roman tradition. But this thesis was questioned, and rightly so, shortly after its publication.[11]

New research has shown that the assumption of a sharp break was more likely the result of the political prejudices of the nineteenth and twentieth centuries than of reality. When the sources are correctly interpreted, the assumed gap in archaeological finds during the fourth and fifth centuries can be closed. The question of continuity is now approached in a quite different way.

Take, for example, the well-investigated Raetic-Alamannic region.[12] All the components of interest are present there. We find a fusing zone in the middle of Europe due to the geography of the Alps, which, on the one hand, was useful for retreat and, on the other hand, was of great importance with respect to traffic once the Alps became the crossroads of power politics. On the routes across the mountain passes the continuity of Roman traffic arrangements can be proved.[13] Thus we find a strong continuity of Roman populations; a well-known example is the Lex Romana Curiensis, the specifically Roman law in the bishopric of Chur in present-day Switzerland.[14] On the other hand, the Alamannic ethnogenesis yielded a tribe with very pronounced features out of "mixed and congested people," which was not altered by the Frankish conquest at the end of the fifth century.[15]

In the first two centuries A.D. the *villa rustica* was the dominant feature of the rural scene. In the third century a tendency toward fortified structures on elevated sites can be noted. These structures did not offer shelter to large numbers of people; they were meant only for the privileged few. The upper classes also left the towns and retreated to the surrounding hills. With that, interestingly, an increase occurred in the amount of imported luxury goods—for example, *terra sigillata* (red-glazed pottery)—despite the unrest.

These examples show the differentiation of society in late antiquity. The rich got richer, the poor got poorer. The military was a main reason for the inflow of money. We find both an increase in the number of gold coins circulating and a decrease in the number of copper coins. This is an indication that money-based economic institutions ceased to exist, a development that began before the Alamannic raids of the late third and fourth centuries.[16]

Slaves became more and more scarce. Since their average life expectancy remained below twenty years, and since a cheap supply from the masses of prisoners of war could no longer be counted on, their numbers decreased markedly. The shortage of workers, however, improved the social situation of the population and created pressure to develop innovations in agriculture.

The continuity of settlements was no more disrupted than was agrarian production. The idea of the "expansionist urge" of the Germans has to be attributed to yesterday's power politics. The first wave of Germans was eager for booty; the work was left to others. Consider an example from the most recent research for the territories within the Roman Empire: Formerly, for example, among the Visigoths and the Burgundians it was said that in the system of hospitality they took over two thirds of the estates: The Romans were confined to one third, the "tertia." Actually it was not the property itself, but the income that was divided. The work was done neither by the Germanic nor by the Roman noblemen, but by servants of different ranks and origins.[17] But in the frontier areas Germanic settlers were taken into the Empire, partly due to a shortage of labor. In many transition zones almost no hostility existed between Roman and Germanic peasants, because of their different lifestyles.[18]

Let us now concentrate on a few examples, first of all, the plough. Joachim Henning of Berlin has come up with striking results. His early works were devoted to tools in the zones of contact between Slavonic peoples, Bulgarians, and Byzantium on the lower Danube. After that he examined the archaeological material from central Europe, which had been buried for some time in the repositories of museums, but had never before been systematically investigated.[19]

Roman fields often were square in shape. Even today, in some regions of Europe such fields can be detected in the landscape. They are square because the plough was moved criss-cross over them. The Roman plough was just a technical improvement of the hook-shaped plough (a scratch ard), which only broke up the ground (see Figure 1). However, the procedure was done several times a year.[20]

During the time of the Roman Empire a new type of ploughshare was developed, with socket and working edges set apart. This type was found particularly in the vicinity of the great villas in Pannonia and Gaul. Along the Rhine and the Danube, where small and medium-sized estates prevailed for a long time, we still find ploughshares fastened to the seat with stick-like extensions. The new type of plough was often larger. The socket was attached more firmly to the ploughseat. We also have hints concerning the use of wheeled frontstools. All this speaks for the use of the plough with heavy teams of oxen that needed more than one person to guide them. This was typical in the case of the large estates (*latifundiae*).[21]

An interesting specimen was found in the hoard of tools in Oster-burken, which was mentioned earlier. This specimen is of the new type, but not as big. Comparable examples existed outside the Roman Empire since late imperial times. The shape of the ploughshare anticipates medieval models, which are often found from the eighth century onward.

Coulters were also found in this hoard. We thus can reconstruct a plough like the one depicted in the ninth century, which at that time was already outdated. The victory of the type of plough that not only broke up the earth but also turned the clods had already begun. The difference is visible only when the ploughshare is examined. The one from Osterburken is worn out almost symmetrically, whereas the Carolingian ones show signs

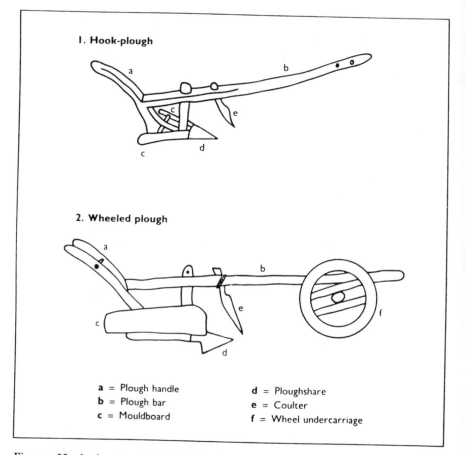

1. Hook-plough

2. Wheeled plough

a = Plough handle   d = Ploughshare
b = Plough bar   e = Coulter
c = Mouldboard   f = Wheel undercarriage

Figure 1. Hook-plough and wheeled plough. After Werner Rösener, *Peasants in the Middle Ages*, fig. 15. By permission of Polity Press.

of use on one side, a result of the use of a moldboard, which forces the ploughshare into an inclination.

The plough of Osterburken—the findings in general suggest a date of approximately the fifth century—thus provides the hitherto missing link between Roman and medieval forms. The importance of the piece is even clearer if we look at other tools found in the same hoard: a collection of scythe blades.[22]

Scythes were known already to the Celts. They were then, as in the Middle Ages, used only for the harvesting of hay. For grain, the danger of losing some of the precious crop was far too high. Only the devil was that crazy. Grain was cut with a sickle, sheaf by sheaf.[23] Only after that was the straw mowed or turned under.

Roman scythes technically were no different from enlarged sickles. But in calling them scythes an important detail is obscured. The scythe blade is fastened to the handle at an angle; this means that if the scythe is moved parallel to the ground the person using it can stand upright. This angle is found in the scythes of Osterburken. There is also another important detail: the use of thinner blades was made possible without a loss in rigidity by the flanging of the back portion of the blade. This made the scythes markedly lighter, without loss of performance.[24]

What is new about these results? First, they are well documented. For the first time the Roman and post-Roman material has been sorted out in a methodologically correct way. The reason for the previous gap in findings is that the great interest in Roman finds swallowed up other things that were found in their vicinity. Precisely because of the continuity, later material, which was of almost the same shape, was dated incorrectly.

Second, the tools are clearly transitional forms. Although they are rooted in Roman culture both in form and in technique, they are technologically advanced, pointing toward the Middle Ages. With respect to these tools, which are of central importance for agriculture, we observe not only the transmission of technology but also creative development and adaption.

This adaptation was one to the immediate society and its circumstances. In that respect, the archaeological evidence further illustrates the observations of social and economic history, and even enhances them. This is an excellent example of the way in which we should think of continuity. Further development is based on a detailed analysis, which sorts out useful features while purposely omitting other details.

This ought to be kept in mind before one deplores the change in lifestyle. The change was related to luxury goods in many cases or to the complex urban civilization as a whole.[25] For example, the Roman technique of heating floors and walls, which seems so comfortable to us, was not retained since it did not, with respect either to comfort or to ease of use, meet the demands of the Germanic nobility. They did know of it. Recently a villa of the Roman type was excavated approximately 40 km north of Vienna.[26] We suspect that this was the dwelling of a retired Roman officer of Germanic origin, who retreated to this place, where he was within the *Germania libera*, but also within sight—on a clear day—of the Roman fort at Vindobona, present-day Vienna. This villa had a floor-heating system, but it probably was rather ineffective because of the cold northerly winds, and in addition it required a large number of servants for its maintenance.

The background for the changes in the plough and the scythe lay in the changed way of life, both for people and for animals. Large ploughs were no longer needed, because the large estates had ceased to exist. The highly specialized forms of agriculture characteristic of late antiquity had lost their markets with the decline of urban civilization. Moreover, there were not enough oxen to make up the large teams needed for the heavy ploughs nor were there enough slaves to drive them. But the benefits of an advanced tool could still be retained if the tool was made smaller. The benefits were, above all, a better anchorage of the ploughshare and, although no examples of it have survived, a special shape of the ploughstool.[27]

At a particular phase of their contact the Germanic peasants were able to plough more deeply than their Roman counterparts, who had preserved the old shape of the plough. Germanic peasants thus were especially good at farming wet soils, which were both fertile and unworked. Thus in some regions of the Alps, for example, south of Salzburg, a mixed settlement of Roman and Germanic people developed in an entirely peaceful way.[28] This has already been shown in regional histories through the analysis of field- and placenames, but now we can imagine how this was achieved practically. These improvements in the plough were of importance for success, up until the development of a plough that turned over the clods.

The Germanic population needed more hay than the Romans did. For this reason barns were erected in addition to the existing granaries.[29] In antiquity the breeding of livestock had been adapted to the climate of the Mediterranean, which permitted the cattle to be left outside throughout the year. Hay was measured in *manipulae* (bundles). The word was retained in regions such as the steep Alpine valleys, where, even today, hay is transported on human backs. In places where flat meadows existed, the measure became the *carradae*, the cart.[30]

North of the Alps more hay was needed for fodder during the winter, probably because the number of cattle kept was larger. As it once had been among the early Romans, so it was now with the Germanic peoples: ownership of cattle was prestigious. We are familiar with the Latin word *pecunia* for money, which is derived from *pecus* (cattle). Cattle in Old High German is *fehu*; from this "fief" is derived. Everything that was valuable to the medieval nobleman thus came from cattle.

And so, it turns out that the changed circumstances of life and society were the motive for changes in the use of tools. The new scythe must have had consequences, the extent of which we are not yet able to determine. The scythe probably changed the landscape; perhaps an analysis of pollen can improve the understanding of this process.

How many head of cattle could be fed during the winter and for what period of time cattle were kept in the stable are important questions. The answers to these questions influence, among other things, the whole question of manure, which is very important with regard to continuity. In the Roman treatises on agriculture there are elaborate discussions of the treatment and use of manure. An important role was played by composting, and even today there is something to be learned from these works. In written sources from the early Middle Ages, evidence for systematic manuring can be found only rarely before the tenth century. After that there is more frequent mention of labor services concerning the collection of stable manure.[31]

Archaeology as yet is confined to a narrow band of the spectrum, because the excavations of settlements have been too few to permit generalized statements. We do not know where the cattle whose bones have been counted were kept. The cattle could have come from the pastures as well. Very recently, phosphate analysis has been applied to solve the problem, but the interpretation is not secure.[32]

We know that the Germanic peoples did not understand the art of systematic manuring. They continued to exhaust the soil. In several cases soils were exhausted in about three generations, and settlements had to be moved. It would be of the utmost importance to know where and when the collection of manure began, since this is a basic factor in permanent settlements in more densely populated regions. The number of people who can be fed from a certain area of land depends on the use of manure, and this, in turn, is a decisive factor for political, cultural, and military development.

We see the connection most markedly in the rise of the Frankish realm under the Carolingians after the seventh century. We have to consider this period—in modification of Pirenne's thesis—as a last impulse of continuity and not simply a "renaissance." The Franks, being mostly of Germanic origin themselves, along with the expansion of their empire brought the world of Roman tradition that existed inside the *limes* to those outside it. With that a shift of the political center of Europe toward the middle of the continent occurred.

A key role in that was played by the spreading of the "bipartite" agricultural system, which influenced the social system of Europe for centuries. By "bipartite" agriculture we mean the division of estates formerly managed using slave labor into a demesne and dependent holdings. In antiquity we find the beginnings of bipartite agriculture in the letting out of parts of the estate to a number of *coloni* for independent cultivation. These *coloni* possessed their own houses and were able to establish their own families. Tacitus, in the first century A.D., believed that the *servi* of the Germans had their own houses and homes, considering this an example for his fellow Roman citizens.[33]

Although the way in which the course of events developed is a matter of disagreement among researchers, the result is clear: This was the successful rural concept of the early Middle Ages. Bipartite agriculture stimulated higher performance through the prospect of individual profit, without a need to change the system of rulership to any major extent. A society emerged that divided itself into the functions of *bellatores, oratores,* and *laboratores*—to those who fought, those who prayed, and those who worked.[34]

At the latest, such a division of labor became evident with the introduction of armed cavalry. The method of fighting made the training of men and horses necessary, and the equipment was expensive. The clergy saw themselves as the warriors of Christ and came from the same upper ranks in society as the secular warriors. Their numbers increased at the same pace as did agrarian production. The rank of laborers (at first this term was confined to rural workers) was able to feed the other ranks as well. The rank of peasants consisted of various people: there were some who did not want to fight anymore—one might call them dismounted free warriors—and there were others who were not allowed to fight, being unfree people; both had various origins. For some the ties of the manorial system meant security; for others it meant a sort of freedom. It is like a glass of water that you can call half empty or half full. People were unfree with respect to the outside world, but free within their own social surroundings. These dependent persons became within medieval society the most important rank among the unfree strata, and with their development the process of transformation was complete.[35]

The division of society into three orders was a question of ideology from the very beginning and did not reflect the total social spectrum. Later on, there were sections of the population that did not fit into this scheme, for example, artisans or traders. In some of these professions ethnic continuity was high. Long-distance traders were Jews and Syrians. In some occupations the proportion of persons bearing Roman names was very high, for example, among blacksmiths and carpenters. Some specialized areas of agriculture included a high degree of loan words from Latin. This was true especially for Alpine cattle-herding and viticulture. Cheese, as is often forgotten today, is besides clarified butter the only dairy product that keeps for relatively long periods of time, and was therefore, in addition to cereals and vegetables, one of the staple foods. Wine as a drink had a high social prestige. The knowledge of how to make cheese and process grapes was certainly not learned from books.[36]

Other kinds of knowledge *were* taken from books. An astonishing amount of information about agriculture was found in some. What was the reason for this at a time when books were rare and dealt mostly with religious subjects? During antiquity a knowledge of agriculture was considered part of an upper-class education and a component of *otium*, leisure. Even the most noble of the Romans saw themselves as *rusticorum mascula militum proles* (Horace, *Odes* III.6.37), as the offspring of strong and valiant peasants. It was not considered dishonorable to bear the surname of Scrofa, meaning breeding sow. Despite that, the full extent of intellectual ability was applied to science. In the first century B.C. the term *experimentum* emerged to describe a methodically controlled series of tests.[37] Much the same was true for the other end of the known world as well. Expertise on the part of the lords about the basics of life was considered a virtue. The qualities of a warrior as the head of his household were noteworthy to the Scandinavians, as we learn from tombstones erected to their memory. A poet like Virgil could expect an appreciative audience when dealing with agricultural themes, and several excellent reference works were written. These should have enabled the Roman nobleman to invest successfully in real estate and to control his specialized slaves.

The leaders of the religious houses and probably the leading officials at the princes' courts had to be knowledgeable about the economic basis of their estates. Erudition was not seen as a significant handicap to applied, practical expertise. Each reform of the monastic system had to make sure of the economic basis of the religious community as a prerequisite for success. Saint Benedict himself wrote: "The monastery should be planned, if possible, in such a way that all necessities are inside its walls; these are: water, a mill, gardens, and the workshops of the artisans, so the monks can stay inside the walls" (66.6). This is based on the assumption that there were artisans among the monks (57.1), which in later times was not true of the nobility. But Benedict himself, in the first half of the sixth century, had probably gathered around him noblemen who had retreated from society rather than peasant boys used to the harshness of life. Benedict told the monks not to complain if they had to help with the harvest (48.7). Christian morality was fundamentally influenced by the high esteem Benedict had for manual labor, but it was not the labor of the peasant.[38]

The monks of Fulda, during the reign of Charlemagne, insisted on working in the kitchen and the pantry, serving at table, working in the bakery and in the garden, and so on, as it had been laid down in the monastic rule.[39] It should be added in this context, however, that liturgical work was more important than anything else: *ergo nihil Operi Dei praeponatur* (43.3). The largest part of scholarship was devoted to the preparation of the liturgy. A prerequisite for that was the functioning of the economy. Thus the economy was considered important by the learned. Knowledge of agriculture was, both in antiquity and in the early Middle Ages, an integral part of education. It was only with the establishment of schools and universities that the spheres of learning were separated. The religious houses were centers of communication, serving as places of retreat, lodging, and pilgrimage, but they were also feudal centers, and as a result everything that happened there was of the utmost importance to contemporaries. This was true, by the way, with practical consequences not only for the Roman tradition but for the Insular-Celtic tradition as well.[40]

The role of the religious houses was most important in those regions of Europe where towns failed to maintain their position or never played a role, again at the frontiers of the Roman world and on their outer edges, because they served also as missionary centers. Because these establishments were relatively large, with perhaps between thirty and a hundred monks during Merovingian times and several times as large a staff, a division of labor could be employed in agricultural production. More specialized artisan techniques had the best chances of persisting within this framework. Considering themselves advanced religious institutions the monks took care—*ut in omnibus glorificetur Deus*—that the liturgy advanced the crafts. Valuable metal tools had been used not only in the erection and decoration of churches but also on the monastic estates. Thus knowledge about their production and use was preserved. In this context we see again most of the tools from the hoard already discussed, this time in written sources. In the house rules of the abbot Adalard of Corbie, a member of the house of the Carolingians, and in an inventory of a Bavarian church in the first half of the ninth century, we find references to spades, hatchets, chisels, drills, sickles, scythes, and ploughshares.[41]

For the transaction of business on such an estate, some administrative regulations, preferably written ones, were needed for the sake of consistency and continuity. These, too, could be modeled on examples from antiquity. Many of these documents have been lost, however, because perishable media were used. From Carolingian times on we have some registers whose structure is basically no different from those of late antiquity, although the last word about their continuity has not yet been said.[42]

Scholarly knowledge was at least of the same importance as the transmission of practical expertise from person to person. The armarium of the library not only was a religious treasury, it also provided direct access to tradition for worldly matters. A particular group of manuscripts from the early Middle Ages transmitted classical knowledge. From Carolingian times through the Middle Ages, even into the Renaissance, a basic stock of reference works were read. In many cases several authors are preserved in a composite form, constituting a kind of basic canon. In such works we find above all Cato the Elder (ca. 140 B.C.), Varro (ca. 27 B.C.) and Columella (first century A.D.). These manuscripts were more than sustenance for scholars and are to be found in places where we may assume there was a need for literature relevant to the subject—at court and in the religious houses of the reform movement—and there are even some examples of learned disputes and of practical testing.[43] The educated world of the Middle Ages, like that of antiquity, was not confined to the *septem artes liberales*, the seven arts of a free man, but was aware of the sciences of life and nutrition. Only in modern times has education favored the ivory tower of erudition. Thus the continuity of Christianity played its part in the technological transfer of agricultural knowledge.

I have tried to show by means of some examples—the plough, the scythe, manuring—what the process of the creative assimilation of technology between antiquity and the Middle Ages was like. The historical scenarios far away from the centers of the Roman Empire and the social consequences of this kind of continuity have been touched upon only briefly. Even a specifically medieval social structure like the bipartite agricultural system had its precursors in Roman systems.

In addition to this transmission and modification of knowledge, which occurred from man to man and woman to woman in different parts of Europe, we find a treasure of knowledge that was used not only in the early Middle Ages but also in many later periods of European history, and which may have been taken from Roman reference works. We find proof in the written sources that this knowledge was used not only for learned dispute, but for practical application.

The question of continuity between antiquity and the Middle Ages has been raised not only because of interest in this singular and great process in which our cultures are rooted but also because the general structure of the course of events can be observed as well. It is important, even for the present, to understand that all types of cultures interrelate in one way or another; and even more so in cases where they seem to be situated at different cultural levels. A one-sided picture of this process has prevailed for a long time. The so-called advanced civilization gives, the more primitive ones take, with more or less understanding and willingness. We are now approaching the concept of mutual give and take. We have to do justice to the intrinsic value of each culture and not look upon them with the attribute "not yet advanced." This approach can result in a new perspective in analyzing the contact of the so-called barbarians with the developed civilization of the Mediterranean.

If we are to assess the historical importance of such a transmission, we have to do some rethinking. History cannot consist only of what those at the top have said and done, only of what is technically most complicated and advanced. Today, progress should be examined from different aspects. In the early Middle Ages progress meant above all the assimilation of technology into different social structures and circumstances of production. Best at that were the middle ranks of the social spectrum, those who were the most likely to survive in times of crisis.

Our image of antiquity has to be altered of course. It often appears to have been a closed, static society. This was true at best for the upper classes and the military organization, and fundamentally even among them only for a limited period of time. The overwhelming success of Augustan propaganda has made everything else into prehistory or decline. The strength of the Lex Romana is to be found in the fact that under its protection a variety of peoples, religions, and styles of life were possible.

The kind of things that have lasted through the ages may be astonishing at first glance. That it was not only the tools of power and war that survived may leave us with at least a little bit of optimism for the future.

Notes

1. Joachim Henning, "Zur Datierung von Werkzeug- und Agrargerätefunden im germanischen Landnahmegebiet zwischen Rhein und oberer Donau (Der Hortfund von Osterburken)," *Jahrbuch des römisch-germanischen Zentralmuseums Mainz* 32 (1985): 570–94.

2. Edward Gibbon, *History of the Decline and Fall of the Roman Empire* (1776/

88), ed. J. B. Bury, 2nd ed., 7 vols. (London: Methuen, 1926); Lynn White, ed., *The Transformation of the Roman World*, Center for Medieval and Renaissance Studies, University of California at Los Angeles, Contributions 3 (Berkeley: University of California Press, 1966). Cf. Walter Goffart, *Barbarians and Romans*, A.D. *418–584: The Techniques of Accommodation* (Princeton, N.J.: Princeton University Press, 1980); A. H. M. Jones, *The Later Roman Empire (284–602): A Social, Economic, and Administrative Survey*, 3 vols. (Oxford: Basil Blackwell, 1964); Walter Pohl, *Die Awaren: Ein Steppenvolk in Mitteleuropa, 567–822 n. Chr.* (Munich: Beck, 1988); Herwig Wolfram, *History of the Goths*, trans. Thomas J. Dunlap (Berkeley: University of California Press, 1988).

3. Evangelos K. Chrysos and Andreas Schwarcz, eds., *Das Reich und die Barbaren*, Veröffentlichungen des Instituts für österreichische Geschichtsforschung 29 (Vienna: Böhlau, 1989); Herwig Wolfram, "The Shaping of the Early Medieval Kingdom," *Viator* 1 (1970): 1–20, and Wolfram, "The Shaping of the Early Medieval Principality as a Type of Non-royal Rulership," *Viator* 2 (1971): 33–51; cf. Wolfram, "Ethnogenesen im frühmittelalterlichen Donau- und Ostalpenraum (6. bis 10. Jahrhundert)," *Frühmittelalterliche Ethnogenese im Alpenraum*, ed. Helmut Beuman and Werner Schroder, *Nationes* 5 (Sigmaringen: Thorbecke, 1985): 97–151.

4. Isidore of Seville, *Etymologiae sive originum libri XX*, ed. W. M. Lindsay, 2 vols. (Oxford: Clarendon, 1911), V, 31: *Nam tempus per se non intelligitur, nisi per actus humanos memoria digni*; Oswald Spengler: "Der Bauer ist der ewige Mensch, unabhängig von aller Kultur," cited in Wilhelm Abel, *Geschichte der deutschen Landwirtschaft vom frühen Mittelalter bis zum 19. Jahrhundert*, Deutsche Agrargeschichte 2, ed. Gunther Franz, 3rd ed. (Stuttgart: Ulmer, 1978): 9; Karl Brunner and Gerhard Jaritz, *Landherr, Bauer, Ackerknecht: Der Bauer im Mittelalter—Klischee und Wirklichkeit* (Vienna: Böhlau, 1985).

5. John Percival, "Field Patterns in the Albertini Tablets," in *The Ancient Historian and His Materials: Essays In Honour of C. E. Stevens on His Seventieth Birthday*, ed. Barbara Levick (Farnsborough, Hants.: Gregg, 1975), pp. 213 ff.; Percival, "Seigneurial Aspects of Late Roman Estate Management," *English Historical Review* 84 (1969): 449–73; Patrick J. Geary, *Aristocracy in Provence: The Rhône Basin at the Dawn of the Carolingian Age* (Philadelphia: University of Pennsylvania Press, 1985); Jean Durliat, "Du caput antique au manse médiéval," *Pallas* 29 (1982): 67–77; David Herlihy, "Three Patterns of Social Mobility in Medieval Society," *Journal of Interdisciplinary History* 3 (1973): 623–47, reprinted in Herlihy, *The Social History of Italy and Western Europe, 700–1500: Collected Studies* (London: Variorum, 1978).

6. Eugippius, *The Life of Saint Severin*, trans. Ludwig Bieler (Washington, D.C.: Catholic University Press of America, 1965); for that and the following, see Herwig Wolfram, *Die Geburt Mitteleuropas: Geschichte Österreichs vor seiner Entstehung, 378–907* (Vienna: Kremayr & Scheriau, 1987), e.g., pp. 55 ff.

7. Karl Brunner, "Nachgrabungen: Sachkultur und Kontinuitätsfragen am Beispiel der bayerischen Quellen des Frühmittelalters," in *Typen der Ethnogenese unter besonderer Berücksichtigung der Bayern* 1, ed. Herwig Wolfram and Walter Pohl, Denkschriften der österreichischen Akademie der Wissenschaften, phil.-hist. Kl. 201, Veröffentlichungen der Kommission für Frühmittelalterforschung 12 (Vienna:

Verlag der österreichischen Akademie der Wissenschaften, 1990) p. 175; Wolfram, *Die Geburt Mitteleuropas*, p. 325, about the *genealogia hominum de Albina*.

8. But cf. now Rosamond McKitterick, *The Carolingians and the Written Word* (Cambridge: Cambridge University Press, 1989); Michael Richter, *The Formation of the Medieval West* (Dublin: Four Courts Press, 1994).

9. Capitulare de villis (CV), *MGH Capitularia regum francorum* 1, ed. Alfred Boretius (Hannover: Hahn, 1883) no. 32, pp. 82 ff. (ca. 800); *Capitulare de villis: Codex Guelferbytanus 254*, ed. Carlrichard Brühl (Stuttgart: Müller & Schindler, 1971), p. 9 and the literature listed there.

10. I would guess late spring as the date in the poem of Walther von der Vogelweide, 94, 11: "Dô der sumer komen was, / und die bluomen dur daz gras / wünneclîchen sprungen"; Walther von der Vogelweide, *Werke*, ed. Joerg Schaefer (Darmstadt: Wissenschaftliche Buchgesellschaft, 1972), no. 39, p. 100.

11. Peter Brown, *Society and the Holy in Late Antiquity* (Berkeley: University of California Press, 1982), pp. 63 ff.; Henri Pirenne, *Mohammed and Charlemagne*, trans. Bernard Miall (London, 1939; reprinted New York: Barnes & Noble, 1955); Alfons Dopsch, *The Economic and Social Foundations of European Civilization* (London: Stephen Austin, 1937); Erna Patzelt, *Die fränkische Kultur und der Islam* (Vienna: Rohrer, 1932; 2nd ed. Aalen: Scientia-Verlag, 1978).

12. Rainer Christlein, *Die Alamannen: Archäologie eines lebendigen Volkes* (Stuttgart: Theiss, 1978); Bernhard Overbeck, *Geschichte des Alpenrheintals in römischer Zeit auf Grund der archäologischen Zeugnisse* 1, Münchner Beiträge zur Vor- und Frühgeschichte 20 (Munich: Beck, 1982); Gudrun Schneider-Schnekenburger, *Churrätien im Frühmittelalter auf Grund der archäologischen Funde*, Münchner Beiträge zur Vor- und Frühgeschichte 26 (Munich: Beck, 1980); Joachim Henning, "Ökonomie und Gesellschaft Rätiens zwischen Antike und Mittelalter," *Klio* 67 (1985): 625–29.

13. Pankraz Fried, "Alemannien und Italien vom 7. bis 10. Jahrhundert. Die transalpinen Verbindungen der Bayern, Alemannen und Franken bis zum 10. Jahrhundert," *Nationes* 6 (Sigmaringen: Thorbecke, 1987): 34 ff.; Wilhelm Störmer, "Zur Frage der Funktion des kirchlichen Fernbesitzes im Gebiet der Ostalpen vom 8. bis zum 10. Jahrhundert," *Nationes* 6 (Sigmaringen: Thorbecke, 1987): 379 ff.; Störmer, "Engen und Pässe in den mittleren Ostalpen und ihre Sicherung im frühen Mittelalter," *Mitteilungen der Geographischen Gesellschaft in München* 53 (1968): 91 ff.; Otto P. Clavadetscher, "Verkehrsorganisation in Rätien zur Karolingerzeit," *Schweizerische Zeitschrift für Geschichte* 5 (1955): 1 ff.

14. *Lex Romana Curiensis*, ed. Elisabeth Meyer-Marthaler, *Die Rechtsquellen des Kantons Graubunden*, Sammlung schweizerischer Rechtsquellen 15, 1 (Aarau: Sauerländer, 1959).

15. Agathias, *The Histories*, trans. Joseph D. Frendo (Berlin and New York: De Gruyter, 1975), I, 6, 3; cf. Ammianus Marcellinus, trans. John C. Rolfe, 3 vols. (Cambridge, Mass.: Harvard University Press, 1950–52), XVI, 12, 26; Reinhard Wenskus, *Stammesbildung und Verfassung: Das Werden der frühmittelalterlichen Gentes*, 2nd ed. (Köln: Böhlau, 1977), e.g., pp. 495 ff.; Dieter Geuenich and Hagen Keller, "Alamannen, Alamannien, Alamannisch im frühen Mittelalter: Möglichkeiten und Schwierigkeiten des Historikers beim Versuch der Eingrenzung," *Die Bayern und*

ihre Nachbarn 1, ed. Herwig Wolfram and Andreas Schwarcz, Denkschriften der österreichischen Akademie der Wissenschaften, phil.-hist. Kl. 179 (Vienna: Verlag der österreichischen Akademie der Wissenschaften, 1985): 135–57.

16. Henning, "Ökonomie und Gesellschaft," p. 627; cf. K. Poczy, "Historische Übersicht," in *Das römische Budapest: Neue Ausgrabungen und Funde in Aquincum* (Lengerich: Kleins, 1986), pp. 65–69; Dieter Geuenich, "Zur Landnahme der Alamannen," *Frühmittelalterliche Studien* 16 (1982): 25–44.

17. Herwig Wolfram and Andreas Schwarcz, eds., *Anerkennung und Integration: Zu den wirtschaftlichen Grundlagen der Völkerwanderungszeit 400–600*, Veröffentlichungen der Kommission für Frühmittelalterforschung 11, Denkschriften der österreichischen Akademie der Wissenschaften, phil.-hist. Kl. 193 (Vienna: Verlag der österreichischen Akademie der Wissenschaften, 1988).

18. Wolfram, *Die Geburt Mitteleuropas*, pp. 331–40.

19. Joachim Henning, *Südosteuropa zwischen Antike und Mittelalter: Archäologische Beiträge zur Landwirtschaft des 1. Jahrtausends u.Z.*, Akademie der Wissenschaft der DDR Zentralinstitut für alte Geschichte und Archäologie, Schriften zur Ur- und Frühgeschichte 42 (Berlin: Akademie Verlag, 1987); Henning, "Zum Problem der Entwicklung materieller Produktivkräfte bei den germanischen Staatsbildungen," *Klio* 68 (1986): 128–38.

20. "Die Urbare der Abtei Werden a.d. Ruhr," *Rheinische Urbare* 2 (Bonn: Gesellschaft für rheinische Geschichtskunde, 1906) IIA. 3, p. 17: arare = proscindere = gibrakon.

21. Henning, "Zur Datierung," pp. 584–85.

22. Ibid., pp. 581 ff.

23. Brunner and Jaritz, *Landherr, Bauer, Ackerknecht*, pp. 64 f.; *manipuli*: Agrestius, v. 22: *Anexia maturis satiet tua vota maniplis* (Kurt Smolak, ed., *Das Gedicht des Bischofs Agrestius*, Sitzungsberichte der österreichischen Akademie der Wissenschaften, phil.-hist. Kl. 284 [Vienna: Verlag der österreichischen Akademie der Wissenschaften, 1973]); cf. "Die Urbare der Abtei Werden," IIA, sec. 3, p. 17; Karl Ilg, "Die Sense in ihrer Entwicklung und Bedeutung," *Festschrift zu Ehren Hermann Wopfner* 2 (Innsbruck: Wagner, 1948): 179 ff.

24. See pictures in Henning, "Zur Datierung," pp. 577 f.; cf. Cato, *De agricultura*, in *Roman Farm Management: The Treatises of Cato and Varro Done into English*, trans. Fairfax Harrison (New York: Macmillan, 1913), 10–11, and Columella, *On Agriculture*, 3 vols. (Cambridge, Mass: Harvard University Press, 1954–60), IV, 25; Herbert Jankuhn, "Archäologische Beobachtungen zur bäuerlichen Lebens- und Wirtschaftsweise im 1. nachchristlichen Jahrtausend," *Wort und Begriff "Bauer,"* ed. Reinhard Wenskus, Herbert Jankuhn, and Klaus Grinda, Abhandlungen der Akademie der Wissenschaften, Göttingen, phil.-hist. Kl. 3.F. 89 (Göttingen: Vandenhoeck & Ruprecht, 1975), p. 40.

25. Cf. the ample supply of various goods in the edict on maximum prices of the emperor Diocletian from the year 301; Siegfried Lauffer, *Diokletians Preisedikt*, Texte und Kommentare 5 (Berlin: De Gruyter, 1971).

26. Oberleis (GB Mistelbach, Lower Austria) by Herwig Friesinger, Vienna.

27. Brunner and Jaritz, *Landherr, Bauer, Ackerknecht*, pp. 38–39.

28. Fritz Moosleitner, "Die Merowingerzeit," in *Geschichte Salzburgs: Stadt*

und Land 1, ed. Heinz Dopsch and Hans Spatzenegger (Salzburg: Universitäts-
verlag Pustet, 1981), pp. 105 ff.; Heinz Dopsch, "Zum Anteil der Römanen und
ihrer Kultur an der Stammesbildung der Bajuwaren," in *Die Bajuwaren: von Severin
bis Tassilo 488–788*, ed. Hermann Dannheimer and Heinz Dopsch (Munich: Prähist.
Staatssammlung, 1988), pp. 47 ff.; and Wolfram, *Die Geburt Mitteleuropas*, pp. 333 ff.

29. Jean Chapelot and Robert Fossier, *The Village and House in the Middle
Ages*, trans. Henry Cleere (London: Batsford, 1985), p. 110; Peter Donat, *Haus, Hof
und Dorf in Mitteleuropa vom 7.–12. Jahrhundert*, Akademie der Wissenschaften der
DDR Zentralinstitut für Alte Geschichte und Archäologie, Schriften zur Ur- und
Frühgeschichte 33 (Berlin: Akademie Verlag, 1980).

30. *Manipulae*: Isidore of Seville, *Etymologiae*, XVII, 9, 107 and Varro, Res
rusticae, in *Roman Farm Management*, I,49; *carradae*: cf., e.g., *Formulae Marculfi*
1: 11, or *Polyptyque de l'abbé Irminon*, ed. B. Guerard, 2 vols. in 3 (Paris: Imprimerie
Royale, 1844) 2: 1 ff. The Romans had taken most of their vocabulary concerning
vessels from the Celts, the only exception being *plaustrum*, a heavy cart-like vessel:
Dieter Timpe, "Das keltische Handwerk im Lichte der antiken Literatur," in *Das
Handwerk in vor- und frühgeschichtlicher Zeit*, ed. Herbert Jankuhn et al., Abhand-
lungen der Akademie der Wissenschaften in Göttingen, phil.-hist. Kl. 3.F., 122, 1
(Göttingen: Vandenhoeck & Ruprecht, 1981): 51–52.

31. At the royal *curtis* in Staffelsee, Bavaria, MGH *Capitularia* 1: 128, c. 8
(ca. 810): *fimat de terra dominica*; "Die Urbare der Abtei Werden," IIA. 40: *cum ster-
cus in agros deducunt*; Ulrich Bentzien in *Jahrbuch für Volkskunde und Kulturgeschichte*
31, NF 16 (1988): 199; cf. Bentzien, *Bauernarbeit im Feudalismus. Landwirtschaftliche
Arbeitsgeräte und -verfahren in Deutschland von der Mitte des ersten Jahrtausends u.Z.
bis um 1800* (Berlin: Akademie Verlag, 1980). Columella I, 6, 21 on the two types of
manure heaps; II, 10, 1–2 and 14, 5 on lupines used as fertilizer.

32. W. Groenman-van Waateringe and L. H. van Wijngaarden-Bakker, eds.,
*Farm Life in a Carolingian Village: A Model Based on Botanical and Zoological Data
from an Excavated Site*, Studies in Prae- en Protohistorie 1 (Assen/Maastricht, Neth.
and Wolfeboro, N.H.: Van Gorcum, 1987); J. P. Pais, "Reconstruction of Landscape and Plant Husbandry," in *Farm Life*, ed. Groenman-van Waateringe and van
Wijngaarden-Bakker, p. 75, about the planting of rye and manuring, both increasing
in the tenth century; and H. A. Heidinga, *Medieval Settlement and Economy North of
the Lower Rhine: Archeology and History of Kootwijk and the Veluwe (the Netherlands)*,
Cingula 9 (Assen/Maastricht, Neth. and Wolfeboro, N.H.: Van Gorcum, 1987);
both works contain a useful compilation of the literature as well. Cf. Jean Cuisenier
and Rémy Guadagnin, *Un Village au temps de Charlemagne: Moines et paysans de
l'abbaye de Saint-Denis du VIIe siècle à l'An Mil* (Paris: Éditions de la Réunion des
Musées Nationaux, 1988).

33. David Herlihy, "The Carolingian Mansus," *Economic History Review* 13
(1960/61): 79–89; Herlihy, "The Agrarian Revolution in Southern France and Italy,
801–1150," *Speculum* 33 (1958): 23–41; and, in an especially clear manner, Percival,
"Seigneurial Aspects," also about *culturae Mancianae* in P. Ital. 3; cf. Percival,
"Ninth-Century Polyptyques and the Villa System: A Reply," *Latomus* 25 (1966):
134–38, all have collected the wider sources: Varro's dialogue on the *Res rusticae*
from the first century B.C. already alludes, according to Herlihy, to the housing

of slaves. The estate of the poet Horace in the Sabine hills was administered by a *villicus*, but there were also, according to Percival, tenant farms let out to a number of *coloni*: Tacitus, *Germania*, c. 25. Columella wrote in the twelve books of his *De re rustica* above all about the lord's demesne. Inscriptions from the second century in Africa tell of a *Lex Hadriana* or *Lex Manciana*, by which coloni were obligated to cultivate abandoned *(rude)* fields: Salvatore Riccobono et al., eds., *Corpus Inscriptionum Latinarum (CIL)*, *Fontes iuris Romani anteiustiniani* (Florence: S. A. G. Barbère, 1941–43), 1: *Leges*, pp. 484–98, no. 100–103 = *CIL* VIII, 25902, 25943, 26416; 10570; cf. 14428. The Albertini Tablets, the private archives of the Geminiani, which likewise originated in Africa around 500, continued to enumerate *culturae Mancianae*: Christian Courtois et al., eds., *Tablettes Albertini: actes privées de l'époque vandale* (Paris: Arts et Métiers Graphiques, 1952). The famous Papyrus Italicus Three from the mid-sixth century presents many characteristics of medieval agrarian organization: Jan-Olof Tjäder, *Die nichtliterarischen lateinischen Papyri Italiens aus der Zeit 445–700* 1 (Lund: Gleerup, 1955): 184–89; see also Peter Classen, "Fortleben und Wandel spätrömischen Urkundenwesens im frühen Mittelalter," in *Recht und Schrift im Mittelalter*, ed. Classen, Vorträge und Forschungen 23 (Sigmaringen: Thorbecke, 1977): 13–54.

34. Otto Gerhard Oexle, "Tria genera hominum: Zur Geschichte eines Deutungsschemas der sozialen Wirklichkeit in Antike und Mittelalter," in *Institutionen, Kultur und Gesellschaft im Mittelalter: Festschrift für Josef Fleckenstein zu seinem 65. Geburtstag*, ed. Lutz Fenske, Werner Rösener, and Thomas Zotz (Sigmaringen: Thorbecke, 1984), pp. 483–500; Oexle, "Deutungsschemata der sozialen Wirklichkeit im frühen und hohen Mittelalter," in *Mentalitäten im Mittelalter: Methodische und inhaltliche Probleme*, ed. František Graus, Vorträge und Forschungen 35 (Sigmaringen: Thorbecke, 1987): 65–117.

35. Cf., e.g., Karl Bosl, "Gesellschaftsentwicklung, 900–1350," in *Handbuch der deutschen Wirtschafts- und Sozialgeschichte* 1, ed. Hermann Aubin and Wolfgang Zorn (Stuttgart: Union Verlag, 1971): 226–73, or Bosl, *Frühformen der Gesellschaft im mittelalterlichen Europa* (Munich: Oldenbourg, 1964).

36. Karl Brunner, "Wovon lebte der Mensch?" in *Die Bajuwaren*, ed. Dannheimer and Dopsch, pp. 192–97.

37. Even Columella points out in *De re rustica* I, 16: *usus et experientia dominantur in artibus.* Compare his contemporary, Manilius, *Astronomica*, trans. G. P. Goold (Cambridge, Mass: Harvard University Press, 1977), I, 61: *per varios usus artem experientia fecit*, and Cicero, *De Divinatione*, in *Cicero, Selections*, trans. William Armistead Falconer (Cambridge, Mass.: Harvard University Press, 1979), II, 146: *observatio diuturna . . . notandis rebus fecit artem.* On classical agrarian writing see also Isidore, *Etymologiae*, XVII.

38. Cf. Max Weber, *The Protestant Ethic and the Spirit of Capitalism*, trans. Talcott Parsons (New York: Charles Scribner's Sons, 1958), p. 118: "Christian asceticism . . . had developed a systematic method of rational conduct with the purpose of overcoming the *status naturae* . . ."; or Weber, "Staat und Hierokratie," in Weber, *Grundriß Sozialökonomik* 3, *Wirtschaft und Gesellschaft*, 2nd ed. (Tübingen: Mohr, 1925): monasteries were "die ersten rational verwalteten Grundherrschaften und, später, Arbeitsgemeinschaften auf landwirtschaftlichem und gewerblichem Ge-

biete"; Barbara Rosenwein, "Reformmönchtum und der Aufstieg Clunys: Webers Bedeutung für die Forschung heute," in *Max Webers Sicht des okzidentalen Christentums: Interpretation und Kritik*, ed. Wolfgang Schluchter (Frankfurt: Suhrkamp, 1988).

39. "Supplex libellus monachorum Fuldensium Carolo imperatori porrectus," c. 16, *Corpus consuetudinum monasticarum* 1 (Siegburg: Schmitt, 1963): 325: *Ut ipsa monasterii ministeria per fratres ordinentur: id est pistrinum, hortus, bratiarium, coquina, agricultura et cetera ministeria.*

40. Heinz Lowe, ed., *Die Iren und Europa im früheren Mittelalter*, 2 vols. (Stuttgart: Klett-Cotta, 1982); Charles Doherty, "Exchange and Trade in Early Medieval Ireland," *Journal of the Royal Society of Antiquaries of Ireland* 110 (1980): 67–89.

41. Statutum Adalhardi. *Corpus consuetudinum monasticarum* 1: 381: *Et unusquisque ad ortum excolendum sive ad alias necessitates explendas fussorios sex* (cf. Brevium Exempla [BE], *MGH Capitularia* 1, no. 128, c.30, p. 255); *bessos duos, secures tres* (Lex Bawariorum XII, 10, *saigam valentem* = 3 den. = value of a young swine); *dolatorium* (Capitulare de villis [CV], *MGH Capitularia* 1, no. 32, c. 42, BE c. 30 et passim; Heinrich Tiefenbacher, "Bezeichnungen für Werkzeuge aus dem Bauhandwerk im Althochdeutschen," *Das Handwerk in vor- und frühgeschichtlicher Zeit* 2: 721: *barta* or *meizel*); *taratra duo maius et minus* (CV c. 42; cf. Isidore, *Etymologiae* XIX, 20, 14; *terebrum* BE 30 et passim, Columella IV. 29, 15, Tiefenbacher, "Bezeichnungen für Werkzeuge," p. 731: *nabager*); *scalprum unum* (CV c. 42, BE c. 30, Isidore XIX. 20, 13 *scalpellus*); *gulbium unum* (Tiefenbacher, "Bezeichnungen für Werkzeuge," p. 728: *nuogil*); *falcilia duo, falcem unam* (cf. *Die Traditionen des Hochstiftes Freising* 1, ed. Theodor Bitterauf, Quellen und Erörterungen zur bayerischen Geschichte N.F. 4 (Munich: Rieger, 1905), n. 652 (842), Inventory of Bergkirchen: *vomerem I et ligonem I falcem maiorem I carras II catenam II;* Herbert Jankuhn, "Archäologische Beobachtungen," p. 40); *truncos duos, cultrum unum* (*cultellum*, Schmidt-Wiegand, "Der 'Bauer' in der Lex Salica," in *Wort und Begriff "Bauer,"* ed. Wenskus, Jankuhn, and Grinda, p. 135, and Pactus Legis Salicae 7.13); *scerum unum* (perhaps *scar* = ploughshare) and the literature on the Statutum there, p. 357, and esp. Josef Semmler and Adriaan Verhulst, "Les statuts d'Adalhard de Corbie de l'an 822," *Moyen Age* 68 (1962): 91–123 and 233–69; cf. on Roman craftsmanship, K. D. White, *Agricultural Implements of the Roman World* (Cambridge: Cambridge University Press, 1967).

42. Walter Goffart, "Merovingian Polyptichs: Reflections on Two Recent Publications," *Francia* 9 (1982): 57–77; Goffart, "Old and New in Merovingian Taxation," *Past and Present* 96 (1982): 3–21; Goffart, "From Roman Taxation to Medieval Seigneurie: Three Notes," *Speculum* 47 (1972), pt. I: 165–87, pt. II: 373–94; Durliat, "Du caput antique au manse médiéval," pp. 67–77.

43. Wahlafrid Strabo, *Liber de cultura hortorum*, based on Columella X; Max Manitius, *Geschichte der lateinischen Literatur des Mittelalters* 1, Handbuch der Altertumswissenschaft Abt. 9, t. 2, Bd. 1 (Munich: Beck, 1911): 309–10. The architect of Aachen and biographer of Charlemagne, Einhard, knew the *De Architectura* of Vitruvius.

Sándor Bökönyi

3. The Development of Stockbreeding and Herding in Medieval Europe

Medieval animal husbandry played an important role in the evolution of domestic animals. On the one hand, these domestic species were, in fact, the descendants of those that survived from Roman times, which had then merged with ancient European stock and another group introduced by the peoples of the Germanic migrations. On the other hand, these species formed a connection with those post-medieval breeds on which the breeding experiments of early modern times were based.

In the first half of this century, our knowledge of domestic animals in the Middle Ages was obtained largely from contemporary descriptions, written documents, and artistic representations. However, these descriptions can only be considered secondary evidence, easily manipulated or misinterpreted. For several reasons, a real change occurred in the field after World War II. In Germany, the Soviet Union, and Hungary, as a result of heavy bombing, whole quarters of cities and towns were destroyed, and thus a green light was given to excavate the early medieval layers lying under the ruins. Also, some countries that regained their earlier territories after foreign domination decided to carry out large-scale excavations to reveal their own past or to find the first traces of their independent states— for example, in western Poland. In addition, the study of rural archaeology began all over Europe, particularly in Denmark and Great Britain, with the aim of reconstructing the early phases of village development. Finally, with the rebuilding of city centers, especially in Great Britain, a great opportunity arose for large-scale excavations, which yielded large samples of archaeological remains, including animal and plant remains. As a result of all these activities, the study of the development of domestic animal species and its connection with animal husbandry in human settlements developed rapidly.

Domestic Species

The development of domestic animals and animal husbandry went through two phases in the Middle Ages. The earlier one lasted from the fifth century through the thirteenth century; the second phase began in the fourteenth century, stretched through the cultural and economic upswing of the Renaissance, and ended at the beginning of the early modern period. In eastern Europe, the second phase lasted somewhat later, until the withdrawal of the Turks to the Balkan peninsula.

During the first phase, animal husbandry was extremely primitive, showing almost no trace of the Roman practices that had once been so highly developed. In this phase, the size of domestic animals decreased to a point where they were smaller than in the early prehistoric periods (Neolithic, Copper Age, Bronze Age); in fact, the situation was comparable only to the Iron Age, which also saw a significant decrease in the size of domestic animals. The best demonstration can be seen in the curve showing the change in the size of cattle from the Neolithic period through modern times (Figure 2). The improved, highly productive Roman breeds disappeared completely along with deliberate, selective breeding and well-determined breeds. The only exceptions were horses and dogs, both of which were owned only by a privileged, higher social class and had a special place not just in economic life but also in warfare and in leisure pursuits.

The dwarfing of cattle—in comparison with their size during the time of the Roman Empire—was really striking, sometimes reaching a decrease of 20 cm in the withers height. Early breeding practices also may have been a factor in the degeneration as well as the early use of animals as draft power.[1] Nevertheless, this dwarfing was not simply the result of the disappearance of selective breeding, for it occurred most strongly in eastern Europe and Scandinavia, territories that had never been under Roman rule. Thus other factors (for example, climatic deterioration or changing lifestyles) must have played some role. It must be remembered that cattle populations were devastated by frequent wars, because the chief meat consumed by the armies was beef ("Beefeaters"), and what the soldiers did not eat they drove off.

Nevertheless, the effects of selective breeding introduced by the Romans in their colonies lasted long after the fall of the Roman Empire, and although it could no longer save the Roman breeds, it kept medieval cattle stock on a slightly higher level than in the neighboring territories. While the average withers height of cattle was between 95.5 and 102 cm in medi-

withers height in cm

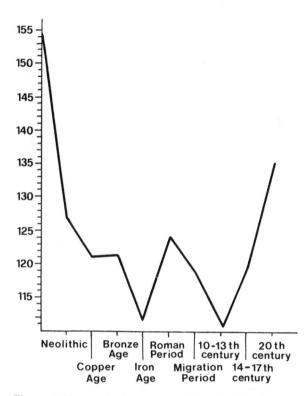

Figure 2. Changes in the average withers height of cattle since the Neolithic period. After Sándor Bökönyi, *History of Domestic Mammals* fig. 9.

eval Russia and less than 110 cm in Poland and non-Roman Germany, cattle on the average taller than 110 cm could be found at early medieval sites in Switzerland, France, and western Hungary and in other former parts of the Roman Empire.[2]

Early medieval cattle, according to the osteological and artistic evidence, were quite uniform all over Europe. They were small, slenderly built, primitive animals with long legs; the trunk was not very deep and had little flesh; and they had a *longifrons* skull with short horns (Figure 3).[3] Hornless (polled) animals are not known from this period, but some longhorns occurred sporadically, mainly in eastern Europe.[4] In this period there is no evidence of the Hungarian gray steppe cattle, a typical longhorn

Figure 3. Representation of small short-horned cattle in a scene showing the Hungarian Conquest. Hungarian Illuminated Chronicle, early fourteenth century. Budapest, National Széchényi Library. After D. Dercsényi et al., eds., *Chronicon pictum* (Budapest, 1964), p. 21.

breed.[5] In fact, the small, short-horned cattle of the early Middle Ages were not an independent breed, nor were they a geographic variant or a breed characteristic of a people or of an ethnic group. They were simply an indication of the very rudimentary conditions of early medieval animal husbandry.[6] As a matter of fact, this type of cattle is still found in many areas of the Balkan Peninsula, Anatolia, and the Near and Middle East, demonstrating that in these regions animal husbandry has remained on the same level as in the early Middle Ages (Figure 4).

Similarly, a rather uniform sheep population lived in early medieval Europe from the Urals eastward to the British Isles. These sheep were about 60 cm high at the withers, about the same size as the German "Heidschnucken."[7] The males had heavy, triangular, twisted horns; the horns of the females were shorter, untwisted, or completely missing. Such sheep were found in Russia and the Ukraine, Bulgaria, Slovakia, Switzerland, Germany, Sweden, and Great Britain.[8] These sheep had, in fact, been living in Europe since the beginning of the Neolithic period and thus can be considered the old European type.

Figure 4. Team of small cows of the medieval type from Bulgaria. The withers of the cows hardly reach the boy's waist. Bökönyi, *History of Domestic Mammals*, fig. 32. By permission of Akadémiai Kiadó.

Such sheep existed in early medieval Hungary, too.[9] Nevertheless, evidence of another type, the so-called "Zackel" group, has been found in the Carpathian Basin.[10] This type probably originated in the southern regions of the Near East including Egypt and the northern Sudan.[11] The "Zackel" have corkscrew-like twisted horns protruding horizontally. Both sexes have horns, although the females are sometimes hornless (Figures 5 and 6). In all probability, they were introduced to central Europe by the Magyars or the Avars; that another route existed via Anatolia and the Balkans is proved both by osteological finds and by artistic evidence (Figure 7). Nevertheless, the "Zackel," with horns standing up in a V shape (Figure 8) and supposed to be the original, ancient sheep breed of the Hungarians,[12] did not exist in the early Middle Ages. The earliest evidence for their appearance dates from the sixteenth and seventeenth centuries.[13] In fact, the original form with horizontal horns still existed around the end of the nineteenth century under the name "Hungarian sheep," while the form with V-shaped horns was called "racka."[14]

Goats were much rarer than sheep in medieval Europe, and different breeds cannot be distinguished among them.[15] Most of them had twisted

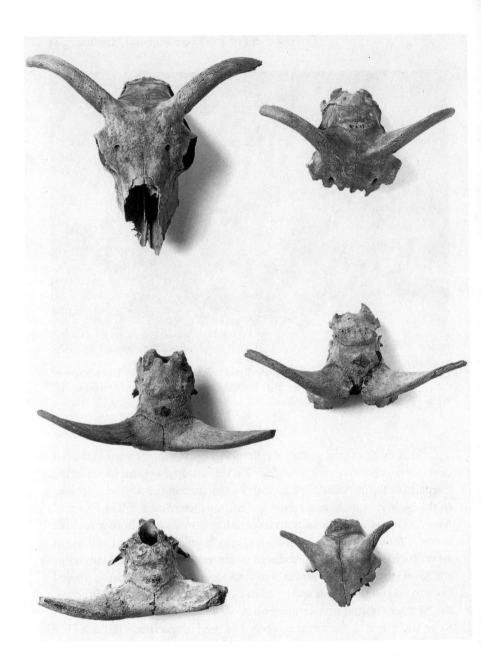

Figure 5. Skull fragments of typical "Zackel" sheep. Skolnok Castle, Hungary, sixteenth and seventeenth centuries. Bökönyi, *History of Domestic Mammals*, fig. 60.
By permission of Akadémiai Kiadó.

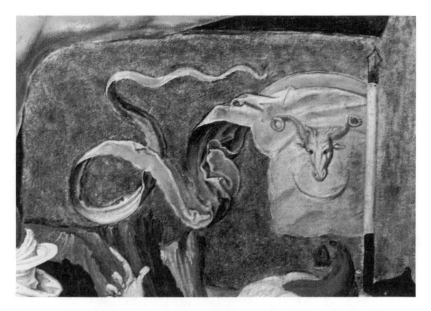

Figure 6. Head and horns of a typical medieval "Hungarian" sheep. Detail of the painting *Calvary*, Master MS., Christian Museum, Esztergom. Bökönyi, *History of Domestic Mammals*, fig. 65. By permission of Akadémiai Kiadó.

Figure 7. Representation of a "Zackel" ram on a Turkish copper bowl from the sixteenth or seventeenth century. Photo: Kálmán Kónya.

Figure 8. Modern "racka" sheep. Note the characteristic V form of the horns. Bökönyi, *History of Domestic Mammals*, fig. 63. By permission of Akadémiai Kiadó.

horns. Hornless goats were considered rarities; the skull of such a goat was found in the Turkish pasha's palace in Buda Castle.[16]

Early medieval pigs, as shown in artistic representations, were small, long-legged, primitive animals. In most cases, the form of the skull is similar to that of the wild boar, probably because pigs were allowed to roam freely and often had the opportunity of crossbreeding with their wild forms (Figure 9). Nevertheless, some pigs were highly domesticated. These were characterized by an extreme shortening of the skull, which is connected with abnormalities in the row of teeth and also with a concave profile.[17]

Among horses the variability was larger than in the domestic species discussed earlier. As was mentioned earlier, there were lasting effects from the selective animal breeding practiced by the Romans. A second influence was the arrival in Europe of large numbers of horses from the East.[18] Finally, one should take into account the effects of the different types of environment, which heavily influenced the development of local horse populations.

The typical early medieval European horse preserved the appearance and size of the eastern strain that reached central Europe first with the

Figure 9. Pigs of the medieval type from southeastern Bulgaria. Note the long muzzle and the hairy crest along the spine. Bökönyi, *History of Domestic Mammals*, fig. 72. By permission of Akadémiai Kiadó.

Scythians and again during the Germanic migrations. In fact, the improved Roman horse basically represented the same type, though some of them were certainly larger and slightly heavier.[19] These horses had an average withers height of 136–37 cm. They had slender legs and a light trunk just as Asian horses have today. In northern and western Europe, the horses were somewhat smaller. The first coldblood (heavy) horses appeared in central Europe (Figure 10). Their wild ancestors were no different from early medieval horses; they were the results of selective breeding carried out to satisfy the need for a strong horse that could carry a knight in heavy armor. The first appearance of these horses seemed to be connected with that of the horseshoe, because their wide, soft hooves needed protection on rocky soil. Although such heavy horses had been bred once before (again, for riders in armor), in the Median Empire, around 500 B.C.,[20] they did not appear afterward in Europe or elsewhere. In fact, coldblood horse breeding started independently in central Europe. In Britain, such heavy horses (the great English war horse) emerged at about the time of the Norman Conquest and may have been related to Scandinavian horses.[21] These horses were very strong even though they were not large: none of them had a withers height exceeding 15 hands (152 cm).

Figure 10. Representation of coldblood horses. Limbourg Frères, *Parade of Hunters*, 1410–16. Chantilly, Musée Condé.

The ass, according to the most recent information, reached Europe via Greece and Italy at the beginning of the second millennium B.C.[22] The ass was found to some extent in the Roman Empire, although, since it is a warmth-loving animal, its distribution in Europe was limited by the cold climate of the northern part of the continent. After the fall of the Roman Empire the ass practically disappeared from Europe and occurred there again only after about 1000 A.D. Most medieval asses were quite small and were mainly used by religious houses and mills as well as on the farms of smallholders.

Mules were even rarer than asses in medieval Europe. Their southern European origin was well known at that time; for example, a Russian record from 1016 mentions a mule imported from Italy. Mules were used as pack animals or by persons who were not skillful riders, such as church dignitaries.

Several breeds of dogs considered luxury animals could be found in the early Middle Ages, pointing to the fact that this fashion of the privileged elite continued after the fall of the Roman Empire. There is evidence of dwarf dachshunds and pomeranians in Opole and Wrocław, in Poland,[23] while greyhounds are represented by skulls[24] and turn up by name in a Hungarian document from 1193.[25] The *kuvasz* was introduced as a sheepdog from the grassy steppe belt of western Asia.[26] Still, the great majority of dogs lacked the characteristics of particular breeds.

Even in the early Middle Ages there were in Europe some foreign domestic species, for example, camels, water buffalo, and probably reintroduced cats. Camels first appeared between the fifth and third centuries B.C. in the south Russian steppes. From there they spread to the Greek colonial towns on the northern coastal region of the Black Sea.[27] They seem to have survived there and moved westward, reaching Romania.[28] At the time of the Roman Empire camels became widespread in Europe, arriving with the eastern military units.[29] After the fall of the Roman Empire, they disappeared but were reintroduced during the Germanic migrations. A pictorial representation of the camel is known from the fourteenth-century Hungarian Illuminated Chronicle (Figure 11). Their rare presence is confirmed by bones found in Hungary and written evidence from western Europe. At the end of the Middle Ages, the Turks brought camels to Europe in larger numbers.

The first water buffalo were brought to Europe (through southern Russia) by the Avars in the sixth century A.D. and by the Bulgarians in the seventh century. It is highly probable that the Hungarians also introduced large numbers of water buffalo to Europe at the end of the ninth century. Nevertheless, the Carpathian Basin was their northernmost area of distribution (water buffalo are still found there); in the western and northern parts of Europe they occurred only as curiosities.

The cat had already been introduced into Europe by the Greeks and particularly by the Romans, although after the fall of the Roman Empire cats either completely disappeared or significantly diminished in numbers.

Figure 11. Representation of camels, in a scene showing the
arrival of the Huns in the Carpathian Basin. Hungarian Illuminated Chronicle, early fourteenth century. Budapest, National
Széchényi Library. After D. Dercsényi et al., eds., *Chronicon
pictum* (Budapest, 1964), p. 7.

They appeared again in central and eastern Europe in the early Middle
Ages, though mainly in religious houses and towns. Their numbers increased only after the fourteenth and fifteenth centuries.

   As for poultry, the chicken was the most common and important bird,
particularly in towns, because it needed little space and a limited amount of
food. Its chief value lay in its eggs, which served as a source of food without the bird's needing to be killed. At the beginning of the Middle Ages
chickens were not numerous and they were small in size, weighing between

1.0 and 1.5 kg. Nevertheless, by the end of the Middle Ages the chicken had attained a leading role among the domestic animals. Other species—for example, geese, ducks, and pigeons—were less common although the numbers of some species, like the mallard duck, could be increased by local domestication. There were also semi-domesticated birds: the twelfth-century *Russkaya Pravda* mentions cranes and swans in the poultry yards of nobles,[30] while in Hungary cranes and herons were often kept in medieval castles and on manors.[31]

With the arrival of the Renaissance, important changes occurred in animal husbandry in Europe. With the reintroduction of conscious animal husbandry (based also on classical sources), animals grew in size, the productivity and quality of domestic animals increased, new breeds were developed and attained considerable economic importance, and new, foreign domestic animal species appeared on the scene.

As for the increase in size, one should look especially at cattle, whose average withers height grew by at least 20 cm. The same trend is found in other domestic species, except probably in Britain, where there is little evidence of any change in the size of domestic animals until after the Middle Ages.[32] In connection with this, there is some evidence that, in some respects, the productivity of animals also grew. For example, while in the early Middle Ages a team of eight oxen was needed to break virgin land, and ten or twelve oxen to pull a plough, in the late Middle Ages four to six oxen were enough for a team even though the ploughs themselves were larger. The quality of wool produced by sheep undoubtedly improved, although we have only limited direct evidence of this. This also was the period of the real development and large scale use of the coldblood horse in warfare.

Milk production by cows in the early Middle Ages has been estimated by Trow-Smith to be about 100 gallons (455 l) a year during a lactation period of 150 days.[33] This yield probably increased during the late Middle Ages, because the same author cites eighteenth-century records that indicate that an average grazing herd produced 450–500 gallons (2050–2275 l) a year and that better fed cows yielded up to 650–800 gallons (2960–3640 l).[34]

Among the new breeds were many varieties of luxury dogs, certain sheep breeds, and improved pig breeds with shorter heads, broken profiles, and better body proportions. But the most important new breed occurred among cattle, namely, the Hungarian gray steppe cattle (Figure 12). This breed first appeared in the fourteenth and fifteenth centuries; before that there were only short-horned, small cattle in the Carpathian Basin.[35] The

Figure 12. Four-ox team of Hungarian gray steppe cattle. Author's photo.

gray steppe cattle quickly became a standard breed not just in Hungary but also in neighboring territories (although the ancient small, short-horned cattle co-existed with them for a long time). The Hungarian cattle had large bodies covered with gray hair (but with black pigment in their skin) and long, vertical white horns with black tips (they may have had some zebu, or humped cattle, blood). They could be driven long distances and, because of their excellent meat, conquered western markets very quickly. The first significant export of cattle from Hungary started with the large scale production of this new breed. The earliest written documentation of this is a decree of King Matthias Corvinus, from 1470, in which the king forbade the export of cattle to Venice before his own cattle had been sold there. From this time until the end of the seventeenth century hundreds of thousands of gray Hungarian cattle were driven to western European markets every year. These cattle were also excellent draft oxen and were used for this purpose until World War II, when most of them perished and the rest were displaced by the use of tractors.

Finally, some of the modern sheep breeds of Britain have their origin in the late Middle Ages, and the German "Heidschnucken" also may go back to this period. In Hungary, at least five sheep breeds appeared at the end or just after the end of the Middle Ages, among them plate-tailed or fat-tailed sheep, which certainly had a Turkish origin.[36] From Spain, the first officially imported merinos arrived in France in 1786, in Saxony in 1765, in Prussia in 1785, and in Lower Austria, Moravia, and Hungary between 1780 and 1790.[37] Thus merinos appeared at least in the best-developed sheep-breeding centers of western and central Europe before the end of the eighteenth century.

Among new animal species the domestic rabbit was the most important. Unborn or newly born rabbits, called *laurices*, had been part of the Roman diet. This habit was taken over by monks in the early Middle Ages, who consumed the fetuses or newborn while they were still blind, during Lent, as Gregory of Tours related with disapproval around 590.[38] In fact their flesh, like that of beaver, was not considered "meat." Cloisters were ideal places to keep rabbits because they could not escape by digging under the deep foundations of the walls. Rabbits reached eastern Europe via France and Germany during the Renaissance. In addition, there is evidence also of peafowl, guinea fowl, and even ferrets.

Animal Husbandry

Animal husbandry in the Middle Ages served three main purposes: it provided meat, fat, and other foodstuffs of animal origin (milk, eggs, honey, etc.); it furnished raw materials (skins, furs, wool, feathers, horns, sinews, etc.); and it produced work animals (draft animals for agriculture and transport, riding animals, and dogs for different purposes).

In central, western, and northwestern Europe, pigs and cattle were the main sources of meat until about 1000. After that, pigs and cattle remained popular until the Renaissance only in central Europe, while in the western and northwestern regions sheep became the main meat animals. The reason was the survival in central Europe of the great forests, which were ideal grazing areas for swine, whereas in western and northwestern Europe there was heavy deforestation. In Hungary, as well as in the southern steppes of eastern Europe, cattle, horses, sheep, and goats provided the bulk of the meat consumed. At the same time, in the northern half of eastern Europe, cattle and pigs were the main food-producing domestic species.

Compared with environment and climate, ethnicity per se does not play as important a role in food habits as do lifestyles. Settled peoples usually ate swine, animals that could not easily be driven but could be fed in towns, with kitchen and table scraps, or could find food in forests during most of the year. Nomadic peoples, on the other hand, ate animals that could be easily driven—primarily cattle, sheep, and horses.

In eastern Europe—in contrast to western Europe (see below)—hunting was sometimes of considerable importance, particularly in settlements located in large, swampy forests. Since the bones of wild and domesticated animals have observable differences, the osteological evidence is important. For several sites for which this evidence has been analyzed, the ratio of hunted animals to domesticated animals was sometimes more than 50 percent, for example, in Grodno between the twelfth and sixteenth centuries[39] and in Voronezh in the ninth and tenth centuries.[40] At Grodno, 98 percent of the wild animals were ungulates hunted for their flesh (red deer, bison, wild swine, roe deer, and elk), and in one of the Voronezh settlements (Kuznetsova dacha) four ungulate species (roe deer, wild swine, elk, and red deer) constituted more than 70 percent of the hunted animals, with the remainder consisting of fur-bearing animals. However, in another settlement of the Voronezh area (Borshevo I) the same ungulates and the saiga antelope made up only 35 percent of the wild animals; the remainder—with the exception of the hedgehog and the vole—consisted of fur-bearing animals, with the valuable beaver constituting nearly 45 percent of the total.

In the early medieval settlements of Latvia, in similar environmental surroundings, wild animals also were frequently found, but their numerical ratio compared with domestic animals never reached 50 percent. As at Borshevo I, fur-bearing animals, in particular, beaver, were predominantly hunted. Thus at Grodno and Kuznetsova dacha the goal of hunting was to provide food; at Borshevo I and in the settlements of Latvia it was to acquire furs.

In eastern Europe pigs and cattle were by far the most common domestic species.[41] In the northern part of medieval Russia (Staraya Ladoga, Kamno, Pskov) pigs were the most frequently found, followed by cattle, sheep, and goats, with horses lagging far behind. In northeastern Russia (Suzdal, Staraya Rjazan) cattle were more common than pigs, and sheep and goats were very rare. In the so-called Tchornaya (Black) Russia (as seen at Grodno), cattle and pigs constituted the bulk of domestic animals; there, too, sheep, goats, and horses were seldom found. On the other hand, in the south, in the region of Kiev and Volin, the raising of sheep and goats took on greater importance. One reason was the drier climate in this region, but another was the fact that numerous peoples or ethnic groups to a greater or lesser extent of Turkish origin had given up their nomadic way of life and settled down in the area. These same peoples were responsible for the fact that the keeping of horses was more important in this region than in the central and northern territories. Up to the fourteenth and fifteenth centuries the raising of domestic poultry was not common anywhere in these eastern territories.

In early medieval Europe everyone was allowed to hunt. After about 1000, hunting became a privilege, the pastime of free people, who enforced laws forbidding serfs to hunt at all. (Fishing, on the other hand, never became a privilege restricted to the nobility, and on or near seacoasts or along rivers fish was a staple food both in peasant villages and in towns.) Later, hunting became a privilege of kings and nobles. In Scotland, for example, King David I established the first hunting preserves in the 1130s, reserving all greater game (red deer, roe deer, and wild swine) to himself. The first forest law was promulgated by the middle of the fourteenth century. In Hungary, this happened much later, in 1504, although the existence of an earlier regulation can be inferred from the fact that King Stephen V gave special permission in 1262 to the inhabitants of Szőlős in northeastern Hungary and in 1272 to the Saxon settlers in Ugocsa County (in the same area) to hunt chamois, wild swine, red deer, and brown bear.[42] The situation may have been very similar in central and western Europe, because in the German peasant insurrection of 1524–25 one of the main demands of the peasants was the free right of hunting.

There was considerable variation in the meat consumed by different social classes. The diet of the upper classes was certainly rich and varied. Besides a lot of venison, meat of newly introduced species—rabbit, peacock, guinea fowl, and perhaps turkey—often appeared on the aristocrat's table. Also common was domestic fowl. The diet of town dwellers was simpler and did not contain the delicacies mentioned above. The lower classes in the suburbs consumed meat parts of lower value, as is proved by the more common occurrence at those sites of animal skulls and foot bones.

In Europe the earliest food taboo dates from the end of the Bronze Age, when the consumption of dog meat ceased.[43] After the widespread conversion to Christianity a second prohibition developed regarding horse meat. In the early Middle Ages most European peoples gave up eating horse meat; the Irish and the Hungarians were the only exceptions. Hungarians, in fact, were secretly eating it in remote settlements, far from royal and religious centers, up to the fifteenth century. In certain villages horse meat made up more than half of all the meat consumed.[44]

Professor Bökönyi unfortunately died while this book was still in preparation. The editor wishes to thank Dr. László Bartosiewicz and Dr. Alice M. Choyke of the Archaeological Institute of the Hungarian Academy of Sciences for their assistance in resolving questions about the manuscript.

Notes

1. Sándor Bökönyi, *History of Domestic Mammals in Central and Eastern Europe*, trans. Lili Halápy (Budapest: Akadémiai Kiadó, 1974), p. 136.
2. Ibid.; Sándor Bökönyi, "Animals, Draft" and "Animals, Food," in *Dictionary of the Middle Ages* 1, ed. Joseph R. Strayer (New York: Charles Scribner's Sons, 1982): 293–98 and 299–302.
3. Sándor Bökönyi, "Die Haustiere in Ungarn im Mittelalter auf Grund der Knochenfunde," in *Viehzucht und Hirtenleben in Ostmitteleuropa*, ed. László Földes (Budapest: Akadé, 1961), pp. 83–111, esp. 86 ff.; Bökönyi, *History of Domestic Mammals*, pp. 136 ff.
4. János Matolcsi, *Állattartás őseink korában* [Animal husbandry in the time of our ancestors] (Budapest: Gondolat, 1982), pp. 216 ff.
5. Bökönyi, "Die Haustiere," pp. 87 ff.; Bökönyi, "Die Entwicklung der mittelalterlichen Haustierfauna Ungarns," *Zeitschrift für Tierzüchtung und Züchtungsbiologie* 77 (1962): 1–15, esp. 3–4; Bökönyi, *History of Domestic Mammals*, pp. 141 ff.
6. Bökönyi, *History of Domestic Mammals*, p. 136.
7. Wolf Herre, "Haustiere im mittelalterlichen Hamburg: Untersuchungen über die Tierknochenfunde in der Kleinen Bäckerstrasse," *Hammaburg* 2 (1950): 7–19, esp. 14.
8. On Russia and the Ukraine, see V. I. Zalkin, "Materialy dlja istorii skotovodstva i okhoty v Drevnej Russi," *Materialy i Issledovanija po Arkheologii* (Moscow) 51 (1956): 109 ff.; Matolcsi, *Állattartás őseink korában*, p. 219. On Bulgaria, see S. Ivanov, "Domašnite i divite životni ot gradiščeto kraj s. Popina, Silistrensko" [Domestic animals and wild animals from a site near Popina in the Silestra region], in Ž. Vožarova, *Slaviano-bulgarskoto selište kraj selo Popina, Silistrensko* (Sofia, 1956), pp. 69–95, esp. 94. On Slovakia, see C. Ambros, "Zvieracie zvysky z Bešeňova a Nitrianskeho Hráduku: Tierknochen aus Bešeňov und Nitrianky Hrádok, Bez. Šurany," *Slovenska Archeologia* (Bratislava) 6(2) (1958): 414–18. On Switzerland, see F. E. Würgler, "Beitrag zur Kenntnis der mittelalterlichen Fauna der Schweiz," *Berichte/Jahrbuch der St. Gallischen Naturwissenschaftlichen Gesellschaft* 75 (1956): 1–89, esp. 73; H. Hartmann-Frick, "Die Knochenfunde der Burg Heitnau," *Thurgauische Beiträge zur vaterländischen Geschichte des Historischen Vereins der Kanton Thurgau* 93 (1957): 53–73, esp. 69. On Germany, see Herre, "Haustiere im mittel-

alterlichen Hamburg," p. 114; H. Requate, "Die Jagdtiere unter den Nahrungs-resten einiger frühgeschichtlicher Siedlungen in Schleswig-Holstein: Ein Beitrag zur Faunengeschichte des Landes," *Schriften des Naturwissenschaftlichen Vereins für Schleswig-Holstein* 28 (1956): 21–41, esp. 28; H. H. Müller, "Die Tierreste von Alt-Hannover," *Hannoversche Geschichtsblätter*, N.F. 12 (1959): 181–259, esp. 234; Otto Gehl, *Gross Raden: Haustiere und Jagdwild der slawischen Siedler*, Beiträge zur Ur- und Frühgeschichte der Bezirke Rostock, Schwerin und Neubrandenburg 13 (Berlin: Akademie Verlag, 1981): 39. On Sweden, see Harry Bergquist and Johannes Lepiksaar, "Medieval Skeletal Remains from Medieval Lund," *Archaeology of Lund: Studies in the Lund Excavation Material* 1 (Lund: Museum of Cultural History, 1957): 11–84, esp. 32. And on Great Britain, see M. L. Ryder, "The History of Sheep Breeds in Britain," *Agricultural History Review* 12(1) (1964): 1–12; (2): 65–82; Ryder, "Livestock," in *The Agrarian History of England and Wales* 1: *Prehistory*, ed. Stuart Piggott (Cambridge: Cambridge University Press, 1981): 301–410, esp. 364 ff.; R. Trow-Smith, *A History of British Livestock Husbandry to 1700* (London: Routledge and Paul, 1957), p. 87.

9. Bökönyi, "Die Haustiere," p. 93; Bökönyi, "Die Entwicklung der mittel-alterlichen Haustierfauna Ungarns," p. 6; Bökönyi, *History of Domestic Mammals*, p. 181.

10. Bökönyi, "Die Haustiere," pp. 93 ff.; Bökönyi, "Die Entwicklung der mittelalterlichen Haustierfauna Ungarns," pp. 7 f.; Bökönyi, *History of Domestic Mammals*, pp. 181 ff.

11. J. U. Duerst and C. Gaillard, "Studien über die Geschichte des aegyp-tischen Hausschafes," *Recueil des travaux relatifs à la philologie et à l'archéologie égyptiennes et assyriennes* 24 (1902): 44 ff.; Joachim Boessneck, *Die Tierwelt des alten Ägypten* (Munich: Beck, 1988), fig. 1, p. 121.

12. B. Hankó, "A magyar juh" [Hungarian sheep], *Természettudományi Közlöny* (1937): 1–9; Hankó, *A magyar háziállatok története* [The history of Hungarian domestic animals] (Budapest: Művelt Nép Könyvkiadó, 1954), pp. 16 ff.

13. Bökönyi, *History of Domestic Mammals*, p. 181.

14. Istvan Balogh, "Pusztai legeltetési rend Debrecenben a XVIII–XIX. száza-ban: Die Weidegerechtigkeit auf den Pussten bei Debrecen im XVIII–IX. Jahr-hundert," *Ethnographia* (Budapest) 69 (1958): 537–66, esp. 553.

15. Bökönyi, "Die Haustiere," p. 97; Bökönyi, *History of Domestic Mammals*, pp. 198 ff.

16. Bökönyi, *History of Domestic Mammals*, fig. 71, p. 200.

17. Bökönyi, "Die Haustiere," table VI, pp. 2–4; Bökönyi, "Die Entwicklung der mittelalterlichen Haustierfauna Ungarns," fig. 10, p. 10; Bökönyi, *History of Domestic Mammals*, fig. 86, p. 222.

18. Sándor Bökönyi, "The Earliest Waves of Domestic Horses in East Eu-rope," *Journal of Indo-European Studies* 6 (1–2) (1978): 17–76, esp. 42 ff.

19. Max Hilzheimer, "Die im Saalburgmuseum aufbewahrten Tierreste aus römischer Zeit," *Saalburg-Jahrbuch* 5 (1924): 105–58, esp. 151; K. H. Habermehl, "Die Tierknochenfunde im römischen Lagerdorf Butzbach," *Saalburg-Jahrbuch* 16 (1957): 67–108, esp. 105; Sándor Bökönyi, "Horse," in *Evolution of Domesticated Animals*, ed. Ian L. Mason (London: Longman, 1984), pp. 162–73.

20. Bökönyi, "Horse," p. 170.

21. Ryder, "Livestock," p. 401.

22. Sándor Bökönyi and G. Siracusano, "Reperti faunistici dell'Età del Bronzo del sito di Coppa Nevigata: un commento preliminare," in *Coppa Nevigata e il suo territorio: Testimonianze archeologiche dal VII al II millennio a.C.*, ed. S. M. Cassano et al. (Rome: Quasar, 1987), pp. 205–10.

23. P. Wyrost, "Badania nad typami psow wczesnosredniowiecznego Opola i Wrocźavia" [Studies on the types of early medieval dogs in Opole and Wrocław], *Silesia Antiqua* 9 (1963): 231–55, esp. 232 ff.

24. Sándor Bökönyi, "Rapport préliminaire sur l'examen des ossements au cours des fouilles de Zalavár," *Acta Archaeologica Hungarica* 4 (1954): 281–86.

25. Hankó, *A magyar háziállatok története*, p. 15.

26. J. Matolcsi and K. Sági, "Épitőáldozat vagy szellemi házőrző?" [Building sacrifice or spiritual watchdog?] *Élet és Tudomany* (Budapest) 38(23) (1983): 716–17.

27. V. I. Zalkin, "Domašnie i dikie životnye Severnogo Pričernomorja v epokhu rannego železa," *Materialy i Issledovanija po Arkheologii* (Moscow) 53 (1960): 50; I. G. Pidopličko, *Materiali do vivćennia minulikh faun URSR* 2 (Kiev, 1956): 92.

28. G. Gheorghiu and S. Haimovici, "Caracteristicile mamiferelor domestice descoperite in asezarea feodala timpurie de la Garvan [Dinogetia]," *Analele Stiintifice ale Universitatii Al .I. Cuza din Iasi* (Jassy) 11 (1965): 175–84, esp. 181.

29. C. Keller, *Geschichte der schweizerischen Haustierwelt* (Frauenfeld, 1919), p. 42; M. Berger and E. Thenius, "Über römerzeitliche Kamelfunde im Stadtgebiet von Wien," *Veröffentlungen des Historischer Museums der Stadt Wien* (1951), 20–22; Sándor Bökönyi, "Camel Sacrifices in Intercisa," *Acta Archaeologica Hungarica* 41 (1989): 399–404.

30. V. P. Lebasheva, *Očerki po istorii russkoj derevni X–XIII vv.* (Moscow, 1956), p. 159.

31. Sándor Takáts, *Macaristan Turk aleminden cizgiler: Rajzok a török világból* [Sketches from Turkish times] (Budapest: Genius, 1917), pp. 67 ff.

32. M. L. Ryder, "Fleece Structure in Some Native and Unimproved Breeds of Sheep," *Zeitschrift für Tierzüchtung und Züchtungsbiologie* 85(2) (1968): 143–70, esp. 160.

33. Trow-Smith, *History of British Livestock Husbandry*, p. 58.

34. Ibid.

35. See the osteological and pictorial evidence in Bökönyi, "Die Haustiere," p. 87; Bökönyi, "Die Entwicklung der mittelalterlichen Haustierfauna Ungarns," pp. 3 ff.; Bökönyi, *History of Domestic Mammals*, pp. 139 ff.; Matolcsi, *Állattartás őseink korában*, pp. 257 f.

36. Bökönyi, *History of Domestic Mammals*, p. 188.

37. B. Denis, "Le peuplement ovin de la France septentrionale avant l'introduction des Mérinos," in *L'Homme, l'animal domestique et l'environnement du moyen âge au XVIIIe siècle*, ed. Robert Durand, *Enquêtes et Documents* 19 (Nantes: Centre de Recherches sur l'Histoire de la France Atlantique, 1993): 177–192; László Gaál, *A magyar állattenyésztés multja* [Hungarian animal husbandry of the past] (Budapest: Akadémiai Kiadó, 1966); Attila Paládi-Kovács, *A magyarországi állattartó kultura*

korszakai [The historical periods of Hungarian animal husbandry] (Budapest: Akadémiai Kiadó, 1993).

38. H. Nachtsheim, *Vom Wildtier zum Haustier* (Berlin, 1949).

39. Zalkin, "Materialy dlja istorii skotovodstva," pp. 178–79.

40. V. Gromova, "Ostatki mlekopitayuščikh iz ranneslavianskikh gorodišč g. Voronezha," *Materialy i Issledovanija po Arkheologii* 8 (1948): 113–23, esp. 122–23.

41. Zalkin, "Materialy dlja istorii skotovodstva," pp. 143 ff; Lebasheva, *Očerki po istorii russkoj*, pp. 178–79.

42. A. Komáromy, "Ugocsa vármegye keletkezése" [The development of Ugocsa County], *Értekezések a Történeti tudományok Köréböl* (Budapest) 16 (1897): 24–26.

43. Bökönyi, *History of Domestic Mammals*, p. 320.

44. Ibid., p. 40.

Andrew M. Watson

4. Arab and European Agriculture in the Middle Ages: A Case of Restricted Diffusion

This is the tale of two worlds that came in contact but did not, to any significant degree, interact. More precisely, it is the story of two distinct agricultural systems—that of the early Islamic world and that of early medieval Europe—which had good opportunities to learn from each other but in fact learned little. Especially it is the story of the failure or relative failure of European agriculture, which by almost any measure was the more "backward," to learn from the more "advanced" agriculture of the Arabs.

I have argued at length elsewhere that the early centuries of Islam witnessed an important agricultural revolution that began in the eastern reaches of the newly formed Islamic world—in Iraq and Iran, the heartland of the former Sassanian Empire—and then moved westward into the Near East, across North Africa, and into Muslim Spain.[1] Central to the revolution were a large number of new crops. Those that I have been able to study in detail include several important cereals, namely rice, sorghum, and hard wheat; sugar cane, which was quickly to become the main source of sweetening; one fiber crop, cotton, which, perhaps more slowly, displaced linen and wool as the principal textile fiber; and a rather large number of fruits and vegetables, including sour oranges, lemons, limes, bananas, watermelons, spinach, artichokes, colocasia, and eggplants. In addition to these useful plants that I have studied, there were certainly many others whose introduction and transmission I have not been able to trace in detail, either for lack of clear evidence or for lack of time (since this is an exceedingly time-consuming task); these include still other fruits and vegetables, as well as a wide variety of medicinal, industrial, and ornamental plants. The agricultural "revolution," however, consisted not merely in the adoption of these new crops almost wherever in the Islamic world they could be grown but also in the use of some of them to spread cultivation into semi-arid regions where sedentary agriculture had previously been unknown and in the incorporation of some crops into new, more intensive rotations that raised considerably the productivity of better, well-watered land. These more intensive rotations required a more careful attention to the fertility of the soil and, most often, more prolonged and heavier irrigation than had previously been available. The totality of the revolution thus embraced new crops, new rotations, new techniques of soil management, and the improvement and extension of irrigation.[2]

The diffusion of this revolution from the eastern reaches of the Islamic world toward the west took at most five hundred years and in places appears to have been much more rapid. By about 1200 the revolution had reached its fullest extent, touching the agriculture, and through agriculture the economy and society, of most of the Islamic West: Africa to the north of the Sahara, much of the east coast of Africa and part of west Africa, and also Muslim Spain, Sicily, and other islands of the Mediterranean. Given the vastness and diversity of the area affected and the many difficulties that the revolution must have encountered in the course of its spread, the process of diffusion was remarkably quick. It tells of a medium whose great powers of conduction were to have notable effects not only on agriculture but also on learning, religion, the arts, popular culture, manners, morals, and much else. The efficiency of this medium gives us much to ponder.

By contrast, Christian Europe seemed extremely unreceptive. When the new crops and the associated farming techniques reached the borders of Christendom, they were not quickly admitted. Rather they were kept languishing at the frontiers like immigrants no one wanted. True, here and there, some small parts of the revolution were accepted without undue delay. Sorghum, for instance, is mentioned as *melega* or *suricum* in an Italian document of 910. In the eleventh century al-Bakrī stated that, along with millet, it was the main crop of Galicia in northern (and by that time Christian) Spain; and in the twelfth century, under the name of *milhoca*, it was being sold in the market of Moissac in the south of France.[3] But the case of sorghum is exceptional. Most of the crops of the Islamic agricultural revolution fared quite differently. They had to wait a long time before being accepted. Only in the thirteenth century did sour oranges, lemons, and limes come to be familiar fruit trees in Christian Spain and Italy. At about the same time hard wheat was spreading into these areas. Spinach also appeared in the thirteenth century and was eventually diffused widely through Europe. Sugar cane was reintroduced into Sicily in the thirteenth century and into Spain in the fourteenth century, becoming a common crop in the fifteenth century. By then it had also appeared in the Portuguese Extremadura, in Madeira, and in the Canaries. At the end of the Middle Ages rice cultivation was appearing in the Po valley and the plains of Pisa, as well as in the Portuguese Extremadura.[4] And so forth. In short, there is an interval of many centuries before Christian Europe takes much, or any, interest in what the Muslims have to offer—at least in agriculture. And even after this long delay, acceptance was at best partial.

In searching for the causes of the relatively slow diffusion of the Islamic agricultural revolution into Europe it would be tempting to see the European climate as a serious, near fatal obstacle—and to look no further. And while it is true that the cold winters of much of Europe did indeed place a northerly limit on many of the new crops, this still left a large area into which the new crops could and eventually *did* move. All the crops could be grown in large parts of the Iberian peninsula, in much of Italy, and in certain areas in southern France. And in time they were successfully introduced into these areas. Some, such as hard wheat, spinach, artichokes, and watermelons, came to be grown over a much wider area. Perhaps they were not always easy to grow in this new physical environment which may, in a number of ways, have been hostile. But for the most part they were not easily grown in the Islamic world either. Since most of the new crops originated in tropical or semi-tropical areas, they required considerable heat and moisture. To be grown in many regions of the Islamic world they needed extensive artificial irrigation, especially in the summer months, when, in earlier agricultural regimes, water had commonly not been available in sufficient quantities. Many required to be hardened or acclimatized, and in some cases new varieties needed to be developed. In the case of cotton a new annual variety, *Gossypium herbaceum*, had to be developed from the ancient perennial tree cotton, *Gossypium arboreum*, before cotton could be introduced into the more northerly reaches of Islamic Asia or into the Mediterranean basin.[5] Though we know little, indeed almost nothing, about the work of adapting these plants to a new environment, it is now becoming clear that the Islamic world from the eighth century onward had a surprising number of what might be called botanical gardens or experimental farms, most of which were patronized by rulers and run by leading botanists and agronomes. Among their activities were the collection, acclimatization, and study of exotic plants, and sometimes the development of new varieties. Europe had nothing of the kind until much later. Its first botanical gardens date from the early fourteenth century—that of Salerno from 1310 and that of Venice from 1330. And not until the sixteenth century were there a number of European botanical gardens that were seriously engaged in work on exotic plants.[6]

That the real problem was something other than climate seems also to be suggested by the fate of the new crops in those parts of Europe that had been under Muslim rule and where the agricultural revolution had been successfully introduced. Ultimately all of Muslim Europe was retaken by the Christians in a reconquest that lasted more than five centuries. The contrasting fate of the agricultural revolution under the two successive regimes, first Muslim and then Christian, is instructive.

In Muslim Spain there can be no doubt that the new crops were widely grown and that, wherever conditions permitted, the new rotations and accompanying intensification of irrigation were widely practiced.[7] Yet Muslim lands that were reconquered by the Christians were, for the main part, soon given over to cereal cultivation or even to ranching and transhumant grazing. In short, a very marked de-intensification of land use occurred. As Thomas Glick has written, "Towns like Seville and Cordoba, which before the [Christian] conquest had enjoyed a mixed agriculture based on wheat, olives and gardens irrigated by *norias*, were now converted to sheepbreeding centres." Around Murcia and Cartagena, many irrigated areas turned to marsh and were abandoned; the Christians continued for a time to grow some grains, sowing the land one year and fallowing it the next; but even this relatively extensive use of the land was abandoned as the area gradually turned into "a vast pasture for sheep."[8] Valencia is something of an exception to the general pattern, for here the Christian conquerors were able to keep the elaborate Muslim system of irrigation intact. But increasingly the lands of its *huerta* were devoted to the production of grains, pulses, and vines—the traditional crops of Christian Europe—at the expense of the new crops introduced by the Arabs.[9] Some rice and citrus fruits did continue to be produced. But most of the new crops either disappeared or became insignificant. Thus in the fourteenth century we find Jaime II of Aragon (1260–1327) sending to Sicily in an effort to reintroduce the cultivation of cotton and sugar cane into his kingdom.[10]

But Sicily, too, was experiencing the same kind of difficulties. In the course of the twelfth century the Muslims who had cultivated the land gradually withdrew from farming or were displaced, and new European lords, ecclesiastical and lay, replaced them. They created large estates that were commonly given over to cereal production. Many, perhaps most, of the new crops disappeared altogether. In the first half of the thirteenth century, Frederick II (1194–1250) had to send to the Holy Land for sugar cane seeds and experts in sugar production, the cultivation of sugar cane having apparently died out in Sicily.[11] His efforts seem to have had no long-term success: by the early fourteenth century the cultivation of sugar cane seems once again to have disappeared. Frederick also attempted to reintroduce two dye crops that had been brought to Sicily by the Arabs, henna and indigo, but again his initiatives were not in the long run successful. With the possible exception of some cotton and citrus fruits, both of which may have continued on a small scale, the new crops brought by the Arabs and so successfully cultivated by them seem completely to have vanished in the century and a half following the coming of the Christians.[12]

In short, in both Spain and Sicily, the Christian conquerors were unable to assimilate the new crops and farming techniques even when they got them on a platter.[13] Thus, to understand the inability of Europe to receive the Arab agricultural revolution we shall have to look beyond climatic factors and probe more deeply into the social, economic, and cultural context of European agriculture. We shall find that this contrasts rather strikingly with the context in which agriculture was practiced in much of the Arab world, and in this contrast we may find the elements of an explanation of the differing performances of these worlds.

To start with we should look at demography—more specifically, at differences in population density. Though no precise figures are available, everything suggests that the Islamic world had relatively high population densities, at least in those parts that were habitable. A number of cities, most notably Baghdad and Cordoba, had populations at least in the hundreds of thousands and perhaps as great, according to some recent estimates, as half a million. There were everywhere many other large cities whose populations were well over 100,000, not to mention a plethora of smaller towns. The countryside, too, gives every appearance of having been very densely settled, especially where irrigated farming was practiced. Even in non-irrigated regions, settlement seems to have been *relatively* dense and to have pushed into areas that in earlier (and later) times were unoccupied. High population densities seem clearly to be bound up, perhaps as both cause and effect, with very intensive agriculture that drew more from the soil.[14]

Perhaps the most important link between demography and agriculture was that which connected the cities of this highly urbanized world with agricultural producers. Interaction occurred through the mediation of markets that allowed money and a money economy to penetrate deeply into the countryside, attracting more and more production toward the cities. This may be a particularly important factor because many of the new crops may have been most efficiently produced in market gardens, as cash crops, rather than on subsistence farms. The agricultural revolution should thus perhaps be seen as fitting into a world that was not only very populous but also highly monetized and commercialized.

By contrast, during the early Middle Ages, western European population densities were very low, as reflected in the total absence of cities of significant size, in its tiny villages, and in the abundance of unused or virtually unused land. Even though densities did rise considerably from around 1000 up to 1300, settlement still remained sparse in comparison to that of the Muslim world. Hand in hand with the low population densities of western Europe went a quite different kind of agriculture: one in which small, relatively isolated communities used land much more extensively. Typically, households had only tiny garden plots, which were often adjacent to dwellings. Fields were sown, usually every second year and sometimes only every third year, with grains and occasionally pulses. And a substantial part of the available land area on a typical estate was given over to the pasturing of animals, which were prized especially as sources of meat and dairy products but also for their wool and hides. Indeed the fact that European men and women had large appetites for meat and dairy products— perhaps especially in areas where Germanic settlement predominated— was perhaps the principal reason for the appearance in Christian Europe of a totally new and highly original type of agriculture: mixed farming, in which the growing of crops, principally grains, which used the land relatively extensively, was integrated with the raising of animals, which used the land still more extensively. Large areas of commons and waste were used for the grazing of animals, and the meadows were used exclusively for the production of hay to feed animals. Even cropland helped to feed animals by producing both fodder crops and the stubble from grains on which village flocks were grazed. Animals were also pastured on croplands during the long periods of fallowing.[15] One final point should be noted: Production in the early Middle Ages was overwhelmingly devoted to providing subsistence for the households of producers and a small surplus for exchange with other members of the community. As there was very little cash in circulation and much of that was used for non-market purposes, a cash nexus typically did not link producers to the outside world.

Not only did the European system make much less intensive use of the soil, as befits a society in which land is abundant and people are few, it was also a system that, once established, resisted change. In the Islamic world, at least in the early centuries of Islam, the greater part of agricultural land was in the hands of owners who had more or less full property rights over it—rights to use land as they saw fit, most often in the most productive and profitable way, and rights to rent or sell it on a market where presumably the highest bidder would be the one who could put it to the best economic use.[16] There was thus a tendency for land to move from less to more productive uses. In the European system, by contrast, the great majority of lords and peasants were caught in the webs of a system in which property rights were shared by lords *and* tenants, and many rights were held by the community at large rather than by individuals. Most areas of land on an estate were exploited in cooperation with others, the cooperative aspects of agriculture being no doubt strongly reinforced by the heavy emphasis placed on animal production. In consequence, there was little opportunity or incentive for enterprising individuals to break loose and innovate. Nor would the sale by either lords or tenants of whatever rights they did enjoy over land lead to more productive use, since those who acquired the rights would be constrained in the same way as those who sold them. When the system was at its most developed, say from the ninth to the eleventh centuries, everything seemed to conspire against innovators. The goal seemed to be not growth but rather survival.

Change did, of course, occur. But it came slowly, and the kind of change that was easiest to effect was the kind that was least threatening—the introduction of another cereal crop, for instance, such as rye, which could be grown in much the same way as wheat and even mixed with wheat.[17] Thus we are not so surprised to learn that sorghum was the first of the crops brought westward by the Muslims to penetrate into Christian Europe. In time, too, slightly more intensive rotations appeared in northern Europe in response to heavier population pressure and greater market opportunities.[18] But these rotations were still a far cry from the farming systems of the early Islamic world. More radical change could come only as individuals—both lords and peasants—began to pursue profit rather than communal survival. They did this by gradually disentangling their property rights from those of others, by dismantling earlier structures.

Another factor that bore heavily on diffusion was the migration or lack of migration of peasants. In the last fifty years we have learned that in the earlier centuries of industrial development the diffusion of difficult technologies occurred most readily through the migration of skilled workers.

More recently, it has become apparent that difficult agricultural technologies, too, were most commonly transmitted by peasants who learned them in one place and carried them elsewhere in the course of migrations. Thus the vectors of the Islamic agricultural revolution seem for the most part to have been peasants, and to a lesser extent landowners, who learned to grow the new crops in their homelands in the eastern reaches of the Islamic world and then, in the course of migrations, took the agricultural revolution, by successive stages, westward. Muslim Spain, for instance, was settled by a great mixture of newcomers—by Berbers from North Africa and by easterners such as Yemeni, Syrians, and Persians. The latter were no doubt the agents of diffusion. But the great movement of peoples across the newly unified Islamic world, and most especially from the eastern to the western parts of this world, stopped when it reached the frontiers of Christendom. There were no carriers available to take the agricultural revolution across the borders and into this still fragmented, almost atomized world that had closed in upon itself.

On the contrary, there was in time to be a reverse flow of people— away from the advancing frontiers of the Christian world. As the Christian reconquest of Spain and Sicily proceeded, and in the centuries following reconquest, landowners and peasants in the reconquered areas gradually moved out of the lands taken by the Christians toward lands still in Muslim hands—toward the southern parts of Spain before these were reconquered, toward North Africa, and toward the heartland of the Islamic world. The retreat began with the reconquest itself, which undoubtedly precipitated a good number of departures. In Spain, it was accelerated after various measures that disadvantaged Muslim cultivators (for example, the reduction of Muslim holdings in Gandía to one quarter their former size by Jaime I), a series of uprisings of conquered Muslim populations, and their subsequent flight. The expulsion of the Muslims from all of Spain from 1501 onward and that of the converted Moriscos after 1609 were only the final stages of a process that had begun much earlier—a process that saw the departure of many, perhaps the great majority, of the most skilled cultivators and their know-how. They were replaced by Christian settlers from northern Spain, Catalonia, and France who were less numerous and who were familiar only with the crops and farming techniques of feudal Europe. With this transfer in populations there was also a pronounced shift in agriculture.[19]

In Sicily, the picture is much the same. Initial tolerance of Muslim peasants, if it ever existed, soon gave way to discrimination, reduction of all Muslim cultivators to serfs of Christian lords, proselytizing, and forced conversions. A great massacre of Muslims took place in Palermo in 1161 and soon spread to the countryside. Further uprisings from 1189 to 1194 resulted in the flight of Muslims from all parts of the island except the Val di Mazara, and Muslims disappeared from there after riots in 1221, 1224, and 1243. By the mid-thirteenth century the Muslim presence on the island had been extinguished. The agricultural revolution that Muslims had brought some centuries earlier was not long in disappearing, too, as the Norman, Frankish, and so-called "Lombard" migrants who streamed in from northern Europe settled the island more sparsely and practiced a kind of agriculture that was more extensive, using the traditional crops of northern Europe.[20]

Should we see in the treatment of the Muslim populations not merely the kind of behavior one might expect conquerors to show to the vanquished but also something more deep-seated—a psychological block on the part of Christians toward *all* things Islamic, which inter alia prevented Christians from learning from Muslim agriculturalists? I do not know. I am at a loss to know what kind of evidence could test such a hypothesis. Certainly this cultural block, if there was one, did not prevent all penetration of Arab culture into western Christendom, since, as is well known, the court in Norman Sicily seems to have enjoyed Arab luxuries, to have adopted (up to a point) Arab lifestyles, and to have taken an interest in Arab poetry and thought. In Christian Spain, too, *mudejar* culture was of some interest to Christian overlords, and in the twelfth and thirteenth centuries many Arabic works of literature, science, and philosophy were translated. But it *may* be true that at the same time, at the level of society that was concerned with the cultivation of the soil, real bigotry was widespread and that this stood in the way of learning.

The role of eating habits in allowing or preventing the diffusion of new crops is also problematic. As is well known, most people and most cultures appear to be firmly attached to their traditional foods and the traditional manner of preparing them. So great are these attachments that they can sometimes effect dramatic changes in the land use of regions into which such peoples migrate. It has been argued, for instance, that the Germanic peoples who overran the western parts of the Roman Empire carried with them a strong and seemingly immutable preference for a diet based mainly on grains and animal products, and that in time this preference brought into being the system of mixed farming that characterized the medieval manor.[21] Was there in the descendants of these same peoples a resistance to the diet of the Arabs, which in areas of sedentary agriculture gave a large place to fruits and vegetables as well as grains and pulses and was conspicuously low in animal proteins? Quite possibly there was. We should note, however, that in most ages of history the very rich, who have the means to be adventurous in their eating, have been interested in exotic foods and that in some periods whole societies, either by predilection or because of changing constraints on food production, have changed their diets in greater or lesser ways. An example of such a change is surely the early Islamic world, where, so it seems, all levels of society were receptive to the new crops and developed a great variety of ways of preparing these. For some of those affected—most particularly for nomadic peoples who became sedentary—the change in eating habits was very great. In the realm of diets, therefore, we may have discovered another area in which early Muslims were very receptive to novelty and medieval Europeans, by and large, were not.

The lesson to be learned from our tale is, I think, the same lesson that we have been learning in the last twenty or thirty years from attempts to export technologies to improve the agriculture of developing countries. It is that agricultural technologies are not easily transferable. What seems to be a superior technology may, as it attempts to move into a new context, encounter economic, social, cultural, and even environmental obstacles that make it inappropriate and, to all intents and purposes, inferior. Agricultural technologies must fit into entire farming systems, and these in turn must harmonize with the context in which farming is practiced. What does not fit in may simply be rejected. Or it may penetrate very slowly into niches that are most favorable to it. To be sure, a very powerful new technology, with great advantages to commend it, may work to effect a gradual change in the context in which agriculture is practiced, a change that eventually will make that context more receptive. But the process will be slow, if it occurs at all.

As it was, these two worlds did not mesh. By and large they went their own ways. Differences in population levels, in urbanization, in the degree of monetization of the economy, in diets, in the skills of cultivators, all seem to have played a part in blocking what could, under other circumstances, have been a highly profitable exchange. Christian distaste for Muslims and their ways may also have played a part. In time, it is true, the context in which the Christians of western Europe practiced agriculture did change, and by the early fourteenth century that context had come much closer to the one in which the Arab agricultural revolution had flourished. At this point a few of the Islamic agricultural innovations did enter into Christendom. By then, however, western Europe was undergoing its own, quite different, agricultural revolution, especially in its temperate zone; and the Islamic world was entering into its long period of agricultural decay. For better or worse the chance was once again lost.[22]

Notes

1. Andrew M. Watson, *Agricultural Innovation in the Early Islamic World: The Diffusion of Crops and Farming Techniques, 700–1100* (Cambridge: Cambridge University Press, 1983). See also Watson, "A Medieval Green Revolution," in *The Islamic Middle East, 700–1900*, ed. A. L. Udovitch (Princeton, N.J.: Darwin, 1981), and Watson, "The Arab Agricultural Revolution and Its Diffusion," *Journal of Economic History* 34 (1974): 8–35.

2. For further information on agricultural practices the reader is referred to my book *Agricultural Innnovation* and to Lucie Bolens, *Les Méthodes culturales au moyen âge d'après les traités d'agronomie andalous: Traditions et techniques* (Geneva: Éditions Médecine et Hygiène, 1974), passim.

3. Al-Bakrī (d. 1094), *Jughrāfiyat al-Andalus wa Ūrūbā* [The geography of Al-Andalus and Europe], ed. 'Abd al-Rahmān 'Alī al-Hajjī (Beirut: Dār al-Irshād, 1968); J.-J. Hémardinquer, "L'introduction du maïs et la culture des sorghos dans l'ancienne France," *Bulletin philologique et historique* (1963): 429–59; Thomas F. Glick, "Agriculture and Nutrition: The Mediterranean Region," in *Dictionary of the Middle Ages*, ed. Joseph R. Strayer (New York: Charles Scribner's Sons, 1982–87), 1: 83.

4. For the diffusion of these crops into Christian Europe see Watson, *Agricultural Innovation*, pp. 82–83 and passim.

5. Andrew M. Watson, "The Rise and Spread of Old World Cotton," in *Studies in Textile History in Memory of Harold B. Burnham*, ed. Veronika Gervers (Toronto: Royal Ontario Museum, 1977), pp. 355–68.

6. See A. M. Watson, "Botanical Gardens in the Early Islamic World," in *The Ronald Smith Festschrift*, ed. E. Robbins and S. Sandahl (Toronto: Tsar, 1994), pp. 105–11. The earliest botanical gardens in Christian Europe appear to have been those planted by Matthaeus Sylvaticus in Salerno ca. 1310 and by Gualterius in Venice ca. 1330. Other European cities and universities did not acquire botanical gardens until the sixteenth and seventeenth centuries: Pisa in 1543; Padua, Parma, and Florence in 1545; Bologna in 1568; Leyden in 1577; Leipzig in 1580; Königsberg in 1581; Paris in 1590; Oxford in 1621. See A. Chiarugi, "Le date di fondazione dei primi orti botanici del mondo," *Nuovo giornale botanico italiano* n.s. 60 (1953): 785–839; A. W. Hill, "The History and Function of Botanical Gardens," *Annals of the Missouri Botanical Garden* 2 (1915): 185–240; F. Philippi, *Los jardines botánicos* (Santiago de Chile, 1878); A. Kerner von Marilaun, *Die botanische Gärten, ihre Aufgabe in der Vergangenheit, Gegenwart und Zukunft* (Innsbruck: Wagnerische Universitäts-Buchhandlung, 1874); G. Kraus, *Über die Bevölkerung Europas mit frem-*

den Pflanzen (Leipzig, 1891), passim; Kraus, *Geschichte der Pfanzeneinführungen in die europäischen botanischen Gärten* (Leipzig, 1894); E. Pavani, "Intorno ai giardini botanici," *Bolletino della Società Adriatica di Scienze Naturali in Trieste* 9 (1886): 51–93; and G. Curset, "Sur les jardins botaniques parisiens au XVIe siècle," *Journal d'agriculture tropicale et de botanique appliquée* 13 (1966): 385–404.

7. Further details on the agriculture of the Muslims in Spain are to be found in Bolens, *Méthodes culturales*; Hermann Lautensach, *Maurische Züge im geographischen Bild der Iberischen Halbinsel* (Bonn: F. Dümmler, 1960); and S. M. Imamuddin, *Some Aspects of the Socio-economic and Cultural History of Muslim Spain, 711–1492 A.D.* (Leiden: E. J. Brill, 1965), pp. 72–96; Pierre Guichard, *L'Espagne et la Sicile musulmanes aux XIe et XIIe siècles* (Lyon: Presses Universitaires de Lyon, 1990), pp. 51–54.

8. Glick, "Agriculture and Nutrition."

9. Thomas F. Glick, *Irrigation and Society in Medieval Valencia* (Cambridge, Mass: Harvard University Press, 1970).

10. J. E. Martínez Ferrando, *Jaime II de Aragón* (Barcelona: CSIC, Escuela de Estudios Medievales, 1948), 2: 19–20.

11. J. A. Huillard-Bréholles, ed., *Historia diplomatica Friderici Secundi* (Paris: Plon, 1852–61), 5: 575.

12. Illuminato Peri, *Città e campagna in Sicilia* (Palermo: Presso l'Accademia, 1953–56); F. Ciccaglione, "La vita economica siciliana nel periodo normanno-svevo," *Archivio storico per la Sicilia orientale* 10 (1913): 321–45; Guichard, *L'Espagne et la Sicile*, pp. 74–80.

13. The revival of the cultivation of sugar cane in Sicily, Spain, and Cyprus after the reconquest of these areas by the Christians may seem to be an interesting exception, but in fact it is not. In Cyprus, sugar cane, which had been introduced by the Arabs during their brief occupation of the island, died out with the Byzantine reconquest and in Sicily it virtually disappeared, surviving only as a marginal garden crop after the Norman invasion. Centuries later, after the loss of the Crusader Kingdom in the Levant, the cultivation of sugar cane and the refining of sugar were reintroduced into Cyprus and Sicily with the active participation of Genoese, Florentine, and Venetian merchants, who were presumably trying to replace Levantine and Egyptian sugar with sugar produced in regions where they could control the production and trade. These same merchants introduced the cultivation of sugar cane into Valencia and Gandía, where it had not been known under Muslim rule. In all these regions the cultivation of sugar cane began to increase during the later part of the fourteenth century and boomed in the early fifteenth century. After a setback resulting from the collapse of world prices of sugar during the second half of the fifteenth century, caused no doubt by the appearance of large quantities of sugar from the Atlantic islands, there was a short revival in the 1540s and 1550s before the crop disappeared from these regions. After the conquest of Granada in 1492, the production of sugar cane continued to prosper through the sixteenth century in much of the coastal plain, especially in the region around Malaga and Salobreña, and declined gradually through the seventeenth century. It is believed that this continuing production depended on Morisco labor supplemented, possibly, by slaves. For Cyprus, see W. Heyd, *Histoire du commerce*

du Levant au Moyen-Age (Leipzig: O. Harrassowitz, 1885–86), 2: 7, and Charles Verlinden, *The Beginnings of Modern Colonization* (Ithaca, N.Y.: Cornell University Press, 1970).

14. Watson, *Agricultural Innovation*, pp. 129–34.

15. See Georges Duby, *The Early Growth of the European Economy* (London: Weidenfeld & Nicholson, 1973), p. 17 ff.; and A. M. Watson, "Towards Denser and More Continuous Settlement: New Crops and Farming Techniques in the Early Middle Ages," in *Pathways to Medieval Peasants*, ed. J. A. Raftis (Toronto: Pontifical Institute of Mediaeval Studies, 1981), pp. 70–71.

16. Watson, *Agricultural Innovation*, pp. 112–16.

17. Marc Bloch, "Les transformations des techniques comme problème de psychologie collective." *Journal de psychologie normale et pathologique* 41 (Jan.–March 1948): 104–15; and Watson, "Towards Denser and More Continuous Settlement," pp. 68–70.

18. Lynn White, Jr., *Medieval Technology and Social Change* (Oxford: Clarendon, 1962), pp. 69–76; D. Faucher, "L'assolement triennal en France," *Etudes rurales* 1 (1961): 7–17; and Watson, "Towards Denser and More Continuous Settlement," pp. 67–73.

19. For Sicily, see C. Trasselli, "Produzione e commercio dello zucchero in Sicilia dal XIII al XIX secolo," *Economia e storia* 2 (1955): 325–42; and C. Trasselli, "Sumario duma historia do açucar siciliano," *Do tempo e da historia* 2 (1968): 49–77. For Spain see M. Birriel Salcedo, *La tierra de Almuñécar en tiempo de Felipe II* (Granada: Universidad de Granada, 1989), pp. 162–65. Further information regarding all these regions is contained in the proceedings of two conferences on the history of sugar cane held in 1990 and 1991 by the Ayuntamiento of Motril: *La caña de azúcar en tiempo de los grandes descubrimientos, 1450–1550* (Motril: Ayuntamiento de Motril, 1990), and *La caña de azúcar en el Mediterraneo* (Motril: Ayuntamiento de Motril, 1991). See also J. H. Galloway, *The Sugar Cane Industry: An Historical Geography from Its Origins to 1914* (Cambridge: Cambridge University Press, 1989), pp. 43–47. The survival of sugar cane cultivation in sixteenth-century Andalusia was probably made possible by the use of Morisco labor. The decline and ultimate collapse of sugar cane production in the south of Spain seems to correspond closely to the departure of the Moriscos.

20. Peri, *Città e campagna in Sicilia*, pp. 40–41; Ciccaglione, "Vita economica siciliana," pp. 322, 340, 343–44.

21. See above, note 15.

22. The diet of people in the early Islamic world has been little studied. See, however, Peter Heine, *Kulinarische Studien: Untersuchungen zur Kochkunst in arabisch-islamischen Mittelatter* (Wiesbaden: O. Harrassowitz, 1988); Rachel Arié, "Remarques sur l'alimentation des Musulmans d'Espagne au cours du Bas Moyen Age," in Arié, *Études sur la civilisation de l'Espagne musulmane* (Leiden: E. J. Brill, 1990); and Catherine Guillaumond, "L'eau dans l'alimentation et la cuisine arabe du IXème au XIIIème siècles," in *L'homme et l'eau en Méditerranée et au Proche Orient* 2, ed. P. Louis (Lyon: Maison de l'Orient, 1986). For discussions of Arab influences on the diet and cooking of Christian Europe during the Middle Ages, see Maxime Rodinson, "Les influences de la civilisation musulmane sur la civilisa-

tion européenne médiévale dans les domaines de la consommation et de la distraction: l'alimentation," in Accademia Nazionale dei Lincei, *Atti dei Convegni* 13 (1971) (Convegno internazionale, 9–15 Aprile, 1969): 477–99; Peter Heine, "Rezeption der arabischen Kochkunst und Getränke in Europa," in *Kommunikation zwischen Orient und Okzident: Alltag und Sachkultur*, Veröffentlichungen des Instituts für Realienkunde des Mittelalters und der frühen Neuzeit 17, Sitzungsberichte der österreichischen Akademie der Wissenschaften, phil.-hist. Kl. 619 (Vienna: Verlag der österreichische Akademie der Wissenschaften, 1994): 379–92.

Bruce M. S. Campbell

5. Ecology Versus Economics in Late Thirteenth- and Early Fourteenth-Century English Agriculture

Environmental issues are topical. Problems of the greenhouse effect, acid rain, deforestation, desertification, and soil erosion serve as stark warnings of the ecological degradation to which mismanagement of the environment can give rise. Likewise, heavy death tolls from famine in Ethiopia and flooding in Bangladesh serve as grim reminders of the demographic price that is sometimes exacted. Yet ecological imbalance and environmental degradation are as old as human exploitation of resources and have sometimes occurred on a scale sufficient to stifle prosperity and occasionally even topple empires. Ancient Jericho and the First Empire of the Mayan civilization of Central America may both have been undermined by over-exploitation of agricultural resources, and a similar explanation has been advanced to account for the crisis to which England in particular, and much of western Europe in general, succumbed during the fourteenth century.[1]

Natural Disaster and Ecological Imbalance in Medieval England

The first half of the fourteenth century was characterized in England by a succession of extreme and disastrous events. In 1315 torrential rain ruined the harvest, and grain prices soared to five or six times normal. By the spring and summer of 1316, England, along with much of the rest of western Europe, was in the grip of a famine of major dimensions—the Great European Famine—and this was accompanied by a virulent and widespread epidemic, perhaps typhus, which greatly increased mortalities. The following harvest was no better, and the two years 1315–16 and 1316–17 mark a rate of inflation in grain prices without parallel in English experience.[2] The hardship and suffering were terrible—twenty-three prisoners in Northampton jail died simply for want of food—and, in the country as a whole, living standards sank to a nadir subsequently matched in misery only by 1597, another year of terrible dearth.[3] To make matters worse, cattle and oxen were ravaged by a severe outbreak of murrain—probably rinderpest—in 1319 and 1320, which depleted plough teams and reduced the ability to cultivate the land, and this was followed in 1321–22 by renewed harvest failure.[4] Dearth recurred in 1330–31, but a run of above-average harvests during the remainder of that decade gave some respite to the hard-pressed rural population, although price deflation made this a difficult time for those who relied on market sales for an income, particularly landlords.[5] Sheep and cattle disease returned in 1334–35, disrupting livestock husbandry, and in 1346–47 renewed harvest failure brought a return to famine conditions in many parts of the country. Then, in 1348–49, bubonic plague struck. The mortality that it precipitated dwarfed that of the Great Famine, large though that had been. Never again was the death rate of a single year to rise so high: within the space of just two years an estimated one-third of the country's population had died, and in the worst-hit communities mortality rates reached 60 percent.[6] Meanwhile, floods had ravaged the low-lying lands around the south and east coasts as a transition to a cooler and wetter climate precipitated a heightened incidence of storm surges within the North Sea.[7] The effects were geographically localized but nonetheless devastating for the communities concerned, as crucial peat cuttings were flooded, salt pans damaged and destroyed, sheep and cattle drowned, and rich pastures and valuable arable land soured with salt or lost to the sea.[8]

Several historians have been tempted to interpret these various natural disasters as a punishment for overexpansion during the previous century or more, as the population grew more rapidly than the economy, impelling the population along a Malthusian trajectory.[9] It was M. M. Postan who first suggested that by 1300 England was possibly supporting a population of 5 to 6 million—a figure not to be exceeded until the Industrial Revolution of the late eighteenth century—and this figure has gained increasing currency among historians.[10] The bulk of that population lived and worked on the land, and the productivity of their labor was fundamental to the vitality of the economy as a whole, since domestic agriculture supplied medieval industry with much of its raw materials, towns with their provisions and fuel, and traders with many of the commodities in which they dealt.[11]

Where agricultural resources were under such mounting pressure, there was an inevitable danger that natural ecological limits would be exceeded. It is precisely this, in J. D. Chambers's damning assessment, that happened. There was

> pressure on agricultural resources to the point at which there was a severe cut-back of agricultural productivity through soil deterioration and falling yields. It was not merely that agricultural techniques were unable to respond to the challenge of increased demand; it was worse than this. The techniques that had sufficed to enable the population to reach the existing limit began to recede owing to the encroachment of arable upon the pasture in order to provide the basic necessity of bread at the expense of the more expendable commodity of meat.[12]

This negative verdict on the productivity and proficiency of the agricultural sector—the mainspring of most wealth in this predominantly agrarian economy—is shared by historians working in very different historiographic traditions. M. M. Postan, for instance, whose broad interpretation of the period still commands attention, believed that by the late thirteenth century yields in medieval agriculture may actually have been falling. As he wrote in 1972: "The hypothesis of declining yields thus stands. Our sources being what they are it must remain no more than a hypothesis. But it is sufficiently well-rooted in evidence to be accepted as a near approximation to an established fact."[13] For Postan the source of the problem lay in the demographically induced colonization of land that was marginal for cultivation and whose productivity could not be sustained in the long term, coupled with the overexpansion of arable at the expense of pasture to the detriment of manure supplies and hence soil fertility.[14]

This "hypothesis" is one of the few parts of Postan's interpretation of the period to have been accepted by the Marxist historian Robert Brenner, who, in 1976, observed: "the crisis of productivity led to demographic crisis, pushing the population over the edge of subsistence."[15] For Brenner and other Marxist historians it was the socio-property relations of feudalism and associated attitudes toward wealth rather than population growth per se that lay behind this productivity collapse. In Brenner's words,

> Because of lack of funds—due to landlords' extraction of rent and the extreme maldistribution of both land and capital, especially livestock—the peasantry were by and large unable to use the land they held in a free and rational manner. They could not, so to speak, put back what they took out of it. Thus the surplus-extraction relations of serfdom tended to lead to the exhaustion of peasant production *per se*.[16]

Nor did landlords employ their superior command over capital and resources to better effect. In R. H. Hilton's view landlords' interests in their estates went little further than the exaction of maximum profit, and the notion of reinvesting profits to raise productivity occurred to very few.[17] The net outcome, as Edward Miller and John Hatcher concluded in 1973, was that "even highly organized and superficially efficient estates were failing in one quite basic requirement of good husbandry: the keeping of the land in good heart."[18]

Whether for demographic, institutional, or a combination of both reasons, there is thus a widely held view that an ecologically induced productivity crisis was the eventual outcome of medieval attempts to expand food production. Compounding the problem were what are perceived to have been relatively low and static levels of technology. To Postan "the inertia of medieval agricultural productivity is unmistakable," with the result that "the land was bound to suffer."[19] Miller and Hatcher agree.[20] J. Z. Titow is convinced that yields were falling on the estates of the bishops of Winchester from the mid-thirteenth century, with "chronic under-manuring" the most likely culprit, and David L. Farmer believes that a shortage of livestock manure was likewise depressing yields on Westminster Abbey's estates.[21] According to Postan, the fact that these great estates obtained such dismal results "makes it all the more probable that yields were falling even more on lands of other and less privileged cultivators."[22] In short, the feudal combination of lymphatic peasants and lackadaisical lords sapped the agricultural sector of dynamism and eroded its ecological base.

The Role of the Market in Determining Production

Implicit in this pessimistic scenario is the assumption that production for consumption predominated over production for exchange. In effect, a state of natural economy is believed to have existed: in the absence of the allocative role of the market, people almost everywhere had to produce the same type of crop and tend the same sort of domesticated animals for meat, wool, and pulling power. Yet, as R. H. Hilton has observed, to view the medieval agrarian economy as a natural economy is to ignore the presence of the exchange economy of the towns.[23]

No one would in fact deny that some trade in agricultural products took place, since towns and industry could not have existed without it, but the former were thought to be small and the latter but little developed;

hence trade remained decentralized, small-scale, and local. It was with the purpose of servicing this trade that, as R. H. Britnell has demonstrated, the thirteenth century witnessed a remarkable proliferation of chartered markets.[24] Part of the impetus behind this movement came from increasing seignorial demands for payments in cash rather than in kind. Peasants therefore found it necessary to produce and market surpluses to get cash to pay rents, fines, amercements, and taxes. As population pressure built up so, too, in Britnell's view, did the number of commercial transactions within rural society increase as economic expediency induced smaller farmers and poorer families to sell their produce and services and buy their food.[25] While the bulk of trade remained at this scale, market demand was unlikely to disrupt or complicate the traditional relationship between farming and the land. Considerations of climate, soil, and topography (mediated by institutional arrangements—field systems, manorial structures, types of tenure—and demographic pressures) consequently remained paramount in determining what crops were grown and what animals were kept. Hence the equation Postan makes between stocking densities and grassland supplies: since livestock were dependent on grassland, livestock were scarce where supplies of grassland were deficient (and vice versa).[26] Hence, too, the verdict of R. E. Glasscock in 1973 that "technology and exchange had not progressed far enough by the early fourteenth century to allow much agricultural specialization."[27]

Yet the medieval exchange economy undoubtedly went further than this, for during the course of the thirteenth century agriculture became increasingly exposed to the influence of expanding urban centers both at home and overseas.[28] As is now becoming clear, the leading cities of the realm were more populous ca. 1300 than has hitherto been recognized. Chief among these was London. By 1300, it was the second largest city north of the Alps and according to Derek Keene had a population approaching 100,000.[29] As Gunnar Persson has argued, its growth, along with that of equivalent cities on the continent and the major provincial capitals (Norwich, for instance, had a population of ca. 25,000 in 1330), presupposes a commensurate rise in agricultural productivity (especially labor productivity), for which increased specialization of production may have been largely responsible.[30] This suggests that, as the metropolis and other leading cities grew and their commercial tentacles reached out, so old patterns of self-sufficiency and localized trading relationships began to break down and a more specialized and spatially differentiated agrarian economy began to emerge.[31] The extent to which individual regions were drawn into this trading nexus varied a great deal, and they were drawn into it in different ways, not least because arable and pastoral production differed quite radically in their transportation requirements. Nor should the impact of overseas demand be underrated. English grain, malt, wool, and hides were all traded internationally, and wool remained the country's biggest foreign-exchange earner throughout this period.[32]

Some indication of the scale and direction of the trade in grain and malt and of the spatial relationship between producer and consumer can be obtained from sheriffs' accounts, customs accounts, and other such sources.[33] These reveal Kings Lynn to have been England's leading grain entrepôt, its excellent river communications via the Great and Little Ouse and the Nene enabling it to tap a rich and extensive hinterland extending deep into East Anglia and the east Midlands.[34] Purveyors' accounts show that river ports such as Lakenheath, Ely, Cambridge, and St. Ives were equipped with commercial granaries that served as collecting houses for grain assembled from the surrounding countryside.[35] This grain was then sent by boat to Kings Lynn, where it was either immediately transshipped onto sea-going vessels that plied a busy trade up and down the east coast and across the North Sea or, again, stored in commercial granaries.[36] Other east-coast ports, such as Yarmouth, Boston, and Ipswich, participated in the same trade and were likewise linked to their hinterlands by cheap and effective river communications.[37] It was the Thames, however, that was probably England's foremost trading artery. At the end of the thirteenth century Henley was an extremely busy river port, replete with granaries and professional boatmen and grain sellers, that functioned as a major collecting point for grain ultimately destined for the London market.[38] The various ports of the Thames estuary also linked farmers on the fertile soils of northern and eastern Kent with the lucrative metropolitan market and the urban concentrations of Flanders and northern France.[39]

Evidence of Agricultural Specialization

What was the impact of regional, national, and international markets upon agricultural production? A partial answer can be obtained for the demesne sector—which contained perhaps 30 percent of all arable land—from information contained in manorial accounts.[40] These record the crops grown and harvested and the livestock kept during the course of a farming year, Michaelmas to Michaelmas. They survive in thousands and afford con-

TABLE I   Demesne Arable-Farming Types, 1250–1349

| | Percentage of Total Sown Acreage | | | | | | |
Type	Wheat	Rye	Barley	Oats	Grain mixtures	Legumes	N
1	24.6	2.6	35.3	12.5	1.0	24.1	61
2	11.1	26.8	31.3	21.7	3.5	5.8	42
3	27.2	1.7	7.1	18.9	37.0	8.2	53
4	34.8	3.0	18.5	28.5	3.8	11.4	113
5	44.8	1.2	4.4	44.8	1.9	2.9	126
6	69.0	0.0	10.7	12.0	4.2	3.9	30
7	13.0	10.0	3.5	68.5	2.6	2.4	40
All	33.5	4.9	14.9	31.5	6.6	8.6	465

siderable potential for future enquiry. An analysis of the acreage devoted to the principal crops from a national sample of accounts drawn from the period 1250–1349 and representing some 465 demesnes, reveals seven basic arable-farming types (Table 1 and Figure 13) differing in the relative range and importance of the crops that were grown and the intensity of their cultivation.[41]

At one extreme there were demesnes on which cultivation of a single crop predominated, usually either oats (arable type 7) or wheat (arable type 6). The former was typical of much of the north, the northwest, and the southwest of England, as well as of certain of the immediate Home Counties, where large-scale oat cultivation was undoubtedly a response to the considerable fodder requirements of the metropolis (oats being a low-value crop incapable of withstanding high transport costs).[42] The latter was very much a specialty of the southwest and shows up particularly in Wilt-shire, Dorset, and Somerset, where wheat was characteristically grown in conjunction with some spring grain in a classic two-course rotation.[43] Such relatively unintensive cultivation systems went hand in hand with these simple crop combinations. More complex and intensive were cropping sys-tems in which wheat and oats were grown in roughly equal proportions (arable type 5). This, in fact, was the most common cropping system of all and was employed on over a quarter of all sampled demesnes, where it was probably associated with a classic three-course rotation of winter grain, spring grain, and fallow. The demesnes in question were widely dis-tributed, but were especially prominent in the northeastern counties of

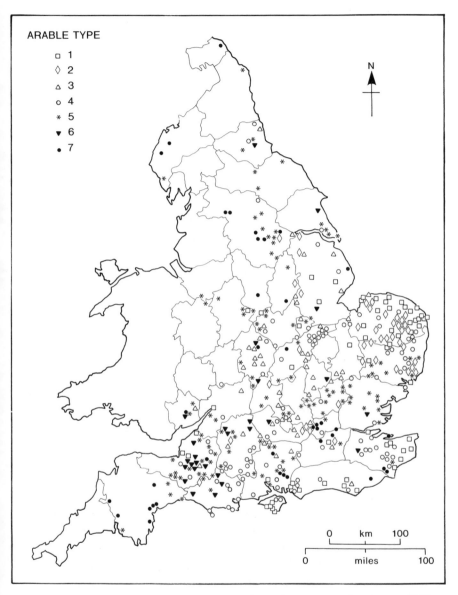

Figure 13. Demesne arable farming types, England, 1250–1349. For explanation, see Table 1.

Durham and Yorkshire, on the heavy clay soils of the Home Counties of Essex, Hertfordshire, and adjacent portions of Suffolk and Middlesex, and, to a lesser extent, in parts of the West Country.[44]

The remaining four cropping types were all much more complex and featured a wider range of crops, with increasing prominence accorded to the malting grains of dredge and barley, together with legumes (peas, beans, and vetches), all of which were spring sown. This transition toward a greater emphasis upon spring crops shows up in arable type 4, concentrations of which occurred in east Suffolk, central Kent, south Wiltshire, and adjacent southern counties, and in the east Midlands, especially northeastern Northamptonshire.[45] A strong spring emphasis also shows up in arable type 3. These demesnes were especially well represented in the Midlands, and substantial acreages were devoted to the cultivation of mixed grains: maslin (wheat and rye) and dredge (barley and oats). In fact, it was in the east Midlands that the greatest range of different cropping patterns was to be found. Lincolnshire and Northamptonshire both furnish examples of all seven different arable types, and even diminutive Rutland furnishes examples of four. Evidently, the counties penetrated by the commercial arteries of the Trent, Welland, Nene, and Ouse contained some of the most varied arable husbandry in England, which is the more remarkable given that this region lay squarely within the bounds of the midland common-field system. Indeed, it is in precisely this area that the system showed signs of evolving more complex forms, with the sowing of *inhoks* from the fallow and the subdivision of furlongs to produce smaller cropping units.[46]

Nevertheless, it was largely outside the limits of the regular common-field system that demesnes employing the most complex and advanced cropping systems—arable types 1 and 2—were to be found. Outstanding in this respect are the coastal counties of Norfolk, Kent, Sussex, and the Isle of Wight, with their unique concentrations of arable type 1 demesnes. What distinguished these demesnes was the exceptional prominence accorded to barley in a four-part combination that also featured winter wheat, some oats, and substantial sowings of legumes. Independent study has shown this specific type of crop combination to have been associated with a particularly intensive and productive rotational regime, in which fallowing was minimized, seeding rates were heavy, and yields per acre were among the highest obtained in medieval England.[47]

On the evidence of the demesne sector, therefore, husbandmen did not everywhere grow the same crops or grow them in the same kinds of way. On the contrary, there was much local and regional variation in the pattern of cropping. Moreover, it was in the more populous parts of the country, and especially in those that were most exposed to commercial influence—notably the east Midlands, East Anglia, and the southeast—that the greatest range of arable-farming types occurred, including several that were rarely encountered in other parts of the country.[48] Away from the south and east the spatial differentiation of arable husbandry diminished, and cropping systems became both simpler and more uniform as methods of cultivation became correspondingly less intensive. In these remoter parts of the country, removed from major markets and often land-locked and handicapped by high costs of overland carriage, there were not the commercial inducements to encourage a greater emphasis upon cash crops and the adoption of more intensive methods.

Such problems of distance and remoteness were not as serious for the livestock producer for the simple reason that animals were capable of walking to market and their products—wool, hides, butter, and cheese—being generally high in value relative to their bulk, were better able than grain to withstand the costs of overland carriage. Trading links could thus exist over remarkably long distances and were less dependent upon access to cheap river and coastal communications. In the mid-thirteenth century the royal larder at Westminster was stocked with beef from Lancashire, and wool produced in the Welsh Marches was ultimately destined for cloth manufacture in Florence.[49]

Analysis of the stock carried on a sample of 660 demesnes reveals that, as with crops, there was much local and regional variation. Seven basic pastoral types may be distinguished (Table 2 and Figure 14). These progress from relatively simple and unintensive regimes dominated by draft animals, via regimes in which the draft and non-draft components were roughly equally balanced, to regimes in which non-working animals predominated. Pastoral husbandry exhibited its least developed and intensive form on those demesnes on which draft animals were of overwhelming importance (pastoral types 5, 6, and 7). These were to be found in all parts of the country but were especially important throughout the north, in parts of the Midlands, and in Monmouthshire and Somerset in the southwest. This is somewhat paradoxical given the supposedly abundant pastoral resources of many of these counties. In part this may represent a genuine emphasis upon the production of surplus oxen for meat and draft power and the breeding of horses for sale. But on the bulk of the lowland demesnes that predominate within the sample, it represents the northern practice of stocking arable demesnes with essential working animals only, while main-

TABLE 2   Demesne Pastoral-Farming Types, 1250–1349

			Percentage of Total Livestock Units*				
Type	Horses	Oxen	Adult cattle	Immature cattle	Sheep	Swine	N
1	11.2	18.0	34.0	29.4	6.1	1.3	100
2	18.7	28.2	22.8	19.7	5.6	5.0	116
3	11.2	20.5	18.3	13.8	33.1	3.1	132
4	9.5	25.5	3.0	2.4	58.0	1.7	75
5	25.0	36.8	5.8	3.4	12.4	16.6	40
6	52.5	37.5	2.1	1.3	5.0	1.7	43
7	15.0	78.6	1.5	1.5	2.3	1.0	154
All	16.7	37.7	14.0	11.6	16.7	3.2	660

*Total livestock units = (horses × 1.0) + ([oxen + adult cattle] × 1.2) + (immature cattle × 0.8) + ([sheep + swine] × 0.1). These weightings correspond to the feed requirements of the different animals. For explanation see Bruce M. S. Campbell, "Land, Labour, Livestock, and Productivity Trends in English Seignorial Agriculture, 1208–1450," in *Land, Labour and Livestock: Historical Studies of European Agricultural Productivity*, ed. Campbell and Overton (Manchester: Manchester University Press, 1991), pp. 156–57.

taining breeding and back-up herds on specialist stock farms—vaccaries and bercaries—on the upland margins.[50] These stock farms are seriously underrepresented in the sample. On the demesnes that make up pastoral types 3 and 4 this emphasis upon draft animals—the tractors and trucks of medieval agriculture—was more muted, and other cattle and sheep were of correspondingly greater importance. Sheep-dominated demesnes were, in fact, relatively few in number and confined to localities with an abundance of suitable pastoral resources: in the vicinity of extensive marshland grazings as in south Yorkshire, southeast Essex, and coastal Sussex, on the light soils of the East Anglian Breckland, and, most notably, on the chalk downlands of southern England, especially in south Wiltshire and north Hampshire.[51]

All the remaining pastoral types—1, 2, and 3—represent more developed and intensive livestock regimes and exhibited a strong geographical bias toward the south and east. Their greater intensity is reflected in the enhanced contribution made by horses to draft power, in the general predominance of cattle over sheep, in a herd structure demographically skewed toward adults (indicative of a specialist interest in dairying), and, ultimately, in pastoral types 1 and 2, in the subordination of working to

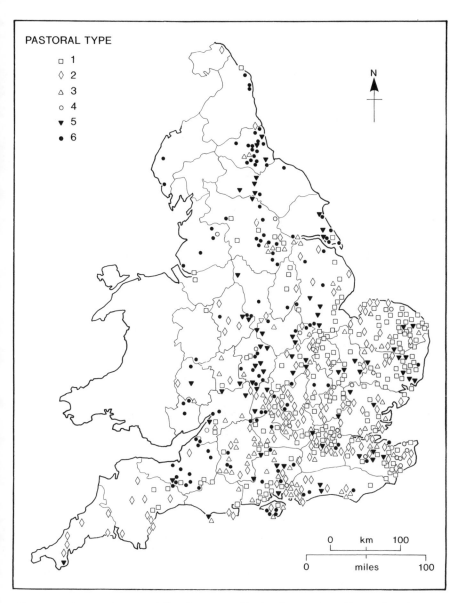

PASTORAL TYPE

□ 1
◇ 2
△ 3
○ 4
▼ 5
● 6

Figure 14. Demesne pastoral farming types, England, 1250–1349. For explanation, see Table 2.

non-working animals.[52] It was, in fact, on the demesnes making up pastoral type 1 that these features exhibited their fullest development and, significantly, notwithstanding the region's strong natural advantage as an arable producer, it was in East Anglia and adjacent parts of the Home Counties that these demesnes were most strongly represented, with notable concentrations in Norfolk, east Suffolk, and the immediate environs of London. Other lesser concentrations occurred around the fen-edge in Lincolnshire and the Soke of Peterborough, in the vicinity of Romney Marsh in southern Kent, in southern Hampshire, and in east Devon.[53] Oswald, Chaucer's reeve from Bawdeswell in Norfolk, managed a demesne with precisely these characteristics. Its stock, as described by Chaucer, comprised "his lord's sheep, his neat [cattle], his dairy, his swine, his horse, his store [fat cattle], and his poultry," a list that compares closely with the stock enumerated on many a Norfolk manorial account.[54] Many of these attributes were shared by the demesnes falling into pastoral type 2, which as a type were more generally distributed throughout the eastern and southern counties.

Overall, as a comparison of Figures 13 and 14 reveals, pastoral husbandry was subject to much greater local variation than was arable husbandry, with the result that examples of most pastoral types were to be found in most parts of the country. Nevertheless, these local variations were superimposed upon a more general trend, whereby the more intensive and developed forms of pastoralism were increasingly well represented toward East Anglia and the southeast.[55] The distinguishing features of these more intensive systems were their greater emphasis upon non-working rather than working animals, horses rather than oxen, cattle rather than sheep, and adult animals rather than immatures. As such they were intensive in their employment of both labor and capital since they were associated with the cultivation of fodder crops, the housing of livestock, and the manufacture of dairy products. Significantly, this spatial progression from less to more intensive systems was not tied to access to cheap river and coastal communications in the same way that applied to crops.

On this evidence, therefore, neither arable nor pastoral husbandry stood in a simple and direct relationship with the physical environment. Varied ecological opportunities obviously explain part of the variation in farming types, yet it is equally plain that climate, soils, and terrain provided farmers with the opportunities to which institutional, social, and economic circumstances determined the response: witness the intensive livestock economy of much of East Anglia, a region with only very limited environmental advantages for pastoralism. To some extent it was field systems and other institutional influences that set limits to the intensity with which arable and pastoral husbandry were conducted. It can be no coincidence that it was largely outside the bounds of the most closely regulated field systems that husbandry assumed its most specialized and intensive forms.[56] Nevertheless, a considerable variety of husbandry types existed even within the confines of the midland system, partly for physical reasons but, more particularly, due to the influence of market forces. While local markets may explain local specializations in particular crops and livestock, the broader trends that have been observed in the character and intensity of both types of husbandry are only explicable with reference to wider patterns of trading. In this respect, farmers were responding to the influence exercised by markets upon prevailing levels of economic rent.

Economic rent represents that part of the revenue above production costs but excluding the farmer's remuneration for capital improvements, his labor, and his managerial skills. In effect it is the return due for the use of the land alone as a factor of production.[57] David Ricardo believed that variations in economic rent were caused by differences in soils and population density and that the margin of cultivation was determined solely by physical factors; it is this view that has hitherto been implicit in most discussion of medieval agriculture.[58] J. H. von Thünen, however, has shown that economic rent was also controlled by distance from the market and transport costs.[59] In a market economy, consequently, economic rent underlies all questions of competition for the use of land and provides the means whereby that competition is resolved to produce patterns of land use. Economic rent varies with distance from a market as well as both the facility with which a commodity can be transported and its perishability, and differed therefore, in a medieval context, between crop and livestock products. Goods that were high in bulk and low in value (e.g., firewood and fodder) and perishable goods (e.g., fresh milk and fat meat) thus yielded an economic rent that declined very sharply with distance from the market. Economic rent also affects production costs and thereby determines the type of farming system employed, so that although the crop being grown or animals tended may remain the same, the manner of cultivation and management may vary. On the whole, the intensity of production tends to decrease with distance from the market, as lower production costs are necessary to offset higher transportation costs.

These ideas help to explain the differentiated patterns of crop and livestock production that characterized the demesne sector. It does not necessarily follow that all demesne production was ultimately destined for the market, for as Kathleen Biddick has demonstrated with reference to Peterborough Abbey, production for consumption continued to prevail over production for exchange on many estates, especially those of major religious institutions with substantial households to provision.[60] But it does mean that the full economic value of those resources was realized only by pitching production for consumption at a level consonant with prevailing levels of economic rent, as does indeed appear to have been the case on the demesnes of Peterborough Abbey. Otherwise it would have been more profitable to lease the land rather than manage it directly unless the security of a constant and assured supply of foodstuffs for the household was sufficient compensation for the sacrifice in both productivity and rent. By implication, therefore, prevailing levels of economic rent must also have exercised some influence upon peasant agriculture.

Peasant farmers were probably responsible for between two-thirds and three-quarters of total agricultural production, and it was this sector that Postan had in mind, in 1959, when he wrote that

> in the main, economic growth and contraction in the medieval economy resulted from activities of countless small agricultural producers who, though not wholly innocent of money and markets, could not be expected to expand or contract their holdings, to take up land or to give it up, to sow more or to work harder in response to the stimuli of prices or under the influence of a pessimistic or optimistic view of future business prospects.[61]

Nevertheless, although direct evidence of peasant agriculture remains elusive, there is a certain amount of circumstantial evidence to suggest that here, too, Postan's view requires revision, at least for the more crowded and commercialized south and east.

The purveyors' accounts of the early fourteenth century that provide such explicit evidence of the grain trade's infrastructure also record large numbers of small "purchases" of grain, bacon, and other necessary provisions, presumably made in local markets from small and middling producers (an impression that is confirmed when, as sometimes occurred, the individuals concerned were named).[62] Such "sales," as is well known, were often made reluctantly due to the difficulties experienced in exacting payment at a fair price.[63] Nevertheless, the monetized peasant economy revealed in East Anglia and the east Midlands from studies of manorial court rolls leaves little doubt that peasant producers were regularly involved in exchange.[64] Land, certainly, was bought and sold with remarkable regularity, and it is clear from analyses of the land market that the propensity to transact land was highly sensitive to harvest conditions and hence the price of grain. Bad harvests evidently induced many peasants to enter the grain market as sellers, whereas good harvests had the opposite effect and allowed some individuals to enlarge their holdings.[65] Amercements for illegal brewing, especially by women, also imply at least a localized trade in the necessary brewing grains.[66]

Where information can be gleaned on the crops and livestock produced on peasant farms—from tithes, tax returns, the multure of mills, and incidental references in court rolls and other manorial sources—it is clear that these were rarely very different from those produced on demesnes in the same locality, with which, of course, peasant cultivators would have been well acquainted.[67] Indeed, it seems likely that demesnes conformed to local husbandry practices in many aspects of their management.[68] That the more substantial peasant farmers were perfectly capable of capitalizing on commercial opportunities when they were offered has been demonstrated by Marjorie McIntosh for the post-1350 period with reference to the royal manor of Havering, fifteen miles east of London.[69] It is also arguable, as A. R. Bridbury has demonstrated, that peasant producers were responsible for the bulk of the wool exported from England.[70] But to what extent did the market, via economic rent, actually structure peasant production? Here the verdict of Biddick in a pioneering, if controversial, analysis of a Bedfordshire lay subsidy roll of 1297 is perhaps most telling:

> The wealthiest taxpayers in both town and country specialized in holding malting grains and sheep, assets more commercial than agrarian. If historians use commercialization, the "production of surpluses for sale," to index development in a primarily agrarian society (as Hilton has done), then the concentration of commercial wealth among such peasants shows considerable engagement in a developing economy.[71]

The Incentives and Disincentives to Agricultural Progress

Once it is recognized that medieval agriculture was more commercialized than has hitherto been appreciated, it follows that agriculture's supposed technological inertia is something of an illusion. Too much of the writing about agricultural technology, both medieval and later, has been preoccupied with novelty: the introduction of new crops, animals, and implements.[72] These may be the most conspicuous forms of change but in economic terms they were not necessarily the most significant. Some changes of this sort certainly took place during the course of the thirteenth and fourteenth centuries. E. M. Veale, John Sheail, and Mark Bailey have documented the introduction and naturalization of the rabbit, which was prized for its meat and fur and its ability to turn to profitable use some of the poorest, sandy soils.[73] John Langdon has demonstrated that horses were increasingly substituted for oxen in haulage—with significant consequences for the economics of road transportation—and, to a lesser extent, in traction.[74] At the same time vetches, a leguminous fodder crop with useful nitrogen-fixing qualities, became more widely grown and contributed to a general increase in the importance of legumes.[75] Toward the end of the twelfth century the windmill was introduced and, as Richard Holt has shown, became generally adopted in those parts of the country that had been deficient in water power.[76] These innovations apart, the crops, livestock, and implements available to medieval cultivators changed relatively little. What did change significantly, however, was the way in which these crops and livestock were combined and integrated into husbandry systems. By the beginning of the fourteenth century there are documented examples of cropping systems with flexible and intensive rotations that obtained a sustained high level of yield notwithstanding the virtual elimination of fallowing; convertible husbandry systems in which land alternated between tillage and temporary grass to the mutual benefit of both; and intensive mixed-farming systems, in which the competing land-use demands of crops and livestock had been reconciled by the cultivation of fodder crops—principally legumes and oats—and such land- and labor-intensive methods of livestock management as the stall-feeding of livestock.[77]

Significantly, those parts of the country that appear to have been most active in adopting these more intensive and progressive farming methods were those with natural environmental advantages, good market access, and few institutional constraints in the form of servile tenures and communal property rights. Where all three of these circumstances coalesced, economic rent, as defined by both Ricardo and von Thünen, was at a maximum, and agricultural systems, as they have been documented in east Norfolk, northeast Kent, and coastal Sussex, attained their most intensive and productive levels.[78] Nevertheless, as Bailey has recently demonstrated with reference to the East Anglian Breckland, it was even possible to triumph over difficult environmental conditions provided that the localities in question were geographically well placed and institutional constraints were few.[79]

The East Anglian Breckland is ecologically one of the most marginal regions for agriculture in lowland England, and even today much of it is devoted to forestry and military use. Its marginality derives from low rainfall and sterile sandy soils, which are retentive neither of nutrients nor of moisture. Postan and others have speculated that under medieval husbandry methods more may have been demanded from these soils than they were capable of sustaining.[80] Nevertheless, although output per unit of land was undoubtedly low, Bailey could find no evidence that it was declining. On the contrary, a flexible agricultural system had evolved based on the close integration of arable and pastoral husbandry and the commercial production of grain—particularly barley—and wool. Breckland's soils may have been poor but they were located within the heart of one of the most prosperous and developed regions in the country and enjoyed convenient river access to Kings Lynn and its thriving grain trade. In short, economic advantages more than offset the region's environmental limitations.[81]

By contrast, it was altogether more difficult to overcome a remote and inconvenient location, especially when it was reinforced by regular field systems and inflexible tenurial structures. Economic development was particularly deterred in regions that were land-locked and far removed from substantial markets.[82] Shropshire and adjacent portions of the west Midlands are a case in point, characterized as they were by arable land values that were among the lowest in the country and by an arable and pastoral husbandry that was neither intensive nor specialized.[83] Here, as in similar situations in northern, western, and southern England, it may have been the lack of substantial market demand that discouraged the adoption of more intensive and productive methods.[84] Indeed, low levels of economic rent—borne of remoteness, transportation difficulties, and imperfections and inefficiencies in the functioning of markets—may have lain like a shadow over much of medieval England, since where economic rent was low there was little incentive to specialize, invest, improve methods, employ more labor, and thereby raise productivity levels. The result may well have been an adherence to farming methods that in their very extensiveness and corresponding deficiency of capital and labor inputs were ultimately inimical to the maintenance of soil fertility. Certainly, Shropshire was one of the English counties most affected by the contraction of arable land after 1291.[85]

Elsewhere, as Persson has effectively argued, in regions exposed to the full force of commercial demand, market specialization was undoubtedly a source of productivity growth within the medieval economy since it enabled producers to maximize their comparative advantage and lent impetus to the process of technological advance.[86] Nevertheless, it was not without its drawbacks, for market dependence incurred its own risks and liabilities. There was a risk to consumers, particularly those in towns, of a failure in supply (witness the heightened mortality of urban populations at times of harvest failure) and a corresponding risk to producers who depended for income on the proceeds of such trade.[87] As Barbara F. Harvey has recently observed, "a market economy and a subsistence level of production—this could be a most unfortunate combination, and those who lived with it lived dangerously."[88] If markets were disrupted and prices fell, indebtedness and, ultimately, economic ruin might result. Moreover, as Andrew B. Appleby has demonstrated for the seventeenth century, shifts in the terms of trade might operate to the disadvantage of particular producers; favoring agriculturalists over artisans and grain producers over pastoralists.[89]

War and Taxation as Sources of Ecological Disequilibrium

War represented one of the most potentially disruptive of all influences. The actual physical destruction of crops, stock, and capital equipment affected only those unfortunate enough to find themselves in the direct path of armies, but the taxes levied to keep these armies in the field fell upon almost everyone. Such taxes, if heavy and repeated, deprived rural producers of crucial investment and working capital, while expropriation and its attendant dislocation of established trade patterns drove up transaction costs to the detriment of all concerned.[90] Warfare, in fact, represented a major threat to the settled pursuit of both agriculture and trade and, significantly, it is from the 1280s—as warfare burgeoned—that signs of contraction become apparent in both.[91]

The 1280s opened with Edward I's costly, but ultimately successful, campaign against Wales.[92] In the 1290s, at much greater cost but with significantly less success, he turned his attention to both France and Scotland. The scale of the forces that he put in the field in his Scottish campaigns were without precedent and were financed by heavy and repeated taxation.[93] The Scottish war dragged inconclusively on until the 1340s, with Edward II actually raising a war tax at the height of the Great Famine, and was accompanied by much economic disruption and physical destruction of stock and property in the northern counties, especially following the disastrous English defeat at Bannockburn in 1314.[94] Then, in 1337, Edward III declared war against France and this was accompanied by renewed tax demands, culminating, in 1341, in the punitive tax equivalent to the tithe on grain, wool, and lambs.[95] This bit more deeply into the country's resources than any previous tax, while to the damage wrought by Scottish raiding parties in the north was added the threat of French attacks along the south coast.[96]

If, by the end of this period, cultivators lacked the necessary livestock and seed to maintain cultivation at its established level, it may well have been, as certain of the returns to the *Nonarum Inquisitiones* of 1342 complained, because of the innumerable taxes and tallages by which the rural population had been impoverished.[97] Certainly, environmental degradation is associated with warfare and excessive military expenditure in many underdeveloped countries today. In these instances ecological breakdown stems not so much from the intrinsic limitations of the agricultural system per se but from agriculture's inability to maintain established levels of biological reproduction in the face of excessive external demands. Such an explanation may well fit the experience of England during the first half of the fourteenth century, when some parts of the country clearly experienced difficulty in weathering the various vicissitudes of the time. The *Nonarum Inquisitiones*, although a notoriously inconsistent source, suggest some of the areas that were most adversely affected.[98] Parts of Yorkshire, economically remote and exposed to attacks from the Scots, show up, along with a concentration of vills in Shropshire, on the Welsh border, likewise remote and no doubt called upon to help provision the crown's Welsh garrisons. On the south coast Sussex had plainly suffered from coastal flooding, French attacks, and the purveyancing activities of royal officials entrusted with the task of provisioning the fleets and armies involved in the French campaigns. Most prominent of all, however, was a group of counties in the east Midlands—southern Cambridgeshire, northeastern Hertfordshire, Bedfordshire, and Buckinghamshire—where, in scores of parishes, arable land lay untilled due to shrinking village populations, impoverished tenants, shortages of seed, and inadequate numbers of plough animals.[99] Substantial portions of these counties lay well within the commercial orbit of London and its east-coast entrepôts, most were dominated by regular commonfield systems, and in a majority of these commonfield townships permanent grassland was in desperately short supply.[100] Even under normal circumstances the ecological equilibrium was probably precarious, and under war conditions the strain may well have become too great.[101]

If the market was more important than the environment in structuring agricultural production, it may therefore have been disruption to the market—partially induced by warfare and excessive taxation—that, indirectly, led to environmental degradation and ecological decline.[102] Purveyancing—the obligatory sale of goods to royal officials for often less than the market price (payment for which was not always honored)—struck at the heart of the nascent market economy, for it was precisely those parts of the country that had become most closely involved in the market and were geared towards the production of surpluses that were most heavily and repeatedly purveyed. As J. R. Maddicott has documented, "throughout the period purveyance fell most often upon the counties of the east coast and the east midlands. . . . The counties afflicted most regularly and heavily were Lincolnshire, Cambridgeshire and Huntingdonshire, while others more fortunate, particularly those of the western and southern midlands, escaped almost entirely."[103] England, in effect, bled from her commercial arteries and the resulting hemorrhage sapped the rural economy of vital reproductive power at precisely the time that the high tide of population growth and a succession of extreme natural events were placing it under increased pressure.[104] It may have been this conjuncture, more than any technological and structural weaknesses inherent within agriculture, that destabilized the ecological status quo and engendered the agrarian crisis of the early fourteenth century.

I am grateful to John Langdon for supplying data, to John Power for research assistance, and to Gill Alexander for drawing the figures. Preliminary versions of this paper were presented to seminars at University College Dublin, the University of Colorado at Boulder, and the University of Saskatchewan at Saskatoon. I am grateful to all those who participated in these meetings, and to Dr. James Masschaele, for their helpful suggestions and comments.

Notes

1. The proposal that soil exhaustion may have been the reason for the decline of ancient Jericho was the subject of a 1990 exchange of letters in the *Times* (London). On the Mayas see J. Eric S. Thompson, *The Rise and Fall of Maya Civilization*, 2nd ed. (Norman: University of Oklahoma Press, 1966); Leslie Bethell, ed., *The Cambridge History of Latin America* 1: *Colonial Latin America* (Cambridge: Cambridge University Press, 1984), pp. 8–10. For medieval England the soil-exhaustion thesis is most fully elaborated in M. M. Postan, "Medieval Agrarian Society in Its Prime, Pt. 7: England," in *The Cambridge Economic History of Europe* 1: *The Agrarian*

Life of the Middle Ages, 2nd ed., ed. Postan (Cambridge: Cambridge University Press, 1966), pp. 549–632.

2. The standard work on the Great Famine remains Ian Kershaw, "The Great Famine and the Agrarian Crisis in England, 1315–1322," *Past and Present* 59 (1973): 3–50; reprinted in *Peasants, Knights and Heretics: Studies in Medieval English Social History*, ed. R. H. Hilton (Cambridge: Cambridge University Press, 1976), pp. 85–132. For a demographic case study of one community during this period see Zvi Razi, *Life, Marriage and Death in a Medieval Parish: Economy, Society and Demography in Halesowen, 1270–1400* (Cambridge: Cambridge University Press, 1980).

3. Barbara A. Hanawalt, "Economic Influences on the Pattern of Crime in England, 1300–1348," *American Journal of Legal History* 18 (1974): 291; E. H. Phelps Brown and Sheila V. Hopkins, "Seven Centuries of the Prices of Consumables Compared with Builders' Wage-Rates," *Economica* 23 (1956): 296–314; reprinted in *Essays in Economic History* 2, ed. E. M. Carus-Wilson (London: Edward Arnold, 1962): 179–96, and Henry Phelps Brown and Sheila V. Hopkins, *A Perspective of Wages and Prices* (London and New York: Methuen, 1981), pp. 13–59; E. A. Wrigley, and R. S. Schofield, *The Population History of England, 1541–1871: A Reconstruction* (London: Edward Arnold, 1981), pp. 402–53.

4. John Langdon, *Horses, Oxen and Technological Innovation: The Use of Draught Animals in English Farming from 1066 to 1500* (Cambridge: Cambridge University Press, 1986), pp. 165–67; Mavis Mate, "The Estates of Canterbury Cathedral Priory Before the Black Death, 1315–1348," *Studies in Medieval and Renaissance History* n.s. 8 (1986): 3–31; Mate, "The Agrarian Economy of South-East England Before the Black Death: Depressed or Buoyant?" in *Before the Black Death: Studies in the "Crisis" of the Early Fourteenth Century*, ed. Bruce M. S. Campbell (Manchester: Manchester University Press, 1991), pp. 79–109.

5. D. G. Watts, "A Model for the Early Fourteenth Century," *Economic History Review* 2nd ser. 20 (1967): 547; Bruce M. S. Campbell, "Population Pressure, Inheritance and the Land Market in a Fourteenth-Century Peasant Community," in *Land, Kinship and Life-Cycle*, ed. Richard M. Smith (Cambridge: Cambridge University Press, 1984), pp. 115–17. But see Mate, "Agrarian Economy of South-East England," pp. 90–103, for a contrasting view of the 1330s.

6. J. F. D. Shrewsbury, *A History of Bubonic Plague in the British Isles* (Cambridge: Cambridge University Press, 1970), pp. 54–125; Graham Twigg, *The Black Death: A Biological Reappraisal* (London: Batsford, 1984), pp. 58–74; Richard M. Smith, "Demographic Developments in Rural England, 1300–48: A Survey," in *Before the Black Death*, ed. Campbell, pp. 25–78; Richard Lomas, "The Black Death in County Durham," *Journal of Medieval History* 15 (1989): 127–40. After 1541 the greatest single mortality peak occurred as a result of the influenza epidemic of 1558: F. J. Fisher, "Influenza and Inflation in Tudor England," *Economic History Review* 2nd ser. 18 (1965): 120–29; Wrigley and Schofield, *Population History of England*, pp. 336–39; John S. Moore, "Jack Fisher's 'Flu': A Visitation Revisited," *Economic History Review* 2nd ser. 46 (1993): 280–307.

7. Mark Bailey, "*Per impetum maris*: Natural Disaster and Economic Decline in Eastern England, 1275–1350," in *Before the Black Death*, ed. Campbell, pp. 184–208.

98   Bruce M. S. Campbell

8. C. T. Smith, "The Historical Evidence," in Smith et al., *The Making of the Broads: A Reconsideration of Their Origin in the Light of New Evidence*, Royal Geographical Society Research Series 3 (London: Royal Geographical Society, 1960), pp. 63–111; P. F. Brandon, "Agriculture and the Effects of Floods and Weather at Barnhorne, Sussex, during the Late Middle Ages," *Sussex Archaeological Collections* 109 (1971): 69–93.

9. This view is most fully articulated in Postan, "Medieval Agrarian Society: England." Key evidence is discussed in M. M. Postan and J. Z. Titow [and Statistical Notes by J. Longden], "Heriots and Prices on Winchester Manors," *Economic History Review* 2nd ser. 11 (1958–59): 392–417; reprinted in Postan, *Essays on Medieval Agriculture and General Problems of the Medieval Economy* (Cambridge: Cambridge University Press, 1973), pp. 150–85. See also J. Z. Titow, *English Rural Society, 1200–1350* (London: Allen & Unwin, 1969). For an early critique of this view see B. F. Harvey, "The Population Trend in England between 1300 and 1348," *Transactions of the Royal Historical Society* 5th ser. 16 (1966): 23–42. For an altogether more optimistic verdict of agrarian trends see H. E. Hallam, *Rural England, 1066–1348* (Brighton, Sussex: Harvester Press, 1981).

10. Postan, "Medieval Agrarian Society: England," pp. 561–62; Smith, "Demographic Developments in Rural England," pp. 48–50. Hallam has suggested a pre-Black Death population as high as 7.2 million: *Rural England*, pp. 246–47. Nevertheless, known patterns of demesne land use and productivity imply that it would have been difficult to feed a population of more than 5.0 million: Bruce M. S. Campbell et al., *A Medieval Capital and Its Grain Supply: Agrarian Production and Distribution in the London Region c. 1300*, Historical Geography Research Series 30 (Historical Geography Research Group, 1993), pp. 37–45.

11. On the economic constraints of a dependence upon inorganic raw materials and animate sources of power, see E. A. Wrigley, "The Supply of Raw Materials in the Industrial Revolution," *Economic History Review* 2nd ser. 15 (1962): 1–16.

12. J. D. Chambers, *Population, Economy, and Society in Pre-industrial England* (New York: Oxford University Press, 1972), pp. 24–25.

13. M. M. Postan, *The Medieval Economy and Society: An Economic History of Britain, 1100–1500* (London: Weidenfeld & Nicolson, 1972), p. 71. But see Mark Overton and Bruce M. S. Campbell, "Productivity Change in European Agricultural Development," in *Land, Labour and Livestock: Historical Studies in European Agricultural Productivity*, ed. Campbell and Overton (Manchester: Manchester University Press, 1991), pp. 29–36.

14. Postan, "Medieval Agrarian Society: England"; M. M. Postan, "Note," *Economic History Review* 2nd ser. 12 (1959–60): 77–82. Postan's views about the colonization of marginal land are derived from those of Wilhelm Abel in *Die Wüstungen des ausgehenden Mittelalters*, 2nd. ed. (Stuttgart: G. Fischer, 1955).

15. Robert Brenner, "Agrarian Class Structure and Economic Development in Pre-industrial Europe," *Past and Present* 70 (1976): 30–75; reprinted as pp. 10–63 in *The Brenner Debate: Agrarian Class Structure and Economic Development in Pre-Industrial Europe*, ed. T. H. Aston and C. H. E. Philpin (New York: Cambridge University Press, 1985), p. 33 (all subsequent references are to this version).

16. Brenner, "Agrarian Class Structure," p. 33.

17. R. H. Hilton, "Rent and Capital Formation in Feudal Society," in Hilton, *The English Peasantry in the Later Middle Ages* (Oxford: Clarendon, 1975), pp. 177–96.

18. Edward Miller and John Hatcher, *Medieval England: Rural Society and Economic Change, 1086–1348* (London: Longman, 1978), p. 217.

19. Postan, *Medieval Economy and Society,* p. 44, and "Medieval Agrarian Society: England," pp. 556–57.

20. Miller and Hatcher, *Medieval England*, p. 217.

21. J. Z. Titow, *Winchester Yields: A Study in Medieval Agricultural Productivity* (Cambridge: Cambridge University Press, 1972), p. 30; David L. Farmer, "Grain Yields on Westminster Abbey Manors, 1271–1410," *Canadian Journal of History* 18 (1983): 331–47. But see Bruce M. S. Campbell, "Land, Labour, Livestock, and Productivity Trends in English Seignorial Agriculture, 1208–1450," and Christopher Thornton, "The Determinants of Land Productivity on the Bishop of Winchester's Demesne of Rimpton, 1208 to 1403," in *Land, Labour and Livestock*, ed. Campbell and Overton, pp. 144–82 and 183–210.

22. Postan, "Medieval Agrarian Society: England," p. 557.

23. R. H. Hilton, "Towns in Societies—Medieval England," *Urban History Yearbook* 9 (1982): 7.

24. R. H. Britnell, "The Proliferation of Markets in England, 1200–1349," *Economic History Review* 2nd ser. 34 (1981): 209–21.

25. Ibid., p. 221.

26. Postan, *Medieval Economy and Society*, p. 59: "We have so far assumed that it was the shortage of pasture that kept the numbers of animals down. That the assumption is right and that the shortage of pasture was great and widespread is revealed by the high and rising rents and by the prices of pastures as given in manorial surveys, custumals and similar manorial valuations of land."

27. R. E. Glasscock, "England *circa* 1334," in *A New Historical Geography of England*, ed. H. C. Darby (Cambridge: Cambridge University Press, 1973), p. 167.

28. Robert S. Lopez, *The Commercial Revolution of the Middle Ages, 950–1350* (Englewood Cliffs, N.J.: Prentice-Hall, 1971); Karl Gunnar Persson, *Pre-industrial Economic Growth: Social Organization and Technological Progress in Europe* (Oxford: Basil Blackwell, 1988), pp. 63–104.

29. Derek Keene, "A New Study of London before the Great Fire," *Urban History Yearbook* 11 (1984): 11–21; Keene, *Cheapside before the Great Fire* (London: Economic and Social Research Council, 1985).

30. Persson, *Pre-industrial Economic Growth*, pp. 12, 31, 66; Karl Gunnar Persson, "Labour Productivity in Medieval Agriculture: Tuscany and the 'Low Countries,'" in *Land, Labour and Livestock*, ed. Campbell and Overton, pp. 124–43; Elizabeth Rutledge, "Immigration and Population Growth in Early Fourteenth-Century Norwich: Evidence from the Tithing Roll," *Urban History Yearbook* 15 (1988): 15–30. Derek Keene estimates the population of Winchester to have been 10–12,000 ca. 1300: *Survey of Medieval Winchester*, Winchester Studies 2 (Oxford: Clarendon, 1985): 366–70.

31. Derek Keene, "Medieval London and Its Region," *London Journal* 14

(1989): 99–111; Campbell et al., *A Medieval Capital*. For the impact of London's growth upon English agriculture in the early modern period, see F. J. Fisher, "The Development of the London Food Market, 1540–1640," *Economic History Review* 2nd ser. 5 (1935): 46–64; reprinted in *Essays in Economic History*, ed. E. M. Carus-Wilson, 1 (London: Edward Arnold, 1954): 135–51; E. A. Wrigley, "A Simple Model of London's Importance in Changing English Society and Economy, 1650–1750," *Past and Present* 37 (1967): 44–70.

32. Eileen Power, *The Wool Trade in English Medieval History* (London: Oxford University Press, 1941; reprint New York: Greenwood, 1987); T. H. Lloyd, *The English Wool Trade in the Middle Ages* (Cambridge: Cambridge University Press, 1977); E. M. Carus-Wilson and Olive Coleman, *England's Export Trade, 1275–1547* (Oxford: Clarendon, 1963); R. A. Pelham, "Medieval Foreign Trade: Eastern Ports," in *An Historical Geography of England before* A.D.*1800*, ed. H. C. Darby (Cambridge: Cambridge University Press, 1963), pp. 314–16; Maryanne Kowaleski, "Town and Country in Late Medieval England: The Hide and Leather Trade," in *Work in Towns, 850–1850*, ed. Penelope J. Corfield and Derek Keene (Leicester: Leicester University Press, 1990), pp. 57–73.

33. N. S. B. Gras, *The Evolution of the English Corn Market from the Twelfth to the Eighteenth Century* (Cambridge, Mass.: Harvard University Press, 1926).

34. Gras, *English Corn Market*, pp. 171–76; Vanessa Parker, "The Economic Development of Lynn and Its Hinterland," in Parker, *The Making of Kings Lynn* (London: Phillimore, 1971), pp. 1–18; H. C. Darby, *The Medieval Fenland* (Cambridge: Cambridge University Press, 1940; 2nd ed. Newton Abbot: David & Charles, 1974); Dorothy M. Owen, ed., *The Making of King's Lynn*, Records of Social and Economic History, n.s. 9 (London: Oxford University Press for the British Academy, 1984); Pelham, "Medieval Foreign Trade," p. 301; James Frederick Edwards and Brian Paul Hindle, "The Transportation System of Medieval England and Wales," *Journal of Historical Geography* 17 (1991): 123–34; John Langdon, "Inland Water Transport in Medieval England," *Journal of Historical Geography* 19 (1993): 1–11.

35. In 1335–36 the sheriff of Cambridgeshire and Huntingdonshire acquired 801 quarters 1 bushel of wheat and 341 quarters 2 bushels of oats for Edward III's military campaign in Scotland and hired eight granaries in the towns of Cambridge, Wisbech, Ely, St. Ives, and Lynn at a charge of £1 13s. 8d. for its storage. The previous year he had acquired 575 quarters 5 bushels of wheat, 136 quarters 2 bushels of peas and beans, and 210 quarters of oats, "bought" in "diverse vills and markets." Seven granaries were then hired in Cambridge and St. Ives to store the grain: London, Public Record Office E358/2.

36. Grain acquired by the sheriff of Cambridgeshire and Huntingdonshire in 1334–36 (see above n. 35) was sent by boat to Kings Lynn at 4d. a quarter, for transshipment to Newcastle at a further charge of 7d. a quarter plus transshipment and storage costs. In 1335–36 it took 14 boats and four ships to shift 1,063 quarters: London, Public Record Office E358/2.

37. In 1299, 245 quarters 2 bushels of wheat, 468 quarters 2 bushels of barley, 266 quarters 7 bushels of oats, and 21 quarters 6 bushels of beans and peas were assembled at Norwich and then transported to Yarmouth by boat via the river Yare

at a total cost of 1 ½d. per quarter: London, Public Record Office E101/574/4. In 1319–20 grain was boated down the river Waveney from Beccles to Yarmouth at 1d. per quarter: London, Public Record Office E101/574/25. In 1345 grain was boated from Wroxham to Yarmouth via the river Bure at 1d. per quarter: London, Public Record Office E101/575/13. A. Saul, "Great Yarmouth in the Fourteenth Century: A Study in Trade, Politics, and Society," unpublished D.Phil. thesis, University of Oxford, 1975. Ipswich in Suffolk and Maldon in Essex were both used as assembly points for grain ultimately destined for Newcastle and Berwick: London, Public Record Office E101/574/33 and E358/2.

38. Keene, "New Study of London," p. 104; David L. Farmer, "Marketing the Produce of the Countryside, 1200–1500," in *The Agrarian History of England and Wales* 3: *1348–1500*, ed. Edward Miller (Cambridge: Cambridge University Press, 1991), pp. 370–72; Campbell et al., *A Medieval Capital*, pp. 47–49.

39. Pelham, "Medieval Foreign Trade," pp. 299–304; R. A. Pelham, "Fourteenth-Century England," in *Historical Geography of England*, ed. Darby, p. 262.

40. The Hundred Rolls of 1279 are our best guide to the relative scale of the demesne sector: E. A. Kosminsky, *Studies in the Agrarian History of England in the Thirteenth Century*, trans. Ruth Kisch and R. H. Hilton (Oxford: Basil Blackwell, 1956), pp. 87–95.

41. For the derivation of this classification see Bruce M. S. Campbell and John P. Power, "Mapping the Agricultural Geography of Medieval England," *Journal of Historical Geography* 15 (1989): 24–39. See also John P. Power and Bruce M. S. Campbell, "Cluster Analysis and the Classification of Medieval Demesne-Farming Systems," *Transactions of the Institute of British Geographers* n.s. 17 (1992): 227–45.

42. Edward Miller, "Farming Techniques: Northern England," and John Hatcher, "Farming Techniques: South-Western England," in *The Agrarian History of England and Wales* 2: *1042–1350*, ed. H. E. Hallam (Cambridge: Cambridge University Press, 1988), pp. 399–411 and 383–98; Campbell et al., *A Medieval Capital*, pp. 115–18.

43. Ian Keil, "Farming on the Dorset Estates of Glastonbury Abbey in the Early Fourteenth Century," *Proceedings of the Dorset Natural History and Archaeological Society* 87 (1966): 234–50; H. S. A. Fox, "Field Systems of East and South Devon, Part I: East Devon," *Transactions of the Devonshire Association* 104 (1972): 88–97.

44. For evidence of the distribution of the three-field system in the Middle Ages see Howard Levi Gray, *English Field Systems* (Cambridge, Mass.: Harvard University Press, 1915), pp. 17–82; Joan Thirsk, "Field Systems of the East Midlands," in *Studies of Field Systems in the British Isles*, ed. Alan R. H. Baker and Robin A. Butlin (Cambridge: Cambridge University Press, 1973); H. S. A. Fox, "The Alleged Transformation from Two-Field to Three-Field Systems in Medieval England," *Economic History Review* 2nd ser. 39 (1986): 526–48. On three-course husbandry in Essex (a county that lacked regular commonfield systems) see R. H. Britnell, "Agriculture in a Region of Ancient Enclosure, 1185–1500," *Nottingham Mediaeval Studies* 27 (1983): 37–55.

45. J. Ambrose Raftis, *The Estates of Ramsey Abbey* (Toronto: Pontifical Insti-

tute of Mediaeval Studies, 1957); Raftis, "Farming Techniques: The East Midlands," in *Agrarian History of England and Wales* 2, ed. Hallam: 325–40; Kathleen Biddick, *The Other Economy: Pastoral Husbandry on a Medieval Estate* (Berkeley: University of California Press, 1989), pp. 65–72.

46. On the sowing of *inhoks* with legumes in Cambridgeshire see Frances M. Page, *The Estates of Crowland Abbey: A Study in Manorial Organisation* (Cambridge: Cambridge University Press, 1934), pp. 118–19; J. R. Ravensdale, *Liable to Floods: Village Landscape on the Edge of the Fens*, A.D. *450–1850* (London: Cambridge University Press, 1974), pp. 116–20. The physical reorganization of furlongs is discussed in David Hall, "The Origins of Open-Field Agriculture: The Archaeological Fieldwork Evidence," in *The Origins of Open Field Agriculture*, ed. Trevor Rowley (London: Croom Helm, 1981), pp. 22–38.

47. Bruce M. S. Campbell, "Agricultural Progress in Medieval England: Some Evidence from Eastern Norfolk," *Economic History Review* 2nd ser. 36 (1983): 26–46; R. A. L. Smith, *The Estates of Canterbury Cathedral Priory* (Cambridge: Cambridge University Press, 1943); Mavis Mate, "Medieval Agrarian Practices: The Determining Factors," *Agricultural History Review* 33 (1985): 22–31; P. F. Brandon, "Demesne Arable Farming in Coastal Sussex during the Later Middle Ages," *Agricultural History Review* 19 (1971): 113–42; Mavis Mate, "Profit and Productivity on the Estates of Isabella de Forz (1260–92)," *Economic History Review* 2nd ser. 33 (1980): 326–34; Campbell et al., *A Medieval Capital*, pp. 135–38.

48. On the distribution of population and wealth in fourteenth-century England see Glasscock, "England *circa* 1334," and Alan R. H. Baker, "Changes in the Later Middle Ages," in *A New Historical Geography of England*, ed. Darby, pp. 137–45 and 190–92.

49. In November 1258 the sheriff of Lancaster was commanded to draft 120 young oxen and cows for delivery to the royal larder at Westminster before December 6: R. Cunliffe Shaw, *The Royal Forest of Lancaster* (Preston, 1956), p. 357. J. H. Munro, "Wool-Price Schedules and the Qualities of English Wools in the Later Middle Ages, c. 1270–1499," *Textile History* 9 (1978): 118–69.

50. G. H. Tupling, *The Economic History of Rossendale* (Manchester: Chetham Society, 1927), pp. 17–41; Shaw, *Royal Forest of Lancaster*, pp. 353–91; R. A. Donkin, *The Cistercians: Studies in the Geography of Medieval England and Wales* (Toronto: Pontifical Institute of Mediaeval Studies, 1978), pp. 68–79; I. S. W. Blanchard, "Economic Change in Derbyshire in the Late Middle Ages, 1272–1540," unpublished Ph.D. thesis, University of London, 1967, pp. 168–74.

51. On seignorial sheep farming in southern England: S. F. Hockey, *The Account-Book of Beaulieu Abbey*, Camden Society, 4th ser. 16 (London: Royal Historical Society, 1975); R. H. Hilton, "Winchcombe Abbey and the Manor of Sherborne," *University of Birmingham Historical Journal* 2 (1949–50): 50–52; Richenda Scott, "Medieval Agriculture," in *Victoria History of the County of Wiltshire* 4, ed. R. B. Pugh, (London: Oxford University Press, 1959): 19–21; J. N. Hare, "Change and Continuity in Wiltshire Agriculture in the Later Middle Ages," in *Agricultural Improvement: Medieval and Modern*, ed. Walter Minchinton, Exeter Papers in Economic History 14 (Exeter: University of Exeter, 1981), pp. 4–9; A. R. Bridbury, "Before the Black Death," *Economic History Review* 2nd ser. 30 (1977): 398–99; M. J.

Stephenson, "Wool Yields in the Medieval Economy," *Economic History Review* 2nd ser. 41 (1988): 368–91. On Breckland sheep farming see Mark Bailey, "Sand into Gold: The Evolution of the Foldcourse System in West Suffolk, 1200–1600," *Agricultural History Review* 38 (1990): 40–57. On the marshland sheep flocks of Essex see H. C. Darby, *The Domesday Geography of Eastern England*, 3rd ed. (Cambridge: Cambridge University Press, 1971), pp. 241–45.

52. John Langdon, "Horse Hauling: A Revolution in Vehicle Transport in Twelfth- and Thirteenth-Century England?" *Past and Present* 103 (1984): 37–66; Bruce M. S. Campbell, "Towards an Agricultural Geography of Medieval England," *Agricultural History Review* 36 (1988): 94–98; Hallam, *Rural England*, p. 248.

53. Pastoral husbandry in the fens and on the fen-edge is discussed in F. M. Page, "'Bidentes Hoylandie': A Mediaeval Sheep Farm," *Economic History* (a supplement to the *Economic Journal*) 1 (1929): 603–13; Raftis, *Estates of Ramsey Abbey*, pp. 129–57; Biddick, *The Other Economy*. On pastoral husbandry in Devon see H. P. R. Finberg, *Tavistock Abbey: A Study in the Social and Economic History of Devon* (Cambridge: Cambridge University Press, 1951), pp. 129–58; N. W. Alcock, "An East Devon Manor in the Later Middle Ages, Part I: 1374–1420, the Manor Farm," *Report and Transactions of the Devonshire Association* 102 (1970): 141–87; K. Ugawa, "The Economic Development of Some Devon Manors in the Thirteenth Century," *Report and Transactions of the Devonshire Association* 94 (1962): 652; Hatcher, "Farming Techniques," pp. 395–98.

54. James Winny, ed., *The General Prologue to the Canterbury Tales* (Cambridge: Cambridge University Press, 1965), p. 70. Chaucer is known to have had first-hand knowledge of Norfolk, and his account of the demesne farm and its stock at Bawdeswell, located in the very heart of the county, may be an attempt at genuine topographical description: Bruce M. S. Campbell, "The Livestock of Chaucer's Reeve: Fact or Fiction?" in *The Salt of Common Life: Individuality and Choice in the Medieval Town, Countryside and Church: Essays Presented to J. Ambrose Raftis on the Occasion of his 70th Birthday*, ed. Edwin Brezette Dewindt (Kalamazoo: Medieval Institute Publications, Western Michigan University, forthcoming). Oswald's shrewd and hard-bargaining character was certainly shared by many of the natives of this county: Campbell, "Population Pressure"; Elaine Clark, "Debt Litigation in a Late Medieval English Vill," in *Pathways to Medieval Peasants*, ed. J. A. Raftis (Toronto: Pontifical Institute of Mediaeval Studies, 1981), pp. 247–79.

55. Campbell and Power, "Mapping the Agricultural Geography," p. 29, fig. 1.

56. For a classification of field systems see Bruce M. S. Campbell, "Common-field Origins: The Regional Dimension," in *Origins of Open Field Agriculture*, ed. Rowley, pp. 112–29. The distribution of field systems is discussed in Gray, *English Field Systems*, and in Baker and Butlin, eds., *Studies of Field Systems*. Debate continues about the efficiency of the commonfields and their influence upon agricultural development: Stefano Fenoaltea, "Transaction Costs, Whig History, and the Common Fields," *Politics and Society* 16 (1988): 171–240.

57. Michael Chisholm, "Johann Heinrich von Thünen," in Chisholm, *Rural Settlement and Land-Use: An Essay in Location* (London: Hutchinson, 1962), pp. 21–35.

58. David Ricardo, *On the Principles of Political Economy and Taxation*, vol. 1 of *The Works and Correspondence of David Ricardo*, ed. Piero Sraffa and M. H. Dobb (Cambridge: Cambridge University Press, 1951); David Grigg, *The Dynamics of Agricultural Change: The Historical Experience* (London: Hutchinson, 1982), pp. 50–59.

59. Johann Heinrich von Thünen, *Der isolierte Staat*, trans. Carla M. Wartenberg and published as *Von Thünen's Isolated State*, ed. Peter Hall (Oxford: Pergamon, 1966); Grigg, *Dynamics of Agricultural Change*, pp. 135–40.

60. Biddick, *The Other Economy*; Kathleen Biddick, "The Link That Separates: Consumption and Production of Pastoral Resources on a Medieval Estate," in *The Social Economy of Consumption*, ed. Henry J. Rutz and Benjamine S. Orlove (Lanham, Md.: University Press of America, 1989); Bruce M. S. Campbell, "Laying Foundations: The Agrarian History of England and Wales 1042–1350," *Agricultural History Review* 37 (1989): 191, n. 15; Campbell et al., *A Medieval Capital*, pp. 145–56, 203–6. The extent of demesne involvement in commercial exchange within ten counties around London in the pre-Black Death period has been the subject of investigation by the research project, "Feeding the City I: The Food Supply of Medieval London," based at the Centre for Metropolitan History, University of London; see Bruce M. S. Campbell, "Measuring the Commercialisation of Seignorial Agriculture c. 1300," in *A Commercialising Economy: England 1086 to c. 1300*, ed. Richard H. Britnell and Bruce M. S. Campbell (Manchester: Manchester University Press, 1995).

61. Postan, "Note," p. 79.

62. In 1352–53 provisions for Calais totaling 152 quarters 6 bushels of wheat, 111 quarters 7 bushels of rye, 70 quarters 4 bushels of oats, 84 quarters 2 bushels of peas, and 599 quarters 7 bushels of malt were obtained from named individuals in 107 different Norfolk vills: London, Public Record Office E101/575/18.

63. John Robert Maddicott, *The English Peasantry and the Demands of the Crown, 1294–1341*, Past and Present Supplement 1 (1975); reprinted in *Landlords, Peasants and Politics in Medieval England*, ed. T. H. Aston (Cambridge: Cambridge University Press, 1987), pp. 285–359, esp. pp. 310–11 (all subsequent references are to this version).

64. J. Ambrose Raftis, *Tenure and Mobility: Studies in the Social History of the Medieval English Village* (Toronto: Pontifical Institute of Mediaeval Studies, 1964); Edwin Brezette DeWindt, *Land and People in Holywell-cum-Needingworth* (Toronto: Pontifical Institute of Mediaeval Studies, 1972); L. R. Poos, "Population and Resources in Two Fourteenth-Century Essex Communities: Great Waltham and High Easter, 1327–1389," unpublished Ph.D. thesis, University of Cambridge, 1983; Campbell, "Population Pressure"; Richard M. Smith, "Families and Their Land in an Area of Partible Inheritance: Redgrave, Suffolk, 1260–1320," in *Land, Kinship and Life-Cycle*, ed. Smith, pp. 135–95; L. Slota, "The Land Market on the St. Albans Manors of Park and Codicote, 1237–1399," unpublished Ph.D. thesis, University of Michigan, 1984; C. Clarke, "Peasant Society and Land Transactions in Chesterton, Cambridgeshire, 1277–1325," unpublished D.Phil. thesis, University of Oxford, 1985.

65. For example, the operation of the land market on the Norfolk manor of Hakeford Hall in Coltishall: Campbell, "Population Pressure."

66. Judith M. Bennett, "The Village Ale-wife: Women and Brewing in Fourteenth-Century England" in *Women and Work in Preindustrial Europe*, ed. Barbara A. Hanawalt (Bloomington: Indiana University Press, 1986), pp. 20–36; Judith M. Bennett, *Women in the Medieval English Countryside: Gender and Household in Brigstock before the Plague* (New York: Oxford University Press, 1987).

67. Lomas, "Black Death in Country Durham"; Campbell, "Agricultural Progress," pp. 39–41.

68. On the regional character of husbandry practices see Bruce M. S. Campbell, "Arable Productivity in Medieval England: Some Evidence from Norfolk," *Journal of Economic History* 43 (1983): 379–404; Mate, "Medieval Agrarian Practices"; Campbell et al., *A Medieval Capital*, pp. 142–43.

69. Marjorie Keniston McIntosh, *Autonomy and Community: The Royal Manor of Havering, 1200–1500* (Cambridge: Cambridge University Press, 1986), pp. 136–52.

70. Bridbury, "Before the Black Death."

71. Kathleen Biddick, "Missing Links: Taxable Wealth, Markets, and Stratification among Medieval English Peasants," *Journal of Interdisciplinary History* 18 (1987): 296–97; also Biddick, "Medieval English Peasants and Market Involvement," *Journal of Economic History* 45 (1985): 823–31.

72. For example, Lynn White Jr., *Medieval Technology and Social Change* (New York: Oxford University Press, 1962); Jean Gimpel, *The Medieval Machine: The Industrial Revolution of the Middle Ages*, 2nd ed. (Aldershot, Hants.: Wildwood House, 1988).

73. E. M. Veale, "The Rabbit in England," *Agricultural History Review* 5 (1957): 85–90; John Sheail, *Rabbits and Their History* (Newton Abbot: David & Charles, 1971); Mark Bailey, "The Rabbit and the Medieval East Anglian Economy," *Agricultural History Review* 36 (1988): 1–20.

74. Langdon, *Horses, Oxen and Technological Innovation*; Langdon, "Horse Hauling."

75. Bruce M. S. Campbell, "The Diffusion of Vetches in Medieval England," *Economic History Review* 2nd ser. 41 (1988): 193–208.

76. Richard Holt, *The Mills of Medieval England* (Oxford: Basil Blackwell, 1988): 17–35.

77. Campbell, "Agricultural Progress"; Mate, "Medieval Agrarian Practices"; Brandon, "Agriculture and the Effects of Floods"; T. A. M. Bishop, "The Rotation of Crops at Westerham, 1297–1350," *Economic History Review* 9 (1938–39): 38–44; Eleanor Searle, *Lordship and Community: Battle Abbey and Its Banlieu, 1066–1538* (Toronto: Pontifical Institute of Mediaeval Studies, 1974), pp. 272–86; Finberg, *Tavistock Abbey*, pp. 86–128. A general survey of medieval farming systems is given in Bruce M. S. Campbell, "People and Land in the Middle Ages, 1066–1500," in *An Historical Geography of England and Wales*, ed. R. A. Dodgshon and R. A. Butlin, 2nd ed. (London: Academic Press, 1990), pp. 89–92; also Power and Campbell, "Cluster Analysis."

78. The profitability of agriculture in these areas of high economic rent is reflected in the exceptionally high value placed upon arable land in manorial extents and *Inquisitiones Post Mortem*. In east Norfolk and northeast Kent these values

commonly exceeded 12d. an acre and sometimes rose as high as 36d.: Campbell, "Agricultural Progress," pp. 28 and 42; Bruce M. S. Campbell, "Medieval Land Use and Land Values," in *An Historical Atlas of Norfolk*, ed. Peter Wade-Martins (Norwich: Norfolk Museum Service, 1993), pp. 48–49; Campbell et al., *A Medieval Capital*, pp. 140–41.

79. Mark Bailey, *A Marginal Economy? East Anglian Breckland in the Later Middle Ages* (Cambridge: Cambridge University Press, 1989).

80. Postan, "Medieval Agrarian Society: England," p. 559; John Saltmarsh, "Plague and Economic Decline in England in the Later Middle Ages," *Cambridge Historical Journal* 7 (1942): 23–41.

81. Mark Bailey, "The Concept of the Margin in the Medieval English Economy," *Economic History Review* 2nd ser. 42 (1989): 1–17.

82. On the constraints to economic growth of institutional factors, and the incentive to growth provided by large and expanding markets, see Douglass C. North, *Structure and Change in Economic History* (New York: Norton, 1981).

83. On many Shropshire demesnes arable land was valued at only 2d. an acre: Campbell, "People and Land," fig. 1A.

84. Notwithstanding significant urban growth during the twelfth and thirteenth centuries the leading urban markets remained too small and their spatial impact too circumscribed. Compare the size of medieval and early modern London: Keene, "Medieval London and Its Region"; E. Anthony Wrigley, "Urban Growth and Agricultural Change: England and the Continent in the Early Modern Period," *Journal of Interdisciplinary History* 15 (1985): 683–728. See also Clyde G. Reed, "Transactions Costs and Differential Growth in Seventeenth Century Western Europe," *Journal of Economic History* 33 (1973): 177–90; R. H. Britnell, "England and Northern Italy in the Early Fourteenth Century: The Economic Contrasts," *Transactions of the Royal Historical Society*, 5th ser. 39 (1989): 168–83. On the circumscribed range of marketing in south Wiltshire in the late-thirteenth and early fourteenth centuries see D. L. Farmer, "Two Wiltshire Manors and Their Markets," *Agricultural History Review* 37 (1989): 1–11.

85. Alan R. H. Baker, "Evidence in the 'Nonarum Inquisitiones' of Contracting Arable Lands in England during the Early Fourteenth Century," *Economic History Review* 2nd ser. 19 (1966): 518–32; reprinted as pp. 85–102 in *Geographical Interpretations of Historical Sources: Readings in Historical Geography*, ed. Baker, John D. Hamshere, and Jack Langton (Newton Abbot: David & Charles, 1970), p. 90 (subsequent references are to this version).

86. Persson, *Pre-industrial Economic Growth*, pp. 31 and 66.

87. The classic study remains Pierre Goubert, *Beauvais et le Beauvaisis de 1600 à 1730*, 2 vols. (Paris: S. E. V. P. E. N., 1960).

88. Barbara F. Harvey, "Introduction: The 'Crisis' of the Early-Fourteenth Century" in *Before the Black Death*, ed. Campbell, p. 15.

89. Andrew B. Appleby, "Disease or Famine? Mortality in Cumberland and Westmorland, 1580–1640," *Economic History Review* 2nd ser. 26 (1973): 403–32.

90. Maddicott, *English Peasantry*. For an up-to-date review of the scale and possible impact of taxation during the first half of the fourteenth century see W. M. Ormrod, "The Crown and the English Economy, 1290–1348," in *Before the Black*

Death, ed. Campbell, pp. 149–83. The adverse consequences of rising transaction costs for the textile industries of northwest Europe in this period are discussed in John H. Munro, "Industrial Transformations in the North-west European Textile Trades c. 1290–c. 1340: Economic Progress or Economic Crisis?" in *Before the Black Death*, ed. Campbell, pp. 110–48.

91. For the adverse consequences for trade of a general escalation in warfare throughout Europe in this period see Munro, "Industrial Transformations." Significantly, the economy of London—very much the economic pulse of the kingdom—appears to have stagnated after 1300: Keene, *Cheapside before the Great Fire*, pp. 19–20.

92. R. R. Davies, *Conquest, Coexistence and Change: Wales, 1063–1415* (Oxford: Oxford University Press, 1987).

93. Maddicott, *English Peasantry*, p. 319.

94. On war taxation see Ormrod, "Crown and English Economy." J. A. Tuck, "War and Society in the Medieval North," *Northern History* 21 (1985): 33–52.

95. Maddicott believes that during the six years 1336–41 "the weight of taxation may have been greater than at any other time in the middle ages, greater even than in the years preceding the revolt of 1381," *English Peasantry*, pp. 329–52.

96. B. Waites, "Medieval Assessments and Agricultural Prosperity in Northeast Yorkshire, 1292–1342," *Yorkshire Archaeological Journal* 44 (1972): 134–45; Mavis Mate, "The Impact of War on the Economy of Canterbury Cathedral Priory, 1294–1340," *Speculum* 57 (1982): 761–78; Mate, "Estates of Canterbury Cathedral Priory."

97. Maddicott, *English Peasantry*.

98. Baker, "Evidence in the 'Nonarum Inquisitiones,'" pp. 88–90, fig. 1. The published edition of the *Nonarum Inquisitiones* (*Nonarum Inquisitiones in Curia Scaccarii*, Record Commission [London: Eyre and Strahan, 1807]) fails to distinguish between the returns from the three separate commissions issued for collection of the tax. Moreover, many other manuscript returns have since come to light (listed individually in Public Record Office, Class E179) which extend the coverage of the published edition by almost 50 percent: Christopher R. Elrington, "Assessments of Gloucestershire: Fiscal Records in Local History," *Transactions of the Bristol and Gloucestershire Archaeological Society* 103 (1985): 5–15.

99. Baker, "Evidence in the 'Nonarum Inquisitiones,'" pp. 91–97. For separate evaluations of the *Nonarum Inquisitiones* evidence for Bedfordshire and Sussex see Alan R. H. Baker, "The Contracting Arable Lands of Bedfordshire in 1341," *Bedfordshire Historical Records* 49 (1970): 7–17; Baker, "Some Evidence of a Reduction in the Acreage of Cultivated Lands in Sussex during the Early Fourteenth Century," *Sussex Archaeological Collections* 104 (1966): 1–5.

100. The raison d'être of these farming systems is discussed further in Campbell, "People and Land."

101. The ecological constraints of commonfield agriculture are debated in William S. Cooter, "Ecological Dimensions of Medieval Agrarian Systems," *Agricultural History* 52 (1978): 458–77; H. S. A. Fox, "Some Ecological Dimensions of Medieval Field Systems," in *Archaeological Approaches to Medieval Europe*, ed. Kathleen Biddick (Kalamazoo: Medieval Institute Publications, Western Michigan University, 1984), pp. 119–58.

102. French medieval historians have long been aware of the debilitating economic consequences of recurrent warfare and taxation: Edouard Perroy, "A l'origine d'une économie contractée: Les crises du XIVe siècle," *Annales: Économies, Sociétés, Civilisations* 3 (1949): 167–82, reprinted in translation as "At the Origin of a Contracted Economy: The Crises of the 14th Century," in *Essays in French Economic History*, ed. Rondo Cameron (Homewood, Ill.: R. D. Irwin, 1970), pp. 91–105; Guy Bois, *The Crisis of Feudalism: Economy and Society in Eastern Normandy, c. 1300–1550* (Cambridge: Cambridge University Press, 1984).

103. Maddicott, *English Peasantry*, p. 301. The correlation with the picture of agricultural contraction conveyed by the *Nonarum Inquisitiones* is not exact. Lincolnshire recorded no vills with uncultivated lands whereas Buckinghamshire, purveyed only three times, recorded many. Nevertheless, as Baker makes clear, the character of the information provided by the *Nonarum Inquisitiones* varies considerably from county to county, nor are returns extant for all counties: "Evidence in the 'Nonarum Inquisitiones,'" pp. 88–89.

104. The adverse consequences of repeated royal purveyancing figure prominently in accounts of the decline of the Anglo-Norman Lordship of Ireland: M. D. O'Sullivan, *Italian Merchant Bankers in Ireland in the Thirteenth Century* (Dublin, 1962); James Lydon, "The Years of Crisis, 1254–1315," in *New History of Ireland* 2: *Medieval Ireland, 1169–1534*, ed. Art Cosgrove (Oxford: Oxford University Press, 1987), pp. 179–204, esp. pp. 195–204.

PART II

Rural Society

6. Thunder and Hail over the Carolingian Countryside

What farmers thought, how they imagined their world to work, and what strategies they consciously adopted to confront a capricious natural world are matters that rarely surface in the records of the Middle Ages. No writer in the ninth century spent much time pondering the thoughts of those who worked and managed the land, probably because it was assumed that those thoughts were unremarkable. Yet what could be more important for our proper appreciation of an age like the Carolingian, in which the countryside and agricultural concerns dominated, than to acknowledge the centrality and, indeed, intelligibility of popular thought? What most worried the Carolingian farmer before his crops were harvested was the weather and what most pressed him afterward were the claims made upon a portion of that harvest by demanding lords. The common Carolingian dread of destructive thunder- and hailstorms not only tells us something about the nature of the rural economy of the period,[1] but also about the way in which people responded to calamity and, on occasion, turned it to their advantage.

One day in 815 or 816, Agobard, the bishop of Lyons, encountered a crowd that was preparing to stone to death three men and a woman who were bound in chains.[2] The bishop learned that the people of his diocese believed that ships traveled in clouds from a region called Magonia in order to retrieve the produce that had been cut down by hail and lost in storms. The people explained to Agobard that the four individuals they had captured were aerial sailors who had fallen out of one of the cloud-ships.[3] After reasoning with the captors, the bishop believed he had finally uncovered the truth behind the story and, with the captors in a state of confusion, the matter was apparently resolved.

Unfortunately, Agobard neglected to tell us what specific and revealing truth he had uncovered. What he did do was write a fascinating tract, "Against the Absurd Belief of the People Concerning Hail and Thunder."[4]

In it he supplies us with a scattering of information about the notion held by the people of his diocese that some individuals could manipulate the weather. The main thrust of his piece, however, was to demonstrate that the weather is controlled by God and not by human beings. Perhaps we shall never know what conspiracy Agobard discovered that day near Lyons, but we can attempt to work our way back into the meaningful world that he and his flock inhabited.

In the treatise Agobard tells us that in the regions around Lyons virtually everyone believed that hail and thunder could be caused by humans. By "virtually everyone," he explained that he meant nobles and common people, city folk and country folk, old and young.[5] As soon as these people heard thunder and saw lightning, they would at once summarily declare that it was a "raised storm."[6] When asked why they described it so, they explained that the storm was raised by the incantations of people called storm-makers (*tempestarii*).[7] Occasionally, when they heard thunder or felt a light puff of wind, people would utter a curse against the evil-speaking tongue of the storm-maker, that it be made parched and silent.[8] Many people believed that the storm-makers had struck a deal with the aerial sailors in which the sailors gave the storm-makers money and received in return grain and other produce which they carried back to Magonia in their ships.[9] Agobard heard incredible accounts of the prowess of the storm-makers, that they could, for instance, control hail so precisely that they could make it fall, if they wished, upon a river and a useless forest or upon a single tub under which one of them hid.[10] On at least one occasion, Agobard heard of someone who had been an eyewitness to one of these events, and he soon tracked him down. Under persistent questioning, however, the man admitted that he had not been personally present when the storm-maker had worked, but he resolutely maintained the truth of his account and named the storm-maker, the time, and the place of the deed.[11] Agobard learned that in many places people claimed that they did not know how to send storms, but, nevertheless, knew how to defend the inhabitants of a place from storms. The people reached an understanding with these defenders about what percentage of their crop they should pay in return for protection, and they called this regular tribute the *canonicus*. Agobard lamented that people who never willingly gave tithes to the church or charitably supported widows, orphans, and the poor freely paid tribute to those who offered to defend them from storms.[12]

In what amounts to an appendix to his tract, Agobard supplies another example of what he thought of as popular foolishness.[13] A few years before

his interest in the storm-makers, when a great number of cattle in the kingdom were struck by disease and died (that is, in 810), a story circulated that
Grimoald, duke of the Beneventans, had sent some of his people to spread
a special dust on the fields, mountains, meadows, and springs of northern
Europe in order to kill the cattle of his enemy, the emperor Charlemagne.
Agobard claims that he both heard and saw that many people were ap-
prehended. Some were killed because of their alleged crime; with plaques
around their necks, they were cast into a river to drown. It amazed Agobard to learn that many of the accused had actually confessed that they had
possessed the poisonous dust and had scattered it, and that neither torture
nor the threat of death had deterred them from giving false witness against
themselves.

Agobard's reaction to these popular beliefs was self-consciously rational and Christian. In the case of the dust-spreaders, for instance, Agobard
thought that the people had not stopped to consider how such dust could
selectively kill cattle and not other animals, how it could be spread over
so wide a territory, and whether, in fact, there were enough Beneventan
men, women, and children to carry out such an immense undertaking. It
saddened him to think that in his time such great foolishness overwhelmed
the wretched world and that such absurdities were believed by Christians.
He was not alone, at least among the powerful and the learned.[14]

The bishop similarly thought that the belief in the power of the storm-
makers to control the weather was another example of popular foolishness
that he could rationally undermine. When the land was too parched to sow,
he wondered why people did not call on the services of the storm-makers
to wet their fields.[15] Why, in fact, should crops ever fail, if the storm-makers
controlled the weather?[16] Why do they not kill their enemies at will with
sudden hailstorms, if they have the power?[17] For Agobard, only God could
control the weather and not the evil-doers whom the people called storm-
makers.[18] Most of his treatise is in fact devoted to the consideration of Old
Testament citations that demonstrate the divine origin of weather. If even
Job could not unlock God's treasure-trove of hail (Job 38: 22–30), then how
could these storm-makers? For these people, "by whom they say violent
winds, crashing thunder, and raised storms can be made, show themselves
to be puny men, devoid of holiness, righteousness, and wisdom, lacking in
faith and truth, and hateful even to their neighbors."[19]

As grateful as we must be for Agobard's fascinating account, we should
not for a minute fall into the trap of thinking that his is an anthropological description of a set of popular beliefs around Lyons in the early ninth

century. The bishop never fully describes the belief in weather-making, in part because his point was to refute and not explain the belief and in part because the believers seem to have resisted talking to him. He admitted that he had great difficulty in locating eyewitnesses, and at least one of these recanted that he had seen such an event, although he stuck fast to the truthfulness of his story. One suspects that Agobard never got inside local belief systems, but operated from the outside as their critic. His rationalism must have met some of the same dead ends that Dr. Livingstone encountered in southern Africa in 1853 as he sought to disprove the local belief in rainmaking.[20] When Livingstone asked a rainmaker if he could make it rain on a specific spot, the rainmaker, apparently puzzled, answered that it had never occurred to him to do so, since all should enjoy the blessings of plentiful rain. In frustration at the rational and persistent questioning of his beliefs, the rainmaker finally said to Livingstone that "your talk is just like that of all who talk on subjects they do not understand. Perhaps you are talking, perhaps not. To me you appear to be perfectly silent."[21] For both Agobard and Livingstone the belief that human beings could control the weather was an important barrier to the Christianization of a rural population.

What we also need to remember about Agobard was that he belonged to the reform movement that was sweeping through the ruling circles of the Carolingian Empire after Charlemagne's death. In 814–815 Louis the Pious cleansed his father's "unwholesome" palace by casting out his "naughty" sisters, ordered the *missi* to restore justice in the kingdom to those who had been arbitrarily mistreated, and introduced a wide-ranging monastic reform.[22] During these years Agobard was the country bishop or suffragan of Lyons and apparently aspired to attract the attention of the reformers.[23] His early writing, which was dominated by works devoted to reform, may well have begun with the treatise against the belief in weather-making. Moreover, as a transplanted Spaniard, Agobard may have brought with him to Francia the deeply ingrained conviction of the higher clergy of Visigothic Spain that weather-making was an evil that should be sharply suppressed.[24] In the battle against perceived paganism Agobard was also cut from that old and venerable mantle worn by the likes of Martin of Tours and Caesarius of Arles, great foes of paganism in Gaul. The tradition had been continued in the Carolingian kingdom by the missionary Boniface and restated by Alcuin, who had urged Arno, the bishop of Salzburg, to imitate Christ by preaching against superstition wherever he found it, be it in castles or countryside.[25] These men were convinced, as one historian has put it, that "there was a barbarian not far below the skin of every Frank."[26]

But Agobard was hardly the first Carolingian to be concerned about the belief in storm-raising, since we find numerous condemnatory references to the custom in official sources. The *Indiculus*, an eighth-century list of popular paganisms, includes a reference to a superstition concerning storms, and Bishop Cathwulf called on Charlemagne as a minister of God to correct, judge, and damn storm-makers and other evil-doers (*malefici*).[27] Indeed, in the famous capitulary, the *Admonitio generalis*, whose chapters were to be repeated so often in other legislation, Charlemagne added to the list of Biblical injunctions against augury and incantation one against the storm-makers.[28] This should alert us to the possibility that storm-raising was considered to be a contemporary problem in Charlemagne's time, perhaps one that worried the king himself.

In the most important legislative reference to weather-making, one of Charlemagne's church councils ordered that those who make storms and do other evil things by incantations, augury, and divination should be captured and turned over to the chief priest of the diocese where they were found. Under his supervision they were to undergo a careful examination to determine whether they would confess to the evil things they had done. Nevertheless, the examination was not to be so harsh that they died as a result; rather storm-makers were to be incarcerated until, with God's help, they promised to reform. But there was a concern that, once caught, these people might escape a strict examination because of rewards given by counts or their subordinates. Hence, it was stipulated that the chief priests should not hide news of comital interference from their bishops who, under such circumstances, should personally take charge of the accused individuals.[29]

The Council of Paris in 829, convened under the auspices of Louis the Pious, specifically condemned storm-making among the other pernicious remnants of paganism: "For some say that by their evil deeds they can stir up the air and send down hail, predict the future, and take away the produce and milk of some and give it to others. And countless other things are said to be done by such people."[30] The Council commanded that, if any men or women were found guilty of practicing storm-making, they were to be severely punished by the lay ruler, since they had openly dared to serve the devil.[31]

Penitentials were equally concerned to curtail the belief in storm-making. Early penitentials, such as the one ascribed to Bede and the so-called Burgundian Penitential, imposed penances of seven years, three of those on bread and water, for the sin.[32] Carolingian penitentials treat the senders of storms (*immissores tempestatum*) as practitioners of magic, and Hrabanus Maurus, in his work *On the Magical Arts*, counted weather divination as a species of magic.[33] The storm-maker, working by means of incantations, was classed by Agobard and the capitularies as a *maleficus*, a word that sometimes means simply evil-doer, but sometimes connotes a witch or magician.[34] The enchanter (*incantator*) was also a *maleficus* and was the pagan worker most closely connected with the enchanting storm-maker by Carolingian sources. Indeed, Agobard's geographical term Magonia might be translated as Magic Land (from *magus* or magician), and, therefore, might be his own satirical coinage.[35] For Regino of Prüm, evil-doers, enchanters, and senders of storms, who cloud the minds of men through the invocation of demons, should be harshly punished.[36] Burchard of Worms, in his tenth-century manual for confessors, the *Corrector*, urged priests to ask their parishioners if they either believed in or had partaken of the perfidy that enchanters and those who call themselves the senders of storms could, through demonic incantations, provoke storms or change people's minds.[37] Both Hrabanus and Agobard were anxious to eliminate the popular support for agents of intercession such as the storm-makers and diviners who stood between the church and its people. Agobard rigorously attacked the storm-makers, not to win them over, since he never seems to have met one, but to discredit their claim to power over the elements and to destroy any popular belief in them. Those who believed that the weather was subject to human control were presented as deluded and superstitious by Agobard, while the penitentials prescribed the means by which the wayward could be reincorporated into Christian society.

Charlemagne and his family may have felt uneasy about the popular belief in storm-making for a complementary set of reasons. From at least the time of Boniface, the Carolingian royal family had tied its own fortunes to the church and to a process of Christianizing Francia. The storm-makers must have posed a perceived threat to official power to the extent that they seemed to belong to the decentralizing face of paganism. What Charlemagne really worried about, if we may judge from his council, was that local priests, counts, and comital agents might conspire to protect the storm-makers. Perhaps he suspected that local counts would establish or always had established regional power bases by forming strong bonds with pagan priests. This was just the sort of ongoing problem Charlemagne had encountered in Saxony; paganism, in short, was a political as well as a religious offense, since it had proved to be a powerful agent of resistance to central authority, both religious and royal.[38] The case of the dust-spreaders is a fascinating example, however opaque, of the perceived mixing of pagan and political interests, for to certain elements of the Carolingian population it seemed to concern a comital enemy of Charlemagne and magic. Moreover, since the Carolingian family had sanctioned one priesthood with a powerful institutional structure, it could ill afford to countenance the presence of another, or rather a diffusion of others, within the kingdom.

But, if the official attitude toward the belief in storm-makers and their claims was critical, rational, and Christian, it was also external. What we would want to recover—were it possible—is the internal attitude, to discover what the people around Lyons actually believed about the storm-makers. Since those individuals Agobard encountered left no record of their own, we shall have to settle for a tentative reconstruction of what they might have believed, why they believed it, and what advantages it brought them.

The popular belief in weather-makers was a very old and, indeed, traditional belief, one that might almost be called western Indo-European, since one can find versions of it among the Greeks, Romans, Germans, and Scandinavians.[39] Seneca made fun of the people of Clenia who appointed hail-officers, and he added that "an older uneducated time used to believe that rains were brought on or driven away by incantations."[40] Six hundred years after Agobard's death, Dr. Johann Hartlieb, who possessed that particularly Renaissance fascination with magic, confronted a confessed hail-maker in Bavaria. When the woman, lying in her cell with one leg in irons, refused to teach the doctor her art unless he spurned Mary and invoked three devils, he turned her over to the inquisitor, who had her burned alive.[41] From antiquity until at least the sixteenth century, it would seem, Europeans believed in storm-makers. In view of this it must have been rigorous orthodoxy that was the newcomer to the countryside of Carolingian Europe, not the popular belief in storm-making. Here, as in so much else, Agobard and an intrusive Carolingian clergy, supported by a centralizing monarchy, were attempting to establish universal standards. It was this very process of intrusion, after all, that brought the belief in storm-making to the attention and scrutiny of officials like Agobard and led to the production of the written record over which we linger.

Of the actual practices of the storm-makers in Carolingian Europe we can say little. All sources agree that they performed their weather magic by incantation, but the words and the actions associated with them are never recorded. The popular curse against the tongue of the storm-maker suggests that people may have thought that the power lay in a secret magical formula spoken by a special individual. The story reported by Agobard that the storm-makers claimed that they could make it hail upon a single tub (*cupa*) under which they hid is interesting, because much later accounts consistently describe the storm-maker as pouring out hail from a tub.[42] As to the sex of the storm-makers, we should note that the church officials at the Council of Paris in 829 thought that a storm-maker might be either male or female. But in the Carolingian sources the words used to describe the storm-maker—*tempestarius*, *immissor*, and *defensor*—are masculine in gender.[43]

Storm-makers seem not to have simply specialized in weather magic, but to have delved into other matters concerning the rural economy, such as the enchantment of crops and foodstuffs and predictions about the farming year. Agobard identified two types of weather-makers: defenders, who prevented storms and received a regular tribute, and the so-called storm-makers, who raised the wind to destroy crops and received a price for them from the sailors from Magonia. The people, in Agobard's account, seem to have thought of these as two faces of the same coin, the benign and malevolent manipulation of the weather respectively. To both of these types we shall return later, but for the moment it is important to think about the popular conception of weather-makers. They were not gods, but the agents of one or of forces that remain unmentioned in the relevant Carolingian literature. In the eleventh century, Adam of Bremen, when writing about Scandinavian religion, said " 'Thor,' they maintain, 'presides over the sky; it is he who rules thunder and lightning, wind and rains, fine weather and crops.' "[44] Though Agobard and the Carolingian penitentials characterized human weather-making as the work of the devil and his agents, the storm-makers themselves may have appealed to some lingering form of that Germanic thunder-god whose name we still honor on Thursday.

Though our knowledge of the actual practices of the storm-makers is bound to remain scanty and speculative, we can still ask why people clung fast to their belief in them and what advantages that belief brought them in their daily lives. The touchstone in both the incidents related by Agobard is the rural economy or, more to the point, crops and cattle. Northern Europe was still covered in the eighth and ninth centuries by old and obstructive forests, so that human habitation clung of necessity to the natural land clearings and easy transport provided by river valleys. But this also made Carolingian villages and small cities particularly vulnerable to storms, to flooding, and to Vikings. Moreover, the Carolingian economy had developed a particular reliance upon cereals and cattle, but the former could be severely damaged by hailstones and the latter by murrain. The Carolingian annals are filled with stories of the destructive power of storms and the ravages of wet weather and with tales of dying cattle. These things mattered to usually laconic annalists because in a subsistence economy they affected a community's very chances for survival. The royal Frankish annalist noted that in 820 constant rain had disastrous economic results, setting off a chain reaction of crop failures, outbreaks of disease among humans and cattle, and delayed planting of the next season's crops.[45] In another year in Frisia 2,437 people were killed by a flash flood,[46] further demonstrating the vulnerability of Carolingian villages to excess and unpredictable rainfall. As the poet put it colorfully when he prayed for good weather:

Look how this wet weather pelts the countryside with violent rain,
And an immense wave of water washes away our fertile fields.[47]

With its low crop yields, this was an economy that had a small margin of success even in good years and could hardly cope with ruinous weather in bad ones.[48]

Some climatologists and historians believe that the climate of Europe deteriorated slightly in the ninth century, becoming marginally colder, wetter, and stormier.[49] Our evidence, however, may be skewed, since a slowly expanding or, at least, concentrated population certainly complained more vociferously about the weather because its economic success was so vulnerable to certain kinds of weather conditions. In fact, today the area around Lyons remains one of the regions in France most susceptible to violent thunder and hailstorms, generally in the months between May and August.[50] At the end of June 1545, Benvenuto Cellini was caught in such a hailstorm just outside Lyons. He claimed that hailstones the size of walnuts and lemons had pummeled his band of travelers, toppled their horses, and broke tree branches. A mile further on they had found fallen trees, the bodies of animals, and dead shepherds.[51] Even if we allow for some measure of exaggeration (which, it must be admitted, abounds in Cellini's autobiography), the storm he suffered through is not atypical. Hailstorms frequently last from fifteen to sixty minutes, are generally the product of severe thunderstorms, and most frequently occur in the summer, often late in the day. Moreover, the size of the hailstones described by Cellini falls within the normal range, and there are frequent reports from India today of both cattle and people dying from hail.[52]

If Cellini's storm was destructive enough to bring down tree branches, sheep, and shepherds, we can imagine what it must have done to fields of standing wheat. Hailstorms tend to drop their stones in a moving swath that may measure 100 by 20 km,[53] and can cause extensive damage. Even today, wheat is the crop most vulnerable to severe damage by hailstorms,[54] though vineyards and orchards may also suffer. The amount of economic damage done by hailstorms, it stands to reason, must be proportionally related to the amount of land under cultivation and to the vulnerability of the crops grown. The Carolingian popular anxiety about hailstorms is, therefore, doubtless a reflection of an economy that had become heavily committed to and dependent upon cereal crops. In the earlier Middle Ages (and despite Cellini's dead shepherds), pastoralists in northern Europe were probably somewhat less worried about the possibility of hailstorms and even floods. In the ninth century, however, the coincidence of grain crops and violent storms led to understandable concern and complaint.

The timing of the hailstorms in the area around Lyons must have also seemed to contemporaries to be most unfortunate, designed in fact to do the maximum amount of damage. If a storm struck in April or May, it would level the unharvested spring wheat so necessary to survival in this subsistence economy after winter reserves of food had run low or been exhausted. In the famous Vienna Labors of the Months illumination made during Agobard's lifetime, a man representing April's work holds a sheaf of spring wheat (Figure 15; see the first scene of the second register). Charlemagne called May *Winnemanoth* or "month of joy" because the spring wheat had been harvested and fodder for the horses and cattle had returned.[55] The king called June *Brachmanoth* ("plowing month") and July *Heuuimanoth* ("hay month"), which are depicted as such in the Labors illumination (second register, scene three and third register, scene one). If a hailstorm struck in those months, it might damage the hay needed to support bovines later in the year. Worse still, even a small hailstorm in late spring or early summer could utterly destroy a field of juvenile wheat before it had had a chance to form ears. The summer must have seemed very long to farmers who had lost their wheat crop in June. If a hailstorm struck in August—Charlemagne's *Aranmanoth* or "the month of the ears," when the heads of cereal plants fully form, shown in the Labors by a farmer harvesting the ripe wheat (third register, second scene)—then mature fields

Figure 15. Labors of the Months (ca. 820). Vienna, Österreichische Nationalbibliothek Cod. 387, fol. 90v.

of grain could be felled. Here the damage was of a different sort as some portion of the flattened crop lying in the field might be collected before it rotted, but even then much of the grain would have been irretrievably scattered by the storm, the hailstones, and the accompanying winds.

Carolingian annalists always associated hailstorms with damage done to crops and subsequent human suffering. The royal Frankish annalist noted that, in 823, in many regions, devastating hail had destroyed crops, and lightning storms had done much damage. After that, disease had raged furiously in the kingdom and had killed people of both sexes and all ages.[56] The Ratisbonne continuator of the Annals of Fulda succinctly stated that, in 889, "with the crops having been destroyed by hail, human beings are suffering wretchedly from want of produce."[57] The annalists of both Xanten and Fulda complained that 872 had experienced a particularly calamitous summer of thunderstorms in which hail had destroyed crops and done much harm to people.[58] The reason why people in the ninth century were so concerned about thunderstorms and hail, therefore, was because they threatened the chief product of the agricultural economy. The loss of much winter or summer wheat to hailstorms could so unbalance local and even regional economies that it might lead to famine. Moreover, the people of the ninth century had come to depend upon bread as their staple food and apparently ate inordinate amounts of it.[59] When wheat was wanting in Charlemagne's world people soon went hungry, or worse. After widespread frost and damage to the harvest in 762, said one chronicler, the following year "many died from lack of bread."[60] In 845, when parts of Gaul were occupied by invaders, people were so desperate for bread that they were said to have eaten loaves made of dirt and a meager amount of flour.[61]

Given the importance of cereal crops and the inevitability of hailstorms in the Carolingian countryside, what were farmers and a society dependent upon cereals to do? There was, it must be said, a tendency at all levels of this society to assign responsibility for the weather to higher powers and to see it in terms of retribution and reward. To the ninth-century mind such events could not, after all, be neutral, mere accidents. Thus the annalist of Fulda understood God's judgment to be at work when, during a terrible storm in September 857, an enormous bolt of lightning shaped like a dragon burst into the Cathedral of Saint Peter at Cologne and struck three men standing at different points in the congregation.[62] In 875 another annalist said that after the appearance of a comet, which gave clear proof of the people's sins, a frightening flash flood had destroyed a village and its eighty-eight inhabitants.[63] God could occasionally cast his retribu-

Figure 16. Hailstorm, Utrecht Psalter, Ps 28 (29): 3–10. Cat. Cod. MS Bibl. Rhenotraiectinae MS 32, fol. 16r. By permission of the University Library, Utrecht.

tion against the enemy, as in 847, when he was thought to have whipped up
a storm to destroy the Saracens who had pillaged Saint Peter's in Rome.[64]
Carolingian iconography, though doubtless dependent on older models,
reinforces the same view of the weather. In the Utrecht Psalter (Figure
16), disembodied heads poke through a cloud to blow up a violent hail-
filled storm over flooding waters and to smash trees, breaking them into
pieces; these heads are personifications of the angry words of God (Ps. 28
[29]: 3–10). In another scene (Figure 17), angels in a cloudy sky cast lances
that become dread bolts of lightning as they approach the earth (Ps. 139: 11
[140: 10]).[65]

The thrust of Agobard's tract was clearly not to argue for a natural
explanation of the weather,[66] but to reassign real responsibility for it to
God, who might be appealed to by special, intercessory agents. Thus, when
frightened by thunder and lightning, he said, the faithful beseech the intercession of the holy prophet Samuel who had once prayed to God for rain
and received it (1 Kings 12: 16–25). His own half-faithful people, on the

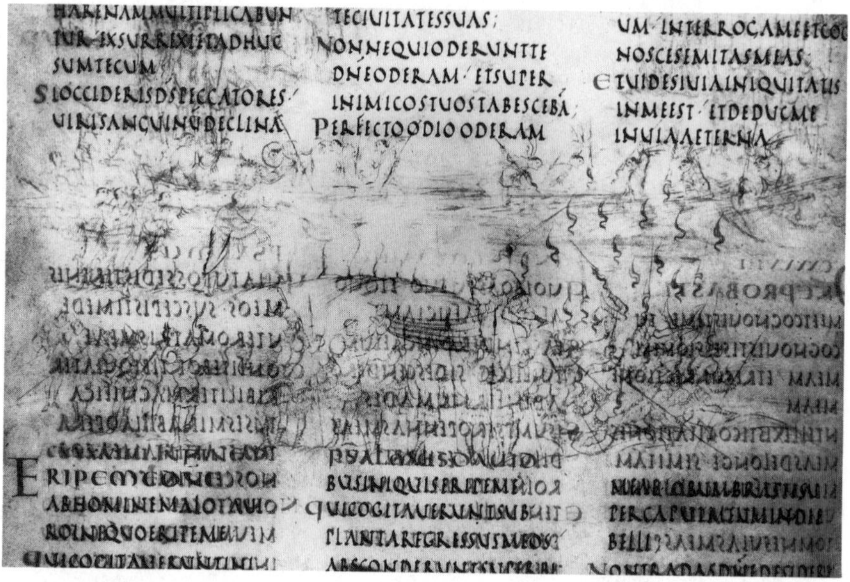

Figure 17. Thunderstorm, Utrecht Psalter, Ps 139: 11 (140: 10). Cat. Cod. MS Bibl. Rhenotraiectinae MS 32, fol. 78v. By permission of the University Library, Utrecht.

other hand, foolishly looked to the trickery of the storm-makers.[67] Since the days of Caesarius of Arles and Pope Gregory the Great, however, the church had sometimes sought to replace paganism by assuming its functions. When the people of the Carolingian countryside turned to the local priest and asked for rain, they found that he had a special prayer designed for the purpose.[68] In the Gregorian Sacramentary used by the Carolingian clergy there were also special masses for driving off thunderstorms.[69] In the account of the life of Abbess Leoba of Bischofsheim (d. 779) written by Rudolf of Fulda in the 830s, we see the general reaction of a rural population to thunderstorms, and the church's intercessory response. An awful storm was said to have once risen in the area around Mainz. Its lightning bolts and thunder claps struck dread into the hearts of all, even the bravest, and the sky suddenly became dark. First the people drove their cattle into their houses, lest they perish in the storm, and then fled themselves—men, women, and children—to Leoba's church. These people probably thought that a stone building was a safer refuge in a thunderstorm than their wood and thatch houses. But the storm rattled even the church, and whimpering families huddled together as the thunder rolled and flashes of lightning cast a strange light through the windows of the darkened church. When all implored Leoba to save them, she cast off her cloak and threw open the doors of the church. She made the sign of the cross into the face of the storm and called on Christ to protect his people. The storm immediately abated, the clouds passed overhead, and the sun shone once again.[70]

Though the church might attribute divine power over the natural world exclusively to God, the people of the Carolingian countryside were not so sure that was their only option; they were prepared to shop around for solutions. Einhard, for instance, tells us of a woman in Niedgau who had dislocated her jaw one morning while yawning. In great pain, she first sought out local women who tried to treat her with herbs and incantations. When that failed, she was taken to the church in Mulinheim, where Einhard had installed the relics of the saints Marcellinus and Peter, which were believed to be able to effect cures. When the woman looked up at the bell-tower of the church, her jaw snapped back into place and credit was given to the saints.[71] The point of the incident, for our purposes, is that the suffering woman had been prepared to seek out a variety of cures successively and, perhaps, saw no contradiction between them. Like most people of the age, her attitude toward the mysteries of the natural world was more open and flexible than Agobard would allow. Hers was a strategy born of sharp necessity; she sought the efficacious, trying whatever might work and perhaps only believing in it when it worked or when it was coincident with a cure.

When a violent thunderstorm struck, people must, therefore, have thought it just as reasonable to look to human agents who, in the name of a deity like Thor, claimed to possess some special control over the elements. The church and people shared an essential conviction that someone controlled the weather. But the farmer's immediate problem was not to determine correct cosmology; it was to cope with bad weather, or the prospect of it. As Evans-Pritchard long ago discerned among the Azande in Africa, one of the functions of the belief in rainmaking and magic is to allow people to explain experience, especially unfortunate events.[72] Such beliefs do more to account for events after they have happened than they do to explain the causes of things before they occur. Notice, in fact, how well the description of the cloud-ships fits the physical nature of a hailstorm. Thunderstorms in the summer often assume the shape of a dark anvil or ship as the updraft of a towering cumulonimbus cloud fashions a frontal projection while hailstones are being formed deep within the storm (see Figure 18).[73] Moreover, after a field of wheat had been thoroughly scattered

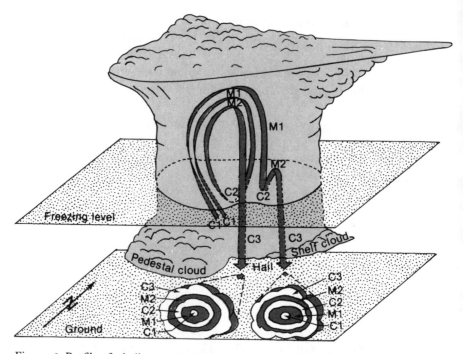

Figure 18. Profile of a hailstorm. Hailstorms are formed by the updraft swirling within a thunderstorm. After Joe R. Eagleman, *Meteorology: The Atmosphere in Action* (New York: Van Nostrand, 1980). By permission of Joe R. Eagleman.

by a violent hailstorm like the one endured by Benvenuto Cellini, it might indeed seem as if the crop had been stolen by the departing cloud, since so little of the cereal would have remained behind. In this light, then, the belief in weather-making could be viewed as a way for people to account for events that were beyond their control.[74]

But the belief in storm-makers also allowed Carolingian farmers to assign responsibility, to blame the malevolent or delinquent agents of the sky-god for the damage done to their crops. Agobard observed that the storm-makers were "hateful even to their neighbors,"[75] and he recorded the curse against the storm-makers that people uttered when a storm was brewing.[76] Moreover, there is the matter of the four aerial sailors who were chained and about to be stoned; whatever Agobard thought was going on, those poor people had certainly been publicly blamed for cooperating in the raising of a recent storm and the destruction of crops. One wonders whether not a few Carolingian storm-makers suffered the fate of the

Ugandan rainmaker who, in 1987, as reported by Reuters, was beaten to death by villagers in the district of Kabale who blamed him for causing the hailstorms and torrential rains that had recently devastated their crops and homes. The unfortunate man had unwisely warned the villagers that, if they did not show him more respect, he would summon up hailstorms to remind them of his power.[77] The people of the Carolingian country-side, too, seem to have been ready to level blame of this sort for natural disasters. When cattle began dying in 810, the story soon spread that they had been poisoned by outsiders, many of whom were promptly executed despite Charlemagne's explicit disapproval.

If Carolingian farmers paid a small tribute to the storm-maker for protection against bad weather, perhaps they thought that this was worth the gamble, since it was the only kind of crop insurance available. Keith Thomas has argued that one of the factors that led to the decline of magic in the early modern world was the rise of insurance.[78] In the ninth century, however, without any protection from the utter devastation of a total crop loss, villages may have found in the storm-maker someone who seemed to offer some small measure of control over potential calamity. But the farmers also knew whom to blame if a violent storm struck anyway, for storm-makers in some regions had been paid to prevent the storm, while in others they were assumed to be bringers of bad weather.

But Agobard thought that he had uncovered something else, not a religious or social reason for the belief in weather-making, but a simple conspiracy, a fraud. He suspected that at its base the belief in weather-making was promoted by greedy people for personal gain. Hence those who claimed only to protect people from storms charged for their service. Agobard lamented, in this case, that the deluded population would will-ingly surrender tribute to a weather-maker while withholding tithes from the church and charity from widows and orphans. In fact, the belief in the existence of storm-makers may have given some people a certain economic leverage in a world where they lacked many such advantages. What we need to bear in mind is that from the farmer's perspective it was not only storms that "robbed" crops, but also lords, both lay and ecclesiastical. To the farmer in the ninth century, formal and informal taxes may have been more predictable than hailstorms, but they were hardly more appreciated. To give just one example, Hincmar of Rheims relates that in Remigius's day, during a time of scarcity, the men and women of one of the saint's vil-lages had become drunk and rebellious. They decided to burn the heaps of grain collected for the bishop rather than surrender them. Saint Remigius responded by calling on God to curse the men with hernias and the women with goiters.[79] One has to wonder, in view of the popular resentment of and occasional resistance to taxation, if the tribute paid to the weather-makers was not effectively an anti-tithe, the means precisely for peasants to avoid paying tithes. Agobard certainly characterizes the *canonicus* as the rival and replacement of the tithes owed by Christians to the church where they received the sacraments.[80] Perhaps peasants occasionally argued that they could not pay their tithes, since there was not enough left of their harvest once their defenders had been paid. Agobard's parish priests, if they heard this excuse, must have wondered—as we do today—whether a weather-tribute had been paid at all or whether this was just another strategy of farmers for fending off demands on their grain in a world of subsistence agriculture.

What are we to make of the story of the four aerial sailors whom the crowd was about to stone to death? Perhaps—to attempt a natural explanation first—the unfortunate four were strangers who had been captured in a field of devastated wheat, attempting to steal its scattered remains, and who had been blamed by a village for the hailstorm itself. The word Magonia, in this case, may refer not to a Magic Land, but rather to a Land of Unscrupulous Dealers (from *mango* or *magono*) in stolen crops.[81] In this light, the people around Lyons would once again have been engaged in assigning blame for crops that had suddenly disappeared.

Agobard's accusation, however, fell not on the four prisoners, but on "those who had produced them in public."[82] After persistent questioning those people became, says Agobard quoting Jeremiah 2: 26, "as confounded as a thief when he is captured." What scam might these people have concocted? Again it is hard not to think that this might have been a way of hiding grain from the scrutiny of those who would tax it. We should not underestimate the ingenuity of those who worked and managed the land, nor underrate their ability to deal with demanding lords. Charlemagne, for instance, had commanded his royal stewards to be alert that "depraved men in no way conceal our seed either under the ground or in other places and, in this way, make the harvest seem scarcer. Likewise let the stewards watch those people for other evil deeds, lest on occasion they be in a position to make mischief."[83] If Charlemagne's own people buried their grain in order to avoid paying their proper dues, attempts to hide grain must have been fairly common in the Carolingian countryside. One suspects that what Agobard learned that day in his diocese was that a group of people had seized the opportunity of a recent storm to hide some ripe wheat from the rest of the village and their lord and to blame its disappearance on a storm and four outsiders. The association of these individuals with the cloud-ships from Magonia would have made some sense to the people around Lyons, because the legend of how storms stole crops was familiar and widely accepted.

Thus the belief in storm-makers persisted because it was traditional, because it was useful in explaining natural phenomena and in assigning blame, and because it supplied some with the means to gain an economic advantage over others. Survival in the Carolingian world demanded different and flexible strategies for coping. In a subsistence economy one wanted to avoid, if possible, the exacting demands made by lords and church on the small surplus of a recent harvest. If the widespread belief in storm-makers and the theft of crops by clouds could be turned to advantage, why would farmers not do it? Like the land, lords needed to be cleverly managed. But storms that threatened a community's very chances for survival, that could bring about hunger and deprivation, were ultimately unavoidable. Carolingians, however, had a weak sense of the existence of purely accidental phenomena; they were convinced that a cause could be found for everything and responsibility could be properly allocated. Storm-makers may have had some prestige in the Carolingian countryside, but they never achieved the central place in their world that African rainmakers did in theirs. They were relegated to the peripheries of power, to the countryside and beleaguered paganism, where they may mostly have been popular tricksters, tipping over tubs to make hail and performing magical acts. Their neighbors, however, were quick to blame the storm-makers for disastrous weather and to revile and curse them. But they also needed such people or, rather, they held fast to the belief in what they represented as agents of intercession who had a special way with the dynamic powers that filled the swirling sky. In a world where men and women were almost completely dependent on the success of their cereal crops for simple survival and where hail and severe thunderstorms threatened even that, the belief that some human being played a role in controlling the weather must have seemed both reasonable and reassuring. Could Agobard have offered more?

Notes

1. For a good general introduction, see Georges Duby, *Rural Economy and Country Life in the Medieval West*, trans. Cynthia Postan (Columbia: University of South Carolina Press, 1968), pp. 3–58, and Wolfgang Metz, "Die Agrarwirtschaft im

karolingischen Reiche," in *Karl der Grosse: Lebenswerk und Nachleben* 1, ed. Helmut Beumann (Düsseldorf: L. Schwann, 1965): 489–500.

2. *Liber contra insulsam vulgi opinionem de grandine et tonitruis* (hereafter *De grandine*) 2, ed. L. Van Acker, in *Agobardi Lugdunensis Opera Omnia*, Corpus Christianorum: Continuatio Mediaevalis 52 (Turnhout: Brepols, 1981), p. 4.7–10 [pp. 3–15]. Also edited in J.-P. Migne, *Patrologia cursus completus, series latina* (hereafter *PL*) 104.148B1–C4 [147A–158C] (the relevant portions of the treatise have been translated by P. E. Dutton in *Carolingian Civilization: A Reader*, ed. Dutton [Peterborough, Ont.: Broadview, 1993], pp. 189–91).

3. Agobard would certainly be disappointed to learn that late in the twentieth century some people have used his report as evidence that the earth was once visited by alien spaceships. See Whitley Strieber, *Communion* (New York: Beech Tree, 1987), pp. 241, 247–48. On Agobard as a rationalist, see Egon Boshof, *Erzbischof Agobard von Lyon* (Cologne: Böhlau, 1969), pp. 8–10, 173, n. 14.

4. The title as it stands in the Migne (*PL*) edition—*Liber contra insulsam vulgi opinionem de grandine et tonitruis*—was taken from an addition to the sole manuscript. Van Acker emended this to "De grandine et tonitruis," which seems less descriptive of the actual content of the work. For a discussion of this text, see, among others, Jacob Grimm, *Teutonic Mythology* 2, trans. J. S. Stallybrass (London: George Bell and Sons, 1883): 638–39; Reginald Lane Poole, *Illustrations of the History of Medieval Thought and Learning*, 2nd rev. ed. (New York: Macmillan, 1920; reprint New York: Dover, 1960), pp. 36–38; J. A. MacCulloch, *Medieval Faith and Fable* (London: George G. Harrap, 1932), p. 20; Allen Cabaniss, *Agobard of Lyons: Churchman and Critic* (Syracuse, N.Y.: Syracuse University Press, 1953), pp. 24–26; Cabaniss, "Agobard of Lyons: Rumour, Propaganda, and Freedom of Thought in the Ninth Century," *History Today* 3 (1953): 128–34; Heinrich Fichtenau, *The Carolingian Empire: The Age of Charlemagne*, trans. Peter Munz (Oxford: Basil Blackwell, 1957; reprint New York: Harper & Row, 1964), pp. 174–75; Boshof, *Erzbischof Agobard*, pp. 170–76; Valerie I. J. Flint, *The Rise of Magic in Early Medieval Europe* (Princeton, N.J.: Princeton University Press, 1991), pp. 111–15.

5. *De grandine* 1, p. 3.1–3 (*PL* 104.147A6–9).

6. *De grandine* 1, p. 3.4 (*PL* 104.147A10): "Aura leuatitia est." While *aura* in classical Latin may mean "a gentle breeze" or "air," in medieval Latin it can also mean "a violent wind" or "storm." *Leuatitia* is a curious word that Agobard may have invented to approximate a vernacular expression. Agobard understood the phrase to mean "aura est levata": see *De grandine* 1, p. 3.7–8 (*PL* 104.147B5–6) and 11, p. 11.23–24 (*PL* 104.154C6–7). See also Charles Du Cange, *Glossarium Mediae et Infimae Latinitatis . . .* , 10 vols. (Paris, 1883–87; reprint Paris: Librairie des Sciences et des Arts, 1937–38), 1: 484: "Galli vulgo dicunt 'Il s'est élevé un Air tempestueux,' pro 'excitata est tempestas.'" Edward B. Tylor, *Primitive Culture: Researches into the Development of Mythology, Philosophy, Religion, Language, Art, and Custom* 1 (London: John Murray, 1871): 84: "The phrase 'raising the wind' now passes as humorous slang, but it once, in all seriousness, described one of the most dreaded of the sorcerer's arts."

7. *De grandine* 1, p. 3.4–8 (*PL* 104.147B1–6). *Tempestarius* literally means "the

one associated with storms" or "stormy one," but a ninth-century council (see note 29 below) speaks of those "qui tempestates et alia maleficia faciunt," so that storm-maker seems a reasonable approximation of the meaning of the word. See also Du Cange, *Glossarium* 8: 49–50.

8. *De grandine* 11, p. 11.19–22 (*PL* 104.154C1–5).

9. *De grandine* 2, p. 4.3–7 (*PL* 104.148B1–8).

10. *De grandine* 7, p. 8.11–14 (*PL* 104.151D2–7).

11. *De grandine* 7, p. 8.15–24 (*PL* 104.151D7–152A9).

12. *De grandine* 15, p. 14.1–12 (*PL* 104.156D11–157A12).

13. *De grandine* 16, pp. 14.1–15.27 (*PL* 104.157C4–158C3).

14. See *Annales regni Francorum* (hereafter *ARF*) 810, ed. G. H. Pertz and rev. F. Kurze, in *MGH: Scriptores rerum Germanicarum in usum scholarum* 6 (Hannover: Hahn, 1895), p. 132; *Capitulare missorum Aquisgranense primum* 4, in *MGH: Capitularia regum Francorum* 1, ed. Alfred Boretius (Hannover: Hahn, 1883), p. 153.11–12: "De homicidiis factis anno praesenti inter vulgares homines, quas propter pulverem mortalem acta sunt"; *Annales Sithienses* 810, ed. G. Waitz, in *MGH: Scriptores* 13 (Hannover: Hahn, 1881): 37: "Boum pestilentia per totam Europam immaniter grassata est, et inde pulverum sparsorum fabula exorta"; and Paschasius Radbertus, *Epitaphium Arsenii* 2.1, ed. E. Dümmler, in *Abhandlungen der Königlich Preußischen Akademie der Wissenschaften* 2 (Berlin: G. Reimer, 1900), p. 61: "Quibus profecto malis precessit prior pulverum fallax adinventio, sub qua tanta fuit vexatio et prodigium mendacii, ut prudentibus daretur intellegi, quod universus orbis ad temptandum esset expositus in manibus inimici."

15. *De grandine* 13, p. 12.1–5 (*PL* 104.155A10–15).

16. See *De grandine* 7, p. 7.4–6 (*PL* 104.151C9–11).

17. *De grandine* 6, p. 7.9–16 (*PL* 104.151A14–B8) and 7, pp. 7.6–8.11 (*PL* 104.151C12–D2).

18. *De grandine* 5, p. 6.1–13 (*PL* 104.150A6–B6).

19. *De grandine* 14, p. 13.9–12 (*PL* 104.156C4–8): "ostendunt nobis homunculos a sanctitate, iustitia et sapientia alienos, a fide et veritate nudos, odibiles etiam proximis, a quibus dicant vehementissimos imbres, sonantia atque tonitrua, et levatitias auras posse fieri." A copy of four of Gregory Nazianzen's sermons, translated into Latin by Rufinus, was given to Lyons by Agobard's predecessor, Bishop Leidrad (d. 815), but the "De grandinis vastatione" was not among them, or at least it is not found in the surviving manuscript, Lyons 599 (515), fols. 10v–60v. Whether or not Agobard was influenced by Gregory in this matter has yet to be established. See *Tyrannii Rufini orationum Gregorii Nazianzeni novem interpretatio*, ed. A. Engelbrecht, Corpus Scriptorum Ecclesiasticorum Latinorum 46 (Vienna: F. Tempsky, 1910): xxxv, 237–261.

20. One can follow the development of Livingstone's thought about the rain-makers in *Livingstone's Missionary Correspondence 1841–1856*, ed. I. Schapera (Berkeley: University of California Press, 1961), pp. 60–65, 102–3, 120–21; *Livingstone's Private Journals, 1851–1853*, ed. I. Schapera (Berkeley: University of California Press, 1960), pp. 239–43; and David Livingstone, *Missionary Travels and Researches in South Africa* . . . (London: John Murray, 1857), pp. 22–25. It should be noted that the pol-

ished dialogue with the rainmaker that appears in *Missionary Travels* does not represent a conversation with a single rainmaker, but one that Livingstone constructed out of the various discussions he had had with the Bakwains over several years.

21. *Livingstone's Private Journals*, p. 243. Cf. Franz Boas, *The Mind of Primitive Man*, rev. ed. (New York: Macmillan, 1938), p. 134.

22. See *ARF* 814, p. 141.5–7; Nithard, *Historiarum libri IV*, 1.2, ed. and trans. Philippe Lauer, in Nithard, *Histoire des fils de Louis le Pieux* (Paris: Société d'Édition "Les Belles Lettres," 1926; reprint 1964), pp. 6–8; and the Astronomer, *Vita Hludowici imperatoris* 21–23, 28, ed. G. H. Pertz, in *MGH: Scriptores* 2 (Hannover: Hahn, 1829): 618–19, 621–22 (trans. by Allen Cabaniss in *Son of Charlemagne: A Contemporary Life of Louis the Pious* [Syracuse, N.Y.: Syracuse University Press, 1961], pp. 54–57, 63–64).

23. On Agobard's career, see Allen Cabaniss, "Agobard of Lyons," *Speculum* 26 (1951): 50–51; Boshof, *Erzbischof Agobard*, pp. 20–37; Stuart Airlie, "Bonds of Power and Bonds of Association in the Court Circle of Louis the Pious," in *Charlemagne's Heir: New Perspectives on the Reign of Louis the Pious (814–840)*, ed. Peter Godman and Roger Collins (Oxford: Clarendon, 1990), p. 194 and n. 21.

24. For the prohibitions against weather magic in Visigothic Spain, see Flint, *Rise of Magic*, pp. 110–111.

25. See Alcuin, *Epistola* 267, ed. E. Dümmler, in *MGH: Epistolae* 4 (1895; reprint Berlin: Weidmann, 1974): 425.29–30 [425–26].

26. D. E. Nineham, "Gottschalk of Orbais: Reactionary or Precursor of the Reformation?" *Journal of Ecclesiastical History* 40 (1989): 12. For a quick tour of Frankish paganism in the Carolingian world, see Pierre Riché, *Daily Life in the World of Charlemagne*, trans. JoAnn McNamara (Philadelphia: University of Pennsylvania Press, 1978), pp. 181–88; Michel Rouche, "The Early Middle Ages in the West," in *A History of Private Life* 1: *From Pagan Rome to Byzantium*, ed. Paul Veyne and trans. Arthur Goldhammer (Cambridge, Mass.: Harvard University Press, 1987): 519–36; and Rosamond McKitterick, *The Frankish Church and the Carolingian Reforms, 789–895* (London: Royal Historical Society, 1977), pp. 119–22.

27. *Indiculus superstitionum et paganiarum* 22, in *MGH: Capitularia* 1: 223.22, and see John T. McNeill and Helena M. Gamer, *Medieval Handbooks of Penance* (New York: Columbia University Press, 1965), pp. 419–21, or Dutton, ed. *Carolingian Civilization*, p. 3. For Cathwulf's letter, ed. Dümmler, see *MGH: Epistolae* 4: 504.12–19 [501–5].

28. *Admonitio generalis* 65, in *MGH: Capitularia* 1: 58.41–59.3: "Omnibus. Item habemus in lege Domini mandatum: 'non auguriamini' (Lev. 19: 26); et in deuteronomio: 'nemo sit qui ariolos sciscitetur vel somnia observet vel ad auguria intendat'; item 'ne sit maleficus nec incantator nec pithones consolatur.' (Deut. 18: 10–11) Ideo praecipimus, ut cauculatores nec incantatores nec tempestarii vel obligatores non fiant; et ubicumque sunt, emendentur vel damnentur." Cf. *Capitulare missorum item speciale* (802) 40, in *MGH: Capitularia* 1: 104.5–7 [102–4]; *Ansegisi abbatis capitularium collectio* 1.62, in *MGH: Capitularia* 1: 402.26–30 [394–450].

29. *Statuta Rhispacensia Frisingensia Salisburgensia* (800) 15, in *MGH: Capitularia* 1: 228.9–17 [226–30], and in *MGH: Concilia* 2.1, ed. Albert Werminghoff (Hannover: Hahn, 1906): 209.18–26.

30. Council of Paris, 829 (69) 2, in *MGH: Concilia* 2.2, ed. Albert Werming-hoff (Hannover: Hahn, 1908): 669.35–37: "Ferunt enim suis maleficiis aera posse conturbare et grandines inmittere, futura praedicere, fructus et lac auferre aliisque dare et innumera a talibus fieri dicuntur."

31. Ibid., p. 669.37–39.

32. For a survey of this material, see McNeill and Gamer, *Medieval Handbooks*, pp. 227 (vi.14) and n. 60, 275 (2.20), 289, and 305 (no. 33).

33. *De magicis artibus*, in *PL* 110.1101D1–4, 1103C11 [1095–1110]. See also Pierre Riché, "La magie à l'époque carolingienne," in *Académie des Inscriptions et Belles-Lettres: Comptes rendus* (Paris, 1973), pp. 134–35.

34. See note 28 above and *Capitula Herardi* 3, in *PL* 121.764B6–10. My pur-pose here is not to pursue the complex definitional and classificatory problem of the separation of religion from magic, in part because it has been treated so well and at such great length by others. See Keith Thomas, *Religion and the Decline of Magic* (New York: Scribner, 1971), pp. 25–50, esp. p. 41, for a series of valuable distinctions. See also Hildred Geertz, "An Anthropology of Religion and Magic: I," *Journal of Interdisciplinary History* 6 (1975): 71–89, and Keith Thomas, "An Anthropology of Religion and Magic: II," *Journal of Interdisciplinary History* 6 (1975): 91–109. For the Middle Ages, see Joseph-Claude Poulin, "Entre magie et religion: recherches sur les utilisations marginales de l'écrit dans la culture populaire du haut moyen âge," in *La Culture populaire au moyen âge*, ed. Pierre Boglioni (Montréal: L'Aurore, 1979), pp. 121–43, and Patrick Geary, "La coercition des saints dans la pratique religieuse médiévale," in *Culture populaire*, ed. Boglioni, pp. 145–61.

35. For this interpretation of the word, see Grimm, *Teutonic Mythology* 2: 639. For another possible interpretation, see page 128 and note 81.

36. Regino of Prüm, *De ecclesiasticis disciplinis et religione christiana libri duo* 2.353, in *PL* 132.350C2–5.

37. Burchard of Worms, *Corrector sive Medicus* 19.5, in *PL* 140.961D3–8. And see Cyrille Vogel, "Pratiques superstitieuses au début au XIe siècle d'après le *Corrector sive medicus* de Burchard, évêque de Worms (965–1025)," in *Études de civilisa-tion médiévale (IXe–XIIe siècles): Mélanges offerts à Edmond-René Labande* (Poitiers: C.E.S.C.M., 1974), pp. 751–61.

38. See *Capitulatio de partibus Saxoniae* (775–790), in *MGH: Capitularia* 1: 68–70, where Charlemagne's frustration with pagan persistence in Saxony is evident. See also J. M. Wallace-Hadrill, *The Frankish Church* (Oxford: Clarendon, 1983), pp. 412–19.

39. See Grimm, *Teutonic Mythology* 2: 636–41; 3: 1086–89; James George Frazer, *The Golden Bough: A Study in Magic and Religion*, 3rd ed. (London: Mac-millan, 1911), 1: 272–74; Ernest J. Moyne, *Raising the Wind: The Legend of Lapland and Finland Wizards in Literature* (Newark: University of Delaware Press, 1981), esp. pp. 13–17.

40. Seneca, *Naturales Quaestiones* 4B.7.3 [4B.7.1–3], in *Seneca in Ten Volumes*, 10: *Naturales Quaestiones*, vol. 2, ed. T. H. Corcoran (Cambridge, Mass.: Harvard University Press, 1972), p. 56: "Rudis adhuc antiquitas credebat et attrahi cantibus imbres et repelli."

41. See Grimm, *Teutonic Mythology* 4: 1769–70, and cf. Jacob and Wilhelm

Grimm, *The German Legends of the Brothers Grimm* 1, trans. Donald Ward (Philadelphia: Institute for the Study of Human Issues, 1981): no. 251, pp. 211–12. See also Carlo Ginzburg, *The Night Battles: Witchcraft and Agrarian Cults in the Sixteenth and Seventeenth Centuries*, trans. John and Anne Tedeschi (Baltimore: Routledge & Kegan Paul, 1983), pp. xx, 22–25.

42. See Grimm, *Teutonic Mythology* 3: 1086–87.

43. The almost exclusive association of women with weather-making in northern lands seems to have taken place in the eleventh century. See Pope Gregory VII, *Registrum* 7.21, ed. E. Caspar, in *MGH: Epistolae selectae* 2.2 (1923; reprint Berlin: Weidmann, 1955): 497–98; Raoul Manselli, "Gregorio VII di fronte al paganesimo nordico: la lettera a Haakon, re di Danimarca (Reg. VII, 21)," *Rivista di storia della chiesa in Italia* 28 (1974): 128–29; Manselli, *La Religion populaire au moyen âge: Problèmes de méthode et d'histoire*, Conference Albert-le-Grand, 1973 (Montréal: Institut d'Études Médiévales Albert-le-Grand, 1975), pp. 46–47 and n. 4.

44. Adam of Bremen, *Gesta Hammaburgensis ecclesiae pontificum* 4.26, 3rd ed., ed. Bernhard Schmeidler, in *MGH: Scriptores rerum Germanicarum in usum scholarum* (Hannover: Hahn, 1917): 258 (also edited in *PL* 146.643A6–8): "'Thor,' inquiunt, 'presidet in aere, qui tonitrus et fulmina, ventos imbresque, serena et fruges gubernat.'"

45. *ARF*, p. 154.

46. *Annales de Saint-Bertin* (hereafter *AB*), ed. F. Grat, J. Vielliard, and S. Clémencet (Paris: Klincksieck, 1964), p. 28. The description in the annals does not allow us to decide whether this disaster was caused by excessive rain or by storm surges. See William H. TeBrake, *Medieval Frontier: Culture and Ecology in Rijnland* (College Station: Texas A&M University Press, 1985), pp. 110–11. See also the account of the 300 villagers in Thuringia who were killed in the flash flood of 889: *Annales Fuldenses* (hereafter *AF*), ed. G. H. Pertz and F. Kurze, in *MGH: Scriptores rerum Germanicarum in usum scholarum* 7 (Hannover: Hahn, 1891): 118.

47. "Ecce nunc aquosus aer imbre rura perluit, / uberes agros, vides, ut uber unda dissipet." Sedulius Scottus, *carmen* 62.3–4, ed. L. Traube, in *MGH: Poetae Latini aevi Carolini* 3 (1886–96; reprint Berlin: Weidmann, 1978): 218 (my translation). Cf. Sedulius Scottus, *On Christian Rulers and the Poems*, trans. Edward G. Doyle, in Medieval and Renaissance Texts and Studies 17 (Binghamton: State University of New York Press, 1983), p. 158.

48. See Duby, *Rural Economy*, p. 27, and Renée Doehaerd, *The Early Middle Ages in the West: Economy and Society*, trans. W. G. Deakin (Amsterdam: North-Holland, 1978), p. 103.

49. See H. H. Lamb, *Climate: Present, Past, and Future* 2: *Climatic History and the Future* (London: Methuen, 1977): 426; Richard Hodges, *Dark Age Economics: The Origins of Towns and Trade*, A.D. *600–1000* (London: Duckworth, 1982), p. 139; Wendy Davies, *Small Worlds: The Village Community in Early Medieval Brittany* (London: Duckworth, 1988), p. 33. For objective indications of summer storminess and heavier precipitation in July and August, see H. H. Lamb, "The Early Medieval Warm Epoch and Its Sequel," *Palaeogeography, Palaeoclimatology, Palaeoecology* 1 (1965): 21–22 and fig. 1. The dendrochronological evidence for northern Europe

does not, however, seem to support the thesis that growing conditions were worse in the ninth century. See the tables printed in Emmanuel Le Roy Ladurie, *Times of Feast, Times of Famine: A History of Climate since the Year 1000*, trans. Barbara Bray (New York: Doubleday, 1971), pp. 386–88.

50. See Narayan R. Gokhale, *Hailstorms and Hailstone Growth* (Albany: State University of New York Press, 1975), pp. 21, 25 (fig. 2–10), 50.

51. See Benvenuto Cellini's autobiography, in *The Life of Benvenuto Cellini: A New Version*, trans. Robert H. Hobart Cust, 2 (London: G. Bell and Sons, 1910): 238.

52. See Joe R. Eagleman, *Severe and Unusual Weather* (New York: Van Nostrand, 1983), p. 138.

53. Gokhale, *Hailstorms*, p. 87.

54. Ibid., p. 13.

55. See Einhard, *Vita Karoli Magni* 29, ed. G. H. Pertz and G. Waitz, in *MGH: Scriptores rerum Germanicarum in usum scholarum* 25 (Hannover: Hahn, 1911; reprint 1965): 33 (trans. Lewis Thorpe in *Einhard and Notker the Stammerer: Two Lives of Charlemagne* [Harmondsworth: Penguin, 1969], p. 82, or trans. S. E. Turner, rev. P. E. Dutton, in Dutton, ed., *Carolingian Civilization*, p. 39.

56. *ARF*, pp. 163–64.

57. *AF*, pp. 117–18: "Grandine vero contritis frugibus mortales inopiam frugum cum miseria patiuntur."

58. See *Annales Xantenses*, ed. B. Simson, in *Annales Xantenses et Annales Vedastini*, in *MGH: Scriptores rerum Germanicarum in usum scholarum* 12 (Hannover: Hahn, 1909): 31, and *AF*, p. 76.

59. See Michel Rouche, "La faim à l'époque carolingienne: essai sur quelques types de rations alimentaires," *Revue historique* 250 (1973): 295–320, and "Les repas de fête à l'époque carolingienne," in *Manger et boire au moyen âge: Actes du Colloque de Nice (15–17 octobre 1982)*, ed. Denis Menjot (Nice: Belles Lettres, 1984), pp. 265–96. Rouche's conclusion, based on his interpretation of monastic provision records, that each monk ate over three pounds of bread per day and had an average dietary intake of approximately 6,000 calories has been called into question by Jean-Claude Hocquet, "Le pain, le vin et la juste mesure à la table des moines carolingiens," *Annales: ESC* 40 (1985): 661–86, answered by Rouche, in *Annales: ESC* 40: 687–88, with a rejoinder by Hocquet, *Annales: ESC* 40: 689–90.

60. *Chronicon Moissiacense*, ed. G. H. Pertz, in *MGH: Scriptores* 1 (Hannover: Hahn, 1826): 294 [282–313]: "multi homines penuria panis perirent."

61. *AB* 843, p. 44.

62. *AF*, p. 48.

63. *AF*, pp. 83–84.

64. *AB*, p. 54.

65. On depictions of wind, rain, and lightning in the Utrecht Psalter, see Suzy Dufrenne, *Les Illustrations du Psautier d'Utrecht: Sources et apport Carolingien* (Paris: Ophrys, 1978), pp. 74–75. For a facsimile edition, see E. T. De Wald, *The Illustrations of the Utrecht Psalter* (Princeton, N.J.: Princeton University Press, 1932).

66. Nor, we should note, was this the purpose of the anonymous author of the tract *De tonitruis Libellus ad Herefridum*, PL 90.609B–614A. Once attributed to

Bede, the work was likely written late in the ninth century, possibly at the court of Charles the Bald. It is, in fact, a work abridged from a Greek text. See Charles W. Jones, *Bedae Pseudepigrapha: Scientific Writings Falsely Attributed to Bede* (Ithaca, N.Y.: Cornell University Press, 1939), pp. 45–47. Interestingly, in the introduction to the work, the author worries that his detractors might charge him with having an interest in magic. The work considers what thunder portends when it comes from one of the four directions or when it occurs in a specific month or on a specific day of the week.

67. *De grandine* 11, p. 10.7–11.16 (*PL* 104.154A14–D2).

68. See "Orationes et Missa ad pluviam postulandam" of the *Supplementum Anianense* 92–93, ed. J. Deshusses, in Deshusses, *Le Sacramentaire grégorien* (Freiburg: Éditions Universitaires, 1971), pp. 448–49.

69. See "Missa ad repellendam tempestatem," of the *Supplementum Anianense* 96, ed. J. Deshusses, *Sacramentaire grégorien*, pp. 450–51.

70. *Vita Leobae abbatissae Biscofesheimensis* 14, ed. G. Waitz, in *MGH: Scriptores* 15.1 (Hannover: Hahn, 1887): 128 [121–31] (trans. C. H. Talbot, in Dutton, ed., *Carolingian Civilization*, p. 321).

71. *Translatio et miracula sanctorum Marcellini et Petri* 3.16, ed. G. Waitz, in *MGH: Scriptores* 15.1: 254 [239–64] (trans. B. Wendell, rev. P. E. Dutton, in Dutton, ed., *Carolingian Civilization*, pp. 226–27).

72. E. E. Evans-Pritchard, *Witchcraft, Oracles, and Magic among the Azande* (Oxford: Clarendon, 1937), pp. 63–83.

73. See Eagleman, *Severe and Unusual Weather*, p. 150.

74. For the plausibility of this kind of general explanation for the occurrence of magical beliefs, but its insufficiency for explaining specific forms and their variation, see Thomas, *Religion and the Decline of Magic*, pp. 647–50. The position that magic and the belief in it are strong when control over the physical environment is weak derives from Bronislaw Malinowski, *Magic, Science, and Religion* (1925), reprinted in Malinowski, *Magic, Science, and Religion, and Other Essays* (Garden City, N.Y.: Doubleday, 1954).

75. *De grandine* 14, p. 13.11 (*PL* 104.156C6).

76. *De grandine* 11, p. 11.19–22 (*PL* 104.154C1–5).

77. *The Globe and Mail* (Toronto), Nov. 14, 1987, p. A2.

78. Thomas, *Religion and the Decline of Magic*, pp. 651–54.

79. *Vita Remigii Episcopi Remensis auctore Hincmaro* 22, ed. Bruno Krusch, in *MGH: Scriptores rerum Merovingicarum* 3 (Hannover: Hahn, 1896): 315.18–316.16 [250–341]. On the resistance to paying tithes, see Giles Constable, "*Nona et Decima*: An Aspect of Carolingian Economy," *Speculum* 35 (1960): 230, 234–35.

80. One of the standard meanings of *canon*, and therefore *canonicus*, was regular tribute or measure of wheat: see *Thesaurus Linguae Latinae* 3 (Leipzig: Teubner, 1906–12): 273, 276.

81. See *Thesaurus Linguae Latinae* 8 (Leipzig: Teubner, 1936): 300, and R. E. Latham, *Revised Medieval Latin Word-List from British and Irish Sources* (London: British Academy, 1965; reprint 1973), p. 288, where "magono" is a variant of "mango."

82. *De grandine* 2, p. 4.14 (*PL* 104.148C2).

83. *Capitulare de villis* 51, in *MGH: Capitularia* 1: 88.5–7: "ut sementia nostra nullatenus pravi homines subtus terram vel aliubi abscondere possint et propter hoc messis rarior fiat. Similiter et de aliis maleficiis illos praevideant, ne aliquando facere possint." See Adriaan E. Verhulst, "Karolingische Agrarpolitik: Das Capitulare de Villis und die Hungersnöte von 792/93 und 805/06," *Zeitschrift für Agrargeschichte und Agrarsoziologie* 13 (1965): 181.

7. Links Within the Village: Evidence from Fourteenth-Century Eastphalia

Historical evidence from the middle of the fourteenth century for the rural area north of the Harz mountains, between the towns of Brunswick, Hildesheim, and Goslar, does not tell us much about the social relationships of the people of the plough. In this respect, it is not comparable to evidence like the great monastic polyptychs of ninth-century France and northern Italy; or the court rolls and manorial accounts of thirteenth- and fourteenth-century England; or the protocols of the Inquisition for the northern Pyrenees in the early fourteenth century; or the famous *Catasto* of Tuscany around 1400. But historians should not limit themselves to the study of source material preserved by chance.

So far, little attention has been paid to the northern Harz region; the few available studies focus mainly on questions concerning rural organization. This essay fits into the framework of broader studies of rural living conditions in the northwest Harz region.[1] The sources for the study are in Latin and middle low German. The following general conditions should be noted: The landscape is a hilly, forested area with small rivers and good soil (see Figure 19). The holding was divided into the farm (Lat. *curia*, mlG. *hoff*: Hof) and the arable land (Lat. *mansus*, mlG. *hove*: Hufe). Each part was rented separately. The size of the Hufe was about 20 to 30 Morgen, that is, 10 to 20 acres.

In the villages there were three types of peasants. At the top there was a small group of rich peasants (*Meier*) who held farms with three to six Hufen, leasing them for life or for a fixed number of years (*Zeitpacht*). The lease generally took the form of share-cropping (one third was due to the lord, usually a member of the nearby town aristocracy, the *Patriziat*), and production was apparently oriented to the town market. The majority of the villagers were peasants who held one or two Hufen. They were hereditary leaseholders and normally paid fixed money rents (*Erbzinsrecht*) to their lords, who belonged to the old ecclesiastical foundations or to the regional nobility. At the bottom of the social strata were the cottagers (*Köt-ner*) who held only half a Hufe or less, or even no arable land, but enjoyed rights of the common; they were often related to the *Meierhöfe*. Besides the rent, the tithes were an important annual payment for villagers; taxes were not levied regularly.

The agricultural system was the three-course rotation (*Dreifelderwirt-schaft*). The fields were divided into strips attached to the Hufen of the holdings; as a consequence, work on the fields had to be organized collectively (*Flurzwang*). Ploughing was done with the heavy, wheeled mold-board plough drawn by oxen or horses.

In the fourteenth century, no severe symptoms of crisis were apparent in the Malthusian or in the ecological sense, but from the time of the first attack of bubonic plague (1350), the villages seem to have become unstable. This can be observed only from the shrinkage in rents in some seignorial accounts.

The subject of this essay is the relationships among the inhabitants, variously called *(agri)cultores*, *buren*, *villani* or *undersaten*, in these villages.[2] The evidence is so sparse that we cannot explicitly use standard terms for the analysis of everyday social forms like "household" and "family." We also must forgo the social science methods so well known to those engaged in family history,[3] which are feasible only on the basis of the source material beginning in the early modern period.[4] Still, we may use such studies as an orientation, always keeping in mind that we cannot transfer the results or the methods applied in them.

The sources used here are lists of dependents; rent registers that link obligations of payments to certain personal names; deeds or charters that settle inheritance disputes or conflicts within the village or that inform us about certain developments; and, in addition, the indispensable seignorial accounts of the cathedral chapter of St. Blasius in Brunswick. It is useful to supplement these with normative sources like Eike von Repgow's custumal (*Sachsenspiegel*) and the Goslar town charter (*Stadtrecht*).

I shall first sketch kinship terminology and the ways in which rural dependents were grouped. Then some remarks will follow on the form of marriage and the transfer of personal goods connected with it. In the third section, I shall deal with signs of gender in respect to rural labor, and with middle low German terms expressing ties to the farm and the land. Finally, I shall examine some indications of economic and social links among the villagers.

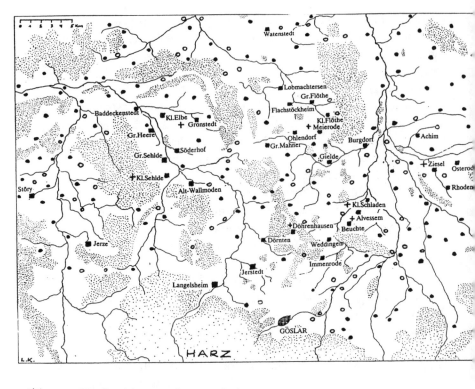

    ⦙⦙⦙⦙ = Woodland (present-day extension)
    ⤸   = Rivers/Streams
    ▮   = Villages with Holdings owned by the
            Neuwerk Monastery/Goslar (1355)
    ●   = Villages (present-day location)
    o / + = Deserted Villages

Figure 19. Medieval settlements northwest of the Harz Mountains.

Kinship Terminology

How were the peasants seen by those who inventoried them? Of course,
this perspective cannot be equated with the way the peasants would have
seen themselves. In trying to analyze documents that reflect the lords' perception, we obtain information mainly about protection and appropria-
tion. The questions are: Who has to pay? or Who is a bondsman? These
questions are posed from above and from the outside. Events within peasant communities are registered only insofar as they are connected with

TABLE 3   Neuwerk Bondsmen (1355)

No.	Village/town	First name	Last name	Kinship term	With wife	With child(ren)	M/C/A*
1	Immenrode	Heneke	Bertholdes		X	X	
2	Immenrode	Johannes	Botels			X	
3	Immenrode	Thidericus	Botels				
4	Immenrode	Ecbertus	(*villicus*)		X	X	16 M
5	Immenrode	Henninghus	de Monte				+4 C
6	Immenrode	(4)	de Wisen	*fratres*			
7	Immenrode	Henninghus	Schure				
8	Immenrode	Conradus	Danhouwere				
9	Immenrode	Conradus +	Oldendorpe	*uxor*			
10	Immenrode	Henricus	Leyen				
11	Immenrode	Deghenhard	Leyen				
12	Weddingen	Meteke	Botels		X	X	7 M
13	Weddingen	Heneke	Grahoves				+2 C
14	Hahndorf	Werneke				X	
15	Hahndorf	Olricus	Teghetmeyer				
16	Hahndorf	Deghen					
17	Dörnten	Thidericus	Teghetmeyer				
18	Dörnten	Thidericus	Teghetmeyer }	*fratres*			20 M
19	Dörnten	Conradus	Teghetmeyer				+6 C
20	Dörnten	Heneke					+4 A
21	Langelsheim	Heneke	Botel				5 M
22	Langelsheim	Greten	Botel	*soror sua*			3 C
23	Bettingerode	Tydericus	Honegher				
24	Ziesel	Johanna		*cum sorore*			10 M
25	Goslar	Hermanni	Minoris	*uxor*			
26	Goslar	Ebelingh	Kinne				

M = mansus; C = curia; A = area.

these questions. Three sources contain a satisfactory amount of information about the personal links within the peasant community: a two-part
"progeny" list from the monastery of Neuwerk (Goslar); an enumeration
of the bondsmen of the noble family of von Heimburg; and a rent roll of
1340 from the cathedral chapter of St. Blasius.

The Neuwerk list contains only 50 names, often without last names or
names related to places (Tables 3 and 4).[5] Only in a few cases is there an
indication of position within the "family" and of kin. Still, the principles
of description can be clearly discerned. First, the majority of the names are

TABLE 4   Neuwerk Bondsmen (1355)

No.	Village/town	First name	Last name	Kinship term	Grain rent			Mone rent
					ch	md	mdl	
1	Hilverdingerode	Hermannus	Mole ⎫		0.5			
2	Hilverdingerode	Henricus	Mole ⎭		0.5			
3	Hilverdingerode	Poltener			0.5			
4	Dörrierode	Conradus	de Ghotighe		1.5			
5	Goslar	Conradus	Doringherod		1.0			
6	Goslar	Berteram	Doringherod		0.5			
7	Goslar	Roghat				6.0		
8	Goslar	Ernestus	?			3.0		
9	Goslar	Ghevehardus			0.5			
10	Goslar	Werpumme			0.5			
11	Goslar	Ghoseke			1.0			
12	Goslar	Marquardus				7.0		
13	Goslar	Ludolfi		soror		6.5		
14	Goslar	Heneke				6.0		
15	Immenrode	Botel			1.5			
16	Immenrode	Ernestus					1.5	15d
17	Immenrode	Bertoldus	Lindeman			1.5		15d
18	Immenrode	Vredeke	Lindemannes			1.5		15d
19	Immenrode	Bertoldus +	Vredeke			1.5		
20	Immenrode	Kines	Boteles			1.5		15d
21	Immenrode	Henricus	Boltoldus					2ℓ
22	Immenrode	Heygerus						2ℓ
23	Immenrode	Ludegerus						2ℓ
24	Immenrode	Henricus	Egghelbrechtes					60d

*ch = chorus; md = modius; mdl = modiolus; ℓ = lot; d = denarius.

those of men. Almost all of them have first names, and the majority have a "family": they are listed "with wife and children" (*cum uxore et pueris*). Some are brothers, either listed only as such or by name. The women listed are in a clear minority. Most of them are listed without names as wife (*uxor*) or sister (*soror*). Whether they are listed with or without names, there is always a personal relationship with a man or another woman (*cum uxore, cum sorore, soror sua*), or else they are the wife of a man not listed, that is, of one who has died or of a free man. The children (*pueri*) are listed as relatives and are unspecified by number, name, or sex.

Despite this narrow basis we can conclude that the principles inferred here are typical. The two other inventories confirm this opinion and offer important additional information. The same basic structure can be found in the list of Heimburg bondsmen, written in low German, which consists of

TABLE 5 Men and Women Listed in the Rent Roll of St. Blasius (1340)

Forms of name/kin terms	With holding	Without holding	Total
Men			
FN + LN	107	13	120
FN	40	11	51
juvenis	1	—	1
sons of men	2	3	5
brothers	3	2	5
sister's husband	1	—	1
Total	154	29	183
Women			
woman's FN	3	—	3
man's FN + woman's FN	1	—	1
woman's FN + LN	8	1	9
widow without name	2	—	2
widow with FN	13	—	13
widow, name with -sche	5	—	5
mater	1	—	1
filia of woman	1	—	1
Total	35	1	36

FN = first name; LN = last name.

44 data units.[6] Here, too, we find the nameless and sexless children (*kinder*, *kindeken*) and also the nameless wife (*husfrouwe*), who is listed together with a man mentioned by name, irrespective of his status as a bondsman or a free man (for example, *hans tymmerman mit syner husvrouwen vn kinderen*). Finally, men are almost always listed by first name and usually also by last name. Moreover, when men are listed as descendants or lateral relatives of men, there is a tendency for them to lose their names; they are nameless sons or brothers of men who are listed by name.[7] A (widowed) woman with children is listed before the siblings of her husband.[8]

The rent roll of St. Blasius of 1340 is very valuable since, in most cases, it gives the size of the holding in addition to the names of 200 men and women.[9] The distribution of the terms used for men and women (Table 5) clearly shows the dominance of men as carriers of names. One can go even further: only men are owners of last names. If a woman has a last name at all, it is that of her deceased husband.[10] Thus a woman casts off second place and her anonymity only when she becomes a widow. In the rent roll, several forms indicate this "appearance": as a *vidua* (still nameless), *vidua Nicolai* (identified in connection with the husband's first name), *vidua Druden* (the husband's last name), *vidua Ludolfi Brules* (the husband's first and last name), and, finally, as *die Bumensche, die Sneysesche,* and so on. All these forms connecting a woman to her husband can be defined even more narrowly: Only the connection with her husband enables the woman to become a bearer of a name at all.

The importance of widowhood for the appearance of women in the source material is evident from the proportion of widows in the total number of women listed by name in the rent roll, roughly 80 percent. And even when men and women are taken together, widows figure significantly as tenants of holdings (13 percent). Thus, a woman becomes independent and "named" to a socially relevant extent through the death of her husband. Still, in "her" name, her husband lives on, and with her name, her role as his representative becomes apparent. Names ending in *-sche*—nouns formed from an attributive adjective—prove this point even more strongly. It would be hard to express more clearly in language the derivative character of a social position.

On the other hand, one should perhaps stress the fact that a woman could attain such a position at all; the holding might have been taken over by guardians (*Vormund*), either from her side of the family or from that of her husband. These, incidentally, would have been guardians of the woman herself and of her children. But we find nothing of the kind; the widow, as we can see here, is so closely linked with her husband that all other blood relations recede into the background, at least from the lord's point of view.

Let us examine the field of kin and conjugal relationships in greater detail. The three inventories reveal a field of kin and conjugal relationships (see Figure 20). The sparseness of the chart is conspicuous. Beginning with the *ego,* there are three generations: that of the *ego* itself, and then an ascending and a descending one. The lateral relations are hardly perceived: clearly, not all possible relationships are considered. No relatives once removed are listed either for the *ego* generation or for that of the parents or children. Only the marriage of the *ego* or the marriages of his siblings or parents are essential and, in addition, the children of the *ego.* The field appears to consist of pairs made up of close relatives. The sister's husband seems to be more important than the cousin. The field narrows even more if we take the distribution of terms for relations into account. Mother and brother-in-law are listed only once each.[11] Wives and sisters are not men-

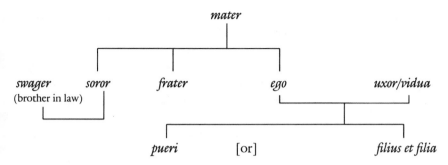

Figure 20. Field of kin and conjugal relationships.

tioned at all where land tenure and payments are concerned. Brothers are rarely listed.

Thus we see that, from the lord's standpoint, everything is centered on the married man and the fact that his wife can act as his representative until she is replaced by the married man of the next generation. This explains why the term *pater* does not exist in the system. In the lord's eyes no married man is exempt from the ownership of a holding or the obligation to make payments—for example, the old peasant who has retired as head of a household and lives apart in a cottage (*Altenteil*). If the *pater* holds this position for a lifetime, this does not necessarily imply that he does it on his own. In the rent roll of 1340, the rents for 12 percent of the holdings were paid by groups of owners. Only in a few cases can the connections between these persons be discerned. We find, for example, two men with the same last name,[12] two men listed by first name and identified as brothers,[13] one man listed by first name "and his brothers" (*et fratres sui*),[14] and finally a widow and her sons (*vidua et filii sui*).[15] This does not appear to be a system in which a patriarch has full power over "rights and obligations."

Thus we can summarize the information that can be gathered from the three registers by the first name, the last name (if applicable), the method of enumeration, the type of relationship, and the holding. The principles according to which the closest links between the peasants are formed are sex, age, and status. Or, to put it differently, masculinity, seniority, and marital status produce forms of ordered and flexible inequality in which (1) the man is ranked above the woman, the brother above the sister, the husband above the wife, the son above the daughter; (2) the father above the son, the mother above the daughter, the elder above the younger brother; and (3) the married man/brother above the unmarried, the widow above the unmarried brother of the husband, the married above the unmarried sister, the legitimate above the illegitimate child.

Clearly the emphasis is on the center of a system in which all three principles are positively correlated—namely, the oldest married man. He is the *ego* of the system. But it is just as characteristic that he has an *alter ego*, his wife, who in the state of widowhood takes his place in matters of land tenure and paying rent. The privileges of age and especially of sex that are gained at birth are broken by the power of marriage, which lives on in the person of the widow in order to preserve rank. This appears to be a fact that should not be underestimated by social historians, but it is unclear whether this alone is constitutive for the lord's perception and what its influence was on the peasant's living conditions, or if it in fact corresponds to them.

Marriage [16]

"Because a man does not want or like to be without a woman, he has to take a legitimate (*echt*) woman [a wife] even if he has been married three or more times and these wives have died. In the same way, a woman takes a husband and has legitimate (*echte*) children with the last as well as the first, and leaves her rights and possessions to them." [17] Legitimacy (*echt*) is the central term in Eike von Repgow's characterization.

Unfortunately, there is only one-sided evidence about marriage in our source material. There is nothing to be found either about preliminary activities (courting, setting up of a marriage contract, engagement) or about the marriage ritual itself. The most distressing blank in this context is the role of the church. This absence may be significant, since in the fourteenth century, in northwestern Germany, the priest was not yet involved in the decisive act of the marriage ceremony. [18] What evidence we have refers to the long-term commitments to the goods (*gude*) and the heirs (*erven*) constituted by the marriage ceremony. Thus the legal act of the marriage is connected to the principle we know from Eike's *Landrecht*. [19]

The wedding ceremony and the *copulatio matrimonii* can be understood as the partial acts that establish such obligations between man and wife. After the consummation of the marriage, described by Eike as the act during which the woman steps into the man's bed (*bedde trit*), the woman becomes the man's companion (*notinne*). [20] It is disputed whether she leaves her father's tutelage (*munt*) through the wedding ceremony or through the *copula carnalis*.[21] In any case, the morning gift (*Morgengabe*) is handed over after the first act of sexual intercourse. A deed from the year 1358 states: "And when they arose after their first night of marriage, the husband, in the presence of many witnesses summoned for the occasion, gave to his wife all his present and future goods for her dowry."[22] That the obligations are confirmed by eyewitnesses is verified in an addition to the Goslar town charter. The widow is supposed to take her morning gift upon the division of the inheritance "in the same form as he had given it to her on the morning after he had lain with her for the first time and they went together to the table before the people."[23] And the bondswoman (*Litonin*), according to a custom from St. Blasius (1450), gives the marriage tax (*beddemunt*) after intercourse (*wen se byslept*).[24]

Marriage thus established a unity of the sexes in which the estate, which belonged to different members of the household, was to be of use to all of them. Marriage unites *man unde wip* as companions (*genoten*); one of them gives the *munt*, the other takes it. Authority and reciprocity are closely integrated. The couple gains (*gewint*) legitimate heirs who, if they survive, will come into their own.

Arbeide: Labor and Work[25]

"And he begot a daughter by her" (*Et genuit ex ea unam filiam*): This is the last sentence of the narrative in the deed of 1358 cited above, concerning a complicated dispute about an inheritance.[26] Economic and social historians sometimes forget what labor meant for the peasant woman during her married life. Eike von Repgow mentions this in a very informative context: "Lend your ear, now, to the case of a wife bearing a child after the death of her husband and being recognized as pregnant on the day of his burial or on the thirtieth day after his death. If the child is born alive and this is witnessed by four men who have heard it and by two women who have helped her during her labor (*arbeide*), the child will inherit from its father."[27]

Soon after the marriage ceremony a large part of a woman's life is spent in pregnancy, the *arbeide* of birth, breastfeeding, and expecting the next pregnancy, until the onset of menopause. Our sources do not tell us anything about the important time in a peasant woman's life spent in pregnancy and as a mother. But even if our search were broadened in time and space, the results would still be meager.[28]

Our knowledge of childhood and youth is just as slight.[29] Eike deals only with questions about the legitimacy of children, about the different stages of life in relation to legal matters, and about guardianship and the right to punish crimes in which children were involved.[30] In the *Latenordnung* from St. Blasius (c. 1450) quoted above, in which the position of children from marriages between parents of different legal status is regulated, it is mentioned only in passing that these children are divided into boys and girls (*knechte* and *megede*).[31]

This fragmented information scarcely constitutes a picture of work. Empirically, little more can be discovered. In the sources, the daily chores during the course of the year are presupposed and therefore never explained. Interest is focused only on creating or preserving conditions for daily livelihood, so as to secure that part to which one is entitled. Thus it is not possible here to construct these activities as an ordered and repeated series of events. Nevertheless, one can outline a central aspect of the social framework of everyday life and formulate its general function as reflected in the use of significant terms. In very different sources, we find a binary orientation in which certain activities and objects are connected to males or to females. This should not be seen, however, as a reduction to a gender-specific division of labor.[32] Rather, it is an extensive form of orientation and organization of life in general, the significance of which should not be underestimated.[33] The threshold that divides males and females—who are often united in everyday life—is evident in the passage from Eike, quoted above, concerning the witnessing of and testimony at the birth of a legitimate child after the death of its father. The women are allowed to testify because they helped the pregnant woman in her labor, the men because they heard it all. The division of the sexes into eye- and ear-witnesses in an important legal matter, the birth of an heir, expresses something fundamental. The men accept their exclusion from the most important proof— the judgment by appearances—so as not to violate the domain of women, that is, the place of pain and birth.

This gender-specific orientation can be found in other aspects of life as well. If the bondsman, in case of death, has to give his best head of cattle (*Besthaupt*), and the woman her best dress (*Bestkleid*), then there must be special bonds that are recognized by the lord between certain objects, especially tools, and the members of the two sexes. The lord at times stresses these ties in the taxation of an inheritance, and even the heirs do this. Certain objects belonging to the inheritance or to the decreased can be passed on only to a member of the same sex: the man's possessions (*[Heer-]Gewäte*)

can only go to the next male relative (*Schwert-Mage*); the woman's possessions (*[Niftel-] Gerade*) only to the next female relative (*Spindel-Mage*).[34]
From the sources we get the following picture:[35] associated with men are
the weapon/knife, riding horse, draft cattle, plough, cart; with women, the
spindle/shears/scissors, poultry, household goods, textiles, jewelry. This
classification enables us—and this is the main point—to draw the conclusion that there is a gender-specific distribution of the most important
activities of the peasants. In connection with other sources—common law,
labor rent rolls, illustrations of the labors of the months, court rolls and
deeds, and literary evidence—the distribution between men and women
can be understood with the help of the following opposing characteristics:
public (male) versus private (female), outside versus inside; heavy (work)
versus light. The Eastphalian sources also suggest that the man as "guardian" is the representative in all public matters (especially in legal conflicts
or requirements concerning the community) and that he shields the house
and home. He performs heavy tasks in the field and meadow, and takes care
of the external economy (*Außenwirtschaft*) of the farm. The woman, on the
other hand, takes care of the physically lighter, domestic tasks.[36] Recent
studies, however, are rather more critical toward such a rigid classification
and emphasize the interconnections between the sexes in the wide range
of peasant activities.[37]

More attention also has been paid to the question of how binding the
gender-specific division of certain activities actually was and what influenced these obligations. The size of the holding and the form of payment
must have had some effect, as did the long-term economic situation,[38] but
the Eastphalian sources provide no empirical evidence to assess this. The
question, for instance, of whether the gender-specific division of activities
was rigid and, considering the wage laborers, stratified on the holdings of
rich villagers, and whether it was necessarily more flexible on the landless—
and thus, ploughless—smallholdings cannot be answered here.

In conclusion, we can make a speculative attempt to account for the
gender-specific ties to certain objects. First, what is characteristic of men
is their closeness to those animals that have a function as tools. Driving
oxen and riding a horse demand the ability to utilize the peculiarities of the
animal and to submit it to one's own will. This means the skillful mobiliza-
tion of force and superiority, not only of physical strength. The common
denominator of the other tools—sword, knife, plough (and axe)—appears
to be the cutting, which, whether we consider ploughing, slaughtering,

lumbering, or wood-cutting, can be regarded as a form of "destruction," a clear initiating and active preparation of *natura* for future usefulness: turning *natura* into *materia*.

As opposed to this, the tie between women and cattle seems to be care for their fertility. The making of raw fibers or hairs into clothing, or of victuals into meals, can be regarded as barely visible changes of "materials" with the aid of fire and water; these conclude the preceding efforts and make the objects ready for use or consumption.

This line of thought can be extended. Man cuts off from the natural "body," such as the earth or plants, what woman then turns into useful things for the human body, such as clothing and food. Things taken by him are made useful by her for living. These differences in behavior are the conditions for living in the peasant household. Man and woman are interconnected in a complementary way, and without such an interconnection, there would be no possibility for the peasants to sustain themselves. Of course these abstractions are not based on reality in the sense of adequate evidence in space and time. And they are one-sided because they do not exhaust the possibilities of interpreting the investigated ties. Their value is limited also by the fact that they cannot be related to the peasant household as the place where they have concrete effects. But, in my opinion, they are useful as an orientation in considering the framework of activities and values within which family, household, and kinship take on different concrete shapes, within which laboring and working are effective.

In trying to determine the basic functions of peasant activities, the way in which the relationships of the peasants to the farm and the holding are expressed in low German sources can be helpful. Here, again, we can see by their language how the lords perceive the peasants. It is astonishing that there is no standardized vocabulary. The verbs used are generally inadequate to express the relationships systematically. In fact, the legal forms of the relationships between people and land are rarely verbalized: neither the free lease limited to a certain number of years (*Meierrecht*) nor the hereditary tenancy *(Erbzinsrecht)* is normally expressed in terms of activity.[39]

In contrast, the mode of description has a tendency toward "factual" terms. These are, above all, the verbs *hebben* (to have), *sitten* (to sit), *wonen* (to dwell), *buwen* (to build), *werken* (to work), *bruken* (to use), and *vruchtigen* (to make fruitful). These words as traced in dictionaries have particular meanings but also interconnections.[40] It is conspicuous that the first three verbs in the sources are very closely connected with the farm: "and cottage belonging to" (*unde I Kothof, den heft Lons*),[41] "and a farm in the steward's possession" (*unde enen hof, dar de meyer uppe sit*),⁴² "a farm in the village of Hary on which Henry M. presently dwells" (*hof, in dem dorpe to H(ary?), dar Heneke Mundeke nu uppe wonet*).⁴³ These verbs indicate the aspect of duration and the subsequent stabilization into "habit" and "custom." It does not seem coincidental that this aspect is stressed especially in the relationship of the peasant to the farm on which he dwells (*haust*), the place that shelters him and his people, the place where they stay. Permanence, habit, steadfastness: these may be the qualities that characterize the relationship of the peasant to house and home.

The verbs of the second group have a quite different meaning; they express the active relationship to the soil. *Buwen* occurs most frequently; *bruken* is much scarcer, *vruchtigen* is rare, and *werken* is used only by Eike.⁴⁴ Most of these verbs refer to the fields, especially in regard to ploughing.⁴⁵ In spite of the variety of verbs used, they seem to convey a homogeneous perception: the connection between constant dwelling and active cultivation. But how this takes place is never expressed in general terms, nor what quality it has for the peasants. Once again, only Eike seems to do this, in describing different forms of rural *buwe* as *arbeide(n)*.⁴⁶

In the dry style of the deeds, labor (*arbeide*), is mentioned only in a remarkably narrow context. Generally, the term is found in leases that specify at whose cost and expense (*cost unde arbeide*) the leasehold will be improved (*beteren*), or who must bear the responsibility in case of damage.⁴⁷ In a lease of 1345 pertaining to a great estate in the vicinity of Alt-Wallmoden that belonged to the cathedral chapter of Riechenberg (Goslar), the knight Johann von Rössing promises that the cost of damage done by mice, foraging soldiers, or hail will be shared by both parties (*dat ek dat buwen unde beteren schal unde wille mit mines sulves cost unde arbeyde*); and when the lease expires, everything will go back to the cathedral chapter "with all increase in yield on the farm and all the seed in the fields."⁴⁸ Calculation, the weighing of profits and losses, has gone into the detailing of the diligent efforts to be undertaken. It is not work in the general sense that characterizes the lives of these peasants; the point is not their livelihood, but a specific attempt to gain profit. It seems clear that there is no comprehensive, generalized word—verb or noun—for the manifold activities of the peasants. Still, we can distinguish between the acts of dwelling on the farm and of working in the fields. That their essential traits are toil and trouble is not expressed at all except perhaps by Eike. *Arbeide* is visible only in reference to a woman's labor in childbirth or when toil is no longer seen simply as one's fate but rather as an integral part of purposeful action for gain or loss.

In other words, labor is seen as a specific calculus in gaining profit. Perhaps it is not too speculative if we see the variety of verbal denominations in relation to the world of the sustenance of the peasant and the income of the lord, but the particular terms of the lease in relation to the world of the profit expectations of the urban patriciate.

Village Organization and Community

For the period considered here, evidence about the village community is scanty.[49] While the period from the tenth century through the thirteenth witnessed local control through seignorial rights over the dependent *familia* (*Hofrecht*), local customs (*Weistümer*) are characteristic of the period after 1400. The transition between the dissolution of the manorial system (*Villikationssystem*) and the formation of the village community (*Dorfgemeinde*) has not been sufficiently studied.[50] We therefore must rely on scattered details from charters and deeds, accounts, and, of course, Eike's *Sachsenspiegel*.

We hear very little in our sources about the division of labor or rural industry.[51] The most important figures—the miller, the blacksmith, and the herdsman—are rarely mentioned since they were not regarded as normal tenants. Mills are listed occasionally in registers (fiefs, holdings), generally without comment. From the rent rolls, we learn that the rents for mills varied from 10 shillings to 1 mark,[52] which was comparable to the rent for a medium-sized or larger holding. Eike tells us about the legal aspects. The mill enjoyed special protection; damage done to it was severely punished. Concerning its use he says proverbially: *"De ok erst to der molen kumt, de scal erst malen"* (that is, first come, first served).[53] Most of these mills were probably powered by water, utilizing the many brooks and dammed ponds.[54] The distribution of mills in the region under study is not known, nor whether each village had its own miller.[55] Only two mill statutes (*Mühlenordnungen*) provide information on how they were run, and they emphasize the honesty of the mill workers and their rate of pay.[56] The ban, that is, the requirement that the peasants use the mill, is not mentioned; the lords, at any rate, do not point it out.

The brief references to blacksmiths and herdsmen cannot be interpreted systematically. Concerning the village herdsman, we can differentiate between the paid community herdsman and the private herdsmen, especially shepherds, hired by rich peasants.[57] Special sites and villages with mineral deposits were particular situations that can only be mentioned marginally here: the salt houses in Gitter/Vöppstedt/Kniestedt (near Salz-gitter)[58] and the pottery works in Ötze and Bengerode.[59]

The taverns should not be omitted. There is evidence for "a small farmstead where the tavern was situated" (*enneke hoveke dar de taverne hadde gewesen*) only in Groß Wehre. But if we agree with Wiswe that the *kopen-pennige* (so-called purchase pennies) should be seen as an ale tax paid by taverners, then there is evidence of taverns in the villages of Jerstedt, Wed-dingen, and Langelsheim, all situated on roads leading to Goslar.[60]

In the sources used here, only the last names of tenants in the rent roll and accounts of St. Blasius and in the Heimburg fief register point to the existence of tradespeople of a small town caliber—baker, weaver, tai-lor, cobbler, joiner, shopkeeper, and so on.[61] Although we can say nothing about their place in the life of the village, still the assumption that poor villagers and cottagers often held these positions may be correct. If we did not have Eike's *Spiegel*, only the outline of village organization would be visible: the leading agent, called *burmester* (*magister civium*), and the village community, the *burscap*.[62] In the lists of court witnesses in deeds dating from the end of the fourteenth century, representatives of thirteen villages in the area are mentioned.[63] In the regional courts (*Godinge*) in Buchladen, near Goslar, and in Liebenburg, where criminal cases and cases concerning the purchase and transfer of land were heard, these represen-tatives acted as jurors.[64] This evidence shows the prevalence in this region of a form of community with the *burmester* at the top. Much strength has been ascribed to this type of organization,[65] although this cannot be con-firmed. We do not know how the burmester was selected, nor what his authority was within the village. What we know about the *meierding*, the seignorial court for bondspeople, is not transferable to court matters con-cerning the *burscap* in general.[66] In the ten charters that deal with disputes between members of the village community (variously denoted as *villani*, *cives*, or *buren*) and local lords, we hear nothing about law-telling, judging, or punishment. The charters testify only to communal actions;[67] the only exceptions concern some privileged villagers (*Erfexen*).[68]

What activities are being described in these charters? In this period the peasants had to cooperate closely. Unfortunately, nothing of substance is said in the deeds about the organization of work in the common fields (*Flurzwang*) and the three-course rotation. Apparently, this was not a mat-ter about which one argued with the lord, whereas the contrary was true concerning the use of the commons, that is, the pasture (*mene*) and the woods (*marke*). Communal activities involved, first, the proper use of the commons; second, their precise boundaries and access to them; and, finally, their allocation. All this concerns the maintenance of the just use of the existing commons. It is important to note that the members of the village community also were jointly engaged in changing prevailing conditions.

In 1311 it was reported that the *cives* of Jerstedt "had converted to arable husbandry some small parcels of fertile land previously given over to grazing for cattle and that, in compensation for those parcels, they had agreed to leave some less fertile land for that purpose."[69] Two aims can clearly be discerned here: the peasants wanted to enlarge the areas of good arable land, but they were unwilling to reduce the size of the pasture. This is a good example of a joint effort to improve the use of local resources without endangering the basic structure. During this period it appears that land cultivation and cattle breeding were in a precise balance and that there was no wish to upset this even slightly. Everyone was responsible for maintaining this balance.

Everyone? Unfortunately, little is known about who in the village community had the right to, and who actually did, participate in the arrangements concerning the village economy though the *Erfexen* were probably the predominant force. However, it is not clear who belonged to this group and who was excluded (perhaps the landless cottagers). The enforcement of legal obligations and the keeping of the peace outside the community on the one hand, and the internal regulation of the use of the commons on the other—these roles attest to the competence and vitality of the village *communitas*.

Two examples will serve to show how the community also looked after the religious and ritual needs of the villagers, for better or for worse.[70] In 1356, the *cives* of Klein Flöthe petitioned the bishop of Hildesheim to raise their newly built church, which had until then only been a daughter church (*Filialkirche*) of the old parish church of Groß Flothe, to the position of a parish church, "so that they might have their own priest and would thus be able to receive the divine sacraments and follow the divine service more easily."[71] Their request was granted, but they had to provide their church with five Hufen and to commit themselves to the perpetual payment of 1 mark per year to the old parish church. Quite an expenditure for a recently cleared hamlet so that the villagers could have their own priest to attend to the cure of souls.[72]

At about the same time, the parish of Mahlum, a village west of Bockenem, had been without any cure of souls or parish law (*ius parrochiale*) for more than twenty years. This absence was a punishment going back to the year 1331, "when in the church of Malem the priest, while celebrating the holy mass, was killed by a certain wicked young fellow who had been instigated by the devil; while those parishioners who had gathered to listen to the divine service neither impeded the miscreant nor made any attempt to detain him after the crime had been committed."[73] The bishop saw the reason in the scandalous fact that the murder and the flight of the attacker were not stopped, because of a conspiracy *(societas occulta)* by the peasants. But why the conspiracy? One can assume that it was directed against that priest who, as earlier deeds show, was appointed by the canons of the cathedral chapter of St. Georgenberg (north of Goslar) without the cooperation of or supervision by the responsible archdeacon.[74] Neither the canon who celebrated there (and who was probably absent most of the time) nor his vicar was seen by the Mahlum peasants as a *clericus ydoneus*, the priest who took care of their liturgical affairs. After this act of collective disobedience, there was no more *solatio* for the parishioners in their own village.[75]

Summary

Despite the paucity of sources for this region of Germany in this period, we can learn something about the relationships within the village. Masculinity, seniority, and marriage are recognized as basic principles that structured social relations, at least as seen from the standpoint of the lords or proprietors who described the relationships in the surviving documents. The social core of the household, which cannot be directly observed, is the conjugal couple. Kinship is consequently understood in a very restricted way. The conjugal couple is perceived as a sexual unit and a community of goods, in which authority and reciprocity are mingled, the man having the tutelage. Besides providing for daily subsistence, it has an important purpose in generating legitimate heirs.

"Labor" and "work" cannot be observed as concrete detailed actions, because the information about peasant labor usually comes from descriptions of obligatory labor services, which are lacking in this case. On the other hand, legal and hence stereotyping sources that comment on gender-related tools and goods allow one to say that everyday "doing" was organized in a binary and complementary way; labor was cooperative, relying on mutual dependency, with the tendency that the man was close to the "beginning" acts, the wife, close to the "ending" ones. On the other hand, the modes of articulating, especially verbalizing, the relationships of people to the farm and the arable show a division between passive dwelling and active growing. The Latin *labor* (mlG. *arbeide*) is not a term with a generalized meaning, like "production" or "work." On the contrary: the words are used in a restricted way to denote the "costs and expenses" of leases, corresponding rather to patrician than to seignorial (or peasant) attitudes.

Only a few hints about the miller, the smith, and the herdsman testify to the so-called division of labor within the local setting. Other specialized crafts were restricted to large villages or to places that had rare mineral resources.

The village community is represented externally by the *burmester* and the court companions (*dingnoten*), about whom little is recorded. Internal disputes about the use of the commons or the balance between cultivated fields and pastures were settled by the *burscap*. Last, but not least, the people of the plough were solicitous about the availability and regular performance of sacred functions by the parish church.

This paper is a revised and abridged version of one chapter of an unpublished work ("Die Neuwerker Bauern und ihre Nachbarn im 14. Jahrhundert," Habilitationsschrift TU, Berlin, 1983). I would like to thank Ylva Eriksson for the translation into English.

Notes

1. Werner Wittich, *Die Grundherrschaft in Nordwestdeutschland* (Leipzig, 1896); O. Teute, "Das alte Ostfalenland. Eine agrar-historisch-statistische Studie," unpublished phil. dissertation, Erlangen University (Leipzig), 1910; R. Hoffmann, "Die wirtschaftliche Verfassung und Verwaltung des Hildesheimer Domkapitels bis zum Beginn der Neuzeit," unpublished phil. dissertation (Münster) 1911; W. Küchenthal, *Bezeichnung der Bauernhöfe und der Bauern—die Klassenteilung der Bauern —im Gebiete des früheren Fürstentums Braunschweig-Wolfenbüttel und des früheren Fürstentums Hildesheim* (Hedeper, 1965); Ernst Döll, *Die Kollegiatstifte St. Blasius und St. Cyriacus zu Braunschweig*, Braunschweiger Werkstücke 36 (Braunschweig: Waisenhaus, 1967); Walter Achilles, "Zur Frage nach der Bedeutung und dem Ursprung südniedersächsischer Hofklassen," *Braunschweiger Jahrbuch* 49 (1968): 86–104; H. D. Illemann, *Bäuerliche Besitzrechte im Bistum Hildesheim. Eine Quellenstudie unter besonderer Berücksichtigung der Grundherrschaft des ehemaligen Klosters St. Michaelis in Hildesheim*, Quellen und Forschungen zur Agrargeschichte 22 (Stuttgart: G. Fischer, 1969); H. Hoffmann, "Das Braunschweiger Umland in der Agrarkrise des 14. Jahrhunderts," *Deutsches Archiv zur Erforschung des Mittelalters* 37 (1981): 162–286; Martin Last, "Villikationen geistlicher Grundherren in Nordwestdeutsch-

land in der Zeit vom 12. bis zum 14. Jahrhundert (Diözesen Osnabrück, Bremen, Verden, Minden, Hildesheim)," in *Die Grundherrschaft im späten Mittelalter* 1, ed. Hans Patze, Vorträge und Forschungen 27 (Sigmaringen: Thorbecke, 1983): 369–450.

2. There are other terms like (Lat.) *coloni, rustici, homines, cives,* and (Eastphalian) *tobehoringe.* See K. Schwarz, "Bäuerliche 'cives' in Brandenburg und benachbarten Territorien: Zur Terminologie verfassungs- und siedlungsgeschichtlicher Quellen Nord- und Mitteldeutschlands," *Blätter für deutsche Landesgeschichte* 99 (1963): 110–17.

3. See, for example, Peter Laslett and Richard Wall, eds., *Household and Family in Past Time* (Cambridge: Cambridge University Press, 1972); David Herlihy, *Medieval Households* (Cambridge, Mass.: Harvard University Press, 1985); Christiane Klapisch-Zuber, *Women, Family, and Ritual in Renaissance Italy*, trans. Lydia Cochrane (Chicago: University of Chicago Press, 1985); Michael Mitterauer and Reinhard Sieder, eds., *Historische Familienforschung* (Frankfurt: Suhrkamp, 1982); Judith M. Bennett, *Women in the Medieval English Countryside: Gender and Household in Brigstock before the Plague* (New York: Oxford University Press, 1987).

4. See especially Lutz K. Berkner, "Inheritance, Land Tenure and Peasant Family Structure: A German Regional Comparison," in *Family and Inheritance: Rural Society in Western Europe, 1200–1800*, ed. Jack Goody, Joan Thirsk, and E. P. Thompson (Cambridge: Cambridge University Press, 1976), pp. 71–95.

5. *Urkundenbuch der Stadt Goslar und der in und bei Goslar belegenen geistlichen Stiftungen* (hereafter *GUB*), ed. Gustav Bode, 4 (1336–65) (Halle, 1905): 397–98, no. 525; for comments on this register see Kuchenbuch, "Die Neuwerker Bauern," pp. 16–17.

6. *Urkundenbuch zur Geschichte der Herzöge von Braunschweig und Lüneburg und ihrer Lande*, ed. H. Sudendorf (hereafter *SUD*), 11 vols. (Hannover: Rumpler, 1859–83), 2: 261–62.

7. For example: "Olrikes sone des meygers mit der husvrowen vnde kinderen"; "hinr(ich) van attenstidde. albrecht sin broder."

8. This is my interpretation of the remarks: "berken husvrouwe vnde kindere. Hartwich bereken broder vnde suster."

9. *Die Vizedominatsrechnungen des Domstifts St. Blasii zu Braunschweig 1299–1450*, ed. Hans Goetting and Hermann Kleinau (hereafter GK), Veröffentlichungen der Niedersächsischen Archivverwaltung 8 (Göttingen: Vandenhoeck & Ruprecht, 1958), pp. 45–48; for critical comments on this rent roll see Hoffmann, "Das Braunschweiger Umland," pp. 171 ff.

10. Some examples (*GK*, p. 46, Twelken, Schöppenstadt; p. 47, Berklingen): "Ghesa Widdekindi, Berta Crighers, Hanna Heydekonis." The last names of the deceased husbands are given in the genitive case.

11. *GK*, p. 47 (Wendessen): "Ghertrudis Widdekindi et swagerus suus de dimidio manso 3 s(olidos) et 3 d(enarios) lini." Is the brother-in-law the guardian or a widower? Ibid. (Ahlum): "Hermannus Derseman et mater sua de uno manso."

12. *GK*, p. 47 (Ührde): "Luderus et Johannes Rivestal de dim(idio) manso."

13. *GK*, p. 47 (Berklingen): "Heydeko et Henricus fratres de uno et dimidio manso 12s."

14. Ibid.

15. *GK*, p. 48 (Bruchmatersen).

16. Useful in general are Herlihy, *Medieval Households*, pp. 98–103, 115–20; Werner Rösener, *Peasants in the Middle Ages*, trans. Alexander Stützer (Urbana: University of Illinois Press, 1992), pp. 178–83; Michael Schröter, "*Wo zwei zusammenkommen in rechter Ehe . . .*": *Sozio- und psychogenetische Studien über Eheschließungsvorgänge vom 12. bis 15. Jahrhundert* (Frankfurt: Suhrkamp, 1985) (based on literary sources).

17. *Das Landrecht des Sachsenspiegels*, ed. Karl August Eckhardt, Germanenrecht 14 (hereafter *SspLaR*) (Göttingen: Musterschmidt, 1955), 2: 23. I refer to the *Sachsenspiegel* because it was certainly used in the Goslar region. See Karl Kroeschell, "Rechtsaufzeichnung und Rechtswirklichkeit: Das Beispiel des Sachsenspiegels," in *Recht und Schrift im Mittelalter*, ed. Peter Classen, Vorträge und Forschungen 23 (Sigmaringen: Thorbecke, 1977): 349–80.

18. Besides Schröter, "*Wo zwei zusammenkommen*," see Ingeborg Schwarz, *Die Bedeutung der Sippe für die Öffentlichkeit der Eheschließung im 15. und 16. Jahrhundert (besonders nach norddeutschen Quellen)*. Schriften zur Kirchen- und Rechtsgeschichte 13 (Tübingen: E. Fabian, 1959), pp. 12–21.

19. See Friedrich-Wilhelm Fricke, *Das Eherecht des Sachsenspiegels: Systematische Darstellung* (Frankfurt: Haag und Herchen, 1978).

20. *SspLaR* 3: 45 §3.

21. Eike Freiherr von Künßberg, *Lehrbuch der deutschen Rechtsgeschichte*, ed. Richard Schröder, 7th ed. (Berlin: De Gruyter, 1932), pp. 803 ff.; in favor of the *Beilager*, K. A. Eckhardt, "Beilager und Muntübergang zur Rechtsbücherzeit," *Zeitschrift der Savigny-Stiftung für Rechtsgeschichte, Germanistische Abteilung* 47 (1927): 174–97.

22. "Et cum surrexissent de lecto prime noctis sui matrimonii, vocatis ad hoc pluribus testibus, donavit . . . sue uxori omnia bona sua presencia et futura in donacionem propter nupcias": *GUB* 4/652. For background on the morning gift see T. Mayer-Maly, "Morgengabe," in *Handwörterbuch zur deutschen Rechtsgeschichte*, ed. Adalbert Erler and Ekkehard Kaufmann (hereafter *HRG*), 3 (Berlin: E. Schmidt, 1984): cols. 678–80.

23. "wu de gegheven heft des morghens, do he erst bi er ghelegen hadde unde se to dische ghinghen vor den luden": *Das Goslarer Stadtrecht* ed. Wilhelm Ebel (hereafter *StRG*) (Göttingen, 1968), BIV 2, p. 190. The affinity of this *Stadtrecht* to Eike's *Sachsenspiegel* is stressed by Wilhelm Ebel, *Rechtsgeschichtliches aus Niederdeutschland* (Göttingen: Schwartz, 1978), pp. 142, 148.

24. Quoted in Döll, *Die Kollegiatstifte*, p. 258 n. 65.

25. As I cannot deny my admiration for the German "Begriffsgeschichte," I can only refer the reader to Werner Conze, "Arbeit," in *Geschichtliche Grundbegriffe* 1, ed. Otto Brunner, Werner Conze, and Reinhart Koselleck (Stuttgart: E. Klett, 1972): 154–215, esp. 160–63; very useful, too, is Raymond Williams, *Keywords: A Vocabulary of Culture and Society* (London, 1988), pp. 176–79 ("labour"), 334–37 ("work"). From the lexicographic standpoint see Robert R. Anderson, Ulrich Goebel, and Oskar Reichmann, "Frühneuhochdeutsch *arbeit* und einige zu-

gehörige Wortbildungen," in *Philologische Untersuchungen, gewidmet Elfriede Stutz zum 65. Geburtstag*, ed. Alfred Ebenbauer (Vienna: Braumüller, 1984), pp. 1–29.

26. See above, note 22.

27. *SspLaR* 1: 33.

28. Even Shulamith Shahar, *The Fourth Estate: A History of Women in the Middle Ages*, trans. Chaya Galai (London: Methuen, 1983), pp. 98–106 ("Woman as Mother"), had to testify to the lamentable state of knowlege in respect to motherhood. Recently Bennett, *Women in the Medieval English Countryside*, has gathered many details, but without grouping them under this heading.

29. An interesting starting point is Klaus Arnold, "Mentalität und Erziehung: Geschlechtsspezifische Arbeitsteilung und Geschlechtersphären als Gegenstand der Sozialisation im Mittelalter," in *Mentalitäten im Mittelalter: Methodische und Inhaltliche Probleme*, ed. František Graus, Vorträge und Forschungen 35 (Sigmaringen: Thorbecke, 1987): 257–88. For concrete details see Bennett, *Women in the Medieval English Countryside*, pp. 65–99.

30. *SspLaR* 1: 36, 37, 42; 2: 65.

31. Quoted in Döll, *Die Kollegiatstifte*, p. 258 n. 65.

32. See the fine survey by Martine Segalen, *Love and Power in the Peasant Family: Rural France in the Nineteenth Century*, trans. Sarah Mathews (Oxford: Basil Blackwell, 1983); a recent comparative anthology is Jochen Martin and Renate Zoepffel, eds., *Aufgaben, Rollen und Räume von Frau und Mann*, 2 vols. (Munich: K. A. Freiburg, 1989). The best local study available is Bennett, *Women in the Medieval English Countryside*, esp. pp. 115–29.

33. An intensive sketch along these lines is Ivan Illich, *Gender* (New York: Pantheon, 1982), esp. chs. 2–6.

34. W. Bungenstock, "Heergewäte und Gerade. Zur Geschichte des bäuerlichen Erbrechts in Nordwestdeutschland," Jur. Diss., Göttingen, 1966. See also the articles "Gerade" and "Heergewäte," *HRG*, 1 (1971), cols. 1527–30; 2 (1978), cols. 29–30. The meaning of *Gewäte* goes back to the warrior's weapons, that of *Gerade* to bride jewelry. *Niftel* (= *Nichte*) means the next female relative (*HRG* 3 [1984], cols. 1007–8). *Mage* means kinship; *Spindelmage*, all female relatives and their descendants (*HRG*, installment 31, cols. 1771–72).

35. The details are to be found in *SspLaR* 3: 76, §3; 2: 58, §2; 59, §2 (plough), and in various deeds (e.g., *GUB* 5/275, 677); Gustav Schmidt, ed., *Urkundenbuch der Collegiatstifter S. Bonifacii und S. Pauli in Halberstadt* (Halle: Hendel, 1884), 6 (1156): draft cattle and cart; the composition of Gewäte and Gerade: *StRG* 1, I & 17, 20, pp. 33–34; detailed composition of the Gerade: *StRG*, BIV 2, pp. 190–91.

36. Jutta Barchewitz, *Von der Wirtschaftstätigkeit der Frau in der vorgeschichtlichen Zeit bis zur Enfaltung der Stadtwirtschaft* (Breslau, 1937; reprint Aalen: Scientia-Verlag, 1982).

37. R. H. Hilton, *The English Peasantry in the Later Middle Ages* (Oxford: Clarendon, 1975), pp. 95–105; Rösener, *Peasants in the Middle Ages*, pp. 183–87; Bennett, *Women in the Medieval English Countryside*, p. 115.

38. Concerning rent, see Christopher Middleton, "The Sexual Division of Labour in Feudal England," *New Left Review* 113–14 (1979): 147–68; concerning the

crisis of the later Middle Ages as a result of the Black Death (1348–50), see Hilton, *English Peasantry*, p. 108.

39. The term "bemeygern" is documented only in *GUB* 3/849/1330, 5/6/1366; "ervetins" (hereditary rent) is used in combination with "hebben" (to have).

40. I have consulted Karl Schiller and August Lübben, *Mittelniederdeutsches Wörterbuch*, 6 vols. (Bremen: J. Kuhtmann, 1875–81); August Lübben and Christoph Walther, *Mittelniederdeutsches Handwörterbuch* (Norden: Soltau, 1888; reprint Darmstadt: Wissenschaftliche Buchgesellschaft, 1979); Agathe Lasch and Conrad Borchling, *Mittelniederdeutsches Handwörterbuch* (Neumünster: K. Wachholtz, 1956–).

41. Walter Deeters, ed., *Quellen zur Hildesheimer Landesgeschichte des 14. und 15. Jahrhunderts*, Veröffentlichungen der Niedersächsischen Archivverwaltung 20 (Göttingen: Vandenhoeck & Ruprecht, 1964): 13 and passim (register of the patrician family Frese/Hildesheim, 1370).

42. *GUB* 4/449/1351, pp. 323–26 (register of the patrician family Dörnten/Goslar); see also *GUB* 4/348/1349, 5/1175/1400.

43. *GUB* 3/585/1321.

44. "Buwen": *GUB* 4/132/1342; 272/1345; 449/1351, pp. 324, 326; 824/1364; 5/550/1384; "bruken": *GUB* 3/188/1308; 4/94/1339; 4/672/1359; 5/237/1373; 5/526/1383; "vruchtigen": *GUB* 4/522–23/1355; 4/544/1356; 5/275/1375; "werken": *SspLaR* 2: 58, §2.

45. Ibid.: "Des mannes sat, de he mit sineme pluge werket, de is verdenet alse se de egede dar over geit"; *GUB* 4/273/1342: "alse se (2.5 Hufen) nu rede beseyet unde mit dem ploghe begrepen sint."

46. *SspLaR* 3: 76, §3; 2: 46, §1 ff.; 2: 58, §3.

47. *GUB* 2/186/1273; 2/301/1282; 2/408/1290; 2/578/1299.

48. "mit aller beteringhe an buwe uppe dem hove unde mit alle der sat uppe dem velde": *GUB* 4/270; 4/778/1363.

49. For a general sketch, see Peter Blickle, "Les communautés villageoises en Allemagne," in *Les Communautés villageoises en Europe occidentale du moyen âge aux temps modernes*, Flaran 4 (Auch: Comité Départemental du Tourisme du Gers, 1984), pp. 129–42; Rösener, *Peasants in the Middle Ages*, pp. 147–68; *Les Communautés rurales/Rural Communities*, pt. 5, Récueils de la Société Jean Bodin 44 (Paris: Dessain et Tolra, 1987).

50. In general, see Peter Blickle, ed., *Deutsche ländliche Rechtsquellen* (Stuttgart: Klett-Cotta, 1977); in detail (lower Saxony), see Last, "Villikationen."

51. For a solid survey see Walter Janssen, "Gewerbliche Produktion des Mittelalters als Wirtschaftsfaktor im ländlichen Raum," in *Das Handwerk in vor- und frühgeschichtlicher Zeit*, 2: *Archäologische und philologische Beiträge*, ed. Herbert Jankuhn et al., Abhandlungen der Akademie der Wissenschaften in Göttingen, phil.-hist. Kl., 3.F. 123 (Göttingen: Vandenhoeck & Ruprecht, 1983): 317–96.

52. *GUB* 4/525/1355, p. 391: "molendinum solvens tres fertones et pullos"; *Urkundenbuch des Hochstifts Hildesheim und seiner Bischöfe*, ed. K. Janicke and H. Hoogeweg (hereafter *UBHHI*) 4 (Leipzig: S. Hirzel, 1908): 726–30: short series of mills (owned by the monastery of St. Michael) with various rents.

53. *SspLaR* 2: 66, §1; 13, §4; 59, §4.

54. W. Kleeberg, *Niedersächsische Mühlengeschichte* (Detmold: H. Böseman, 1964), p. 27.

55. See Kuchenbuch, "Die Neuwerker Bauern," p. 68 n. 120.

56. Richard Doebner, ed., *Urkundenbuch der Stadt Hildesheim* 1 (Hildesheim: Gerstenberg, 1881; reprint Aalen: Scientia-Verlag, 1980): 835/1331; 877/1334.

57. Based on Eike: G. Buchda, "Die Dorfgemeinde im Sachsenspiegel," in *Die Anfänge der Landgemeinde und ihr Wesen* 2, Vorträge und Forschungen 8 (Sigmaringen: Thorbecke, 1964): 21; comparing text and illustration (fourteenth century): Ulrike Lade, "Dorfrecht und Flurordnung in den Illustrationen der Sachsenspiegel-Bilderhandschriften," in *Text-Bild-Interpretation: Untersuchungen zu den Bilderhandschriften des Sachsenspiegels*, ed. Ruth Schmidt-Wiegand (Munich: W. Fink, 1986) 1: 181–82.

58. About this in detail, see Franz Zobel, *Das Heimatbuch des Landkreises Goslar* (Goslar: K. Kraus, 1928), pp. 14–19.

59. *UBHHI* 4/638/1321, p. 346; for the archaeological evidence, see Edith Ennen and Walter Janssen, *Deutsche Agrargeschichte: Vom Neolithikum bis zur Schwelle des Industriezeitalters* (Wiesbaden: Steiner, 1979), p. 159; for glasswork and charcoal burning in or near Königshagen, see Walter Janssen, *Königshagen: Ein archäologisch-historischer Beitrag zur Siedlungsgeschichte des südwestlichen Harzvorlandes*, Quellen und Darstellungen zur Geschichte Niedersachsens 64 (Hildesheim: Lax, 1965), pp. 152 ff.

60. Groß Wehre: *GUB* 4/449/1351, p. 327; "kopenpennige": *GUB* 4/506; E. Wiswe, *Chronik des Dorfes Remlingen* (Remlingen, 1974), pp. 179–80.

61. *GK*, pp. 45–48; *SUD* 2/484/1354.

62. The Eastphalian village community (*Dorfgemeinde*), as especially reflected in Eike's *Sachsenspiegel*, has been carefully studied by Buchda in "Die Dorfgemeinde," pp. 7–24; in addition, see Berent Schwineköper, "Die mittelalterliche Dorfgemeinde in Elbostfalen und in den benachbarten Markengebieten," in *Die Anfänge der Landgemeinde und ihr Wesen* 2: 115–48.

63. *GUB* 5/858/1391: *burmester* of Burgdorf, Schladen, Gielde, Neuenkirchen; *UBHHI* 6/1086/1395: Othfresen, Dörnten, Groß und Klein Döhren, Heissum, Lewe; *GUB* 5/982/1395: Wolfhagen, Langelsheim, Astfeld.

64. Gotz Landwehr, *Die althannoverschen Landgerichte* (Hildesheim: Lax, 1964), pp. 176–80.

65. Schwineköper, "Die mittelalterliche Dorfgemeinde," pp. 143–44.

66. The *meierding* is a court restricted to cases concerning rent and legal status; see Illemann, *Bäuerliche Besitzrechte*, p. 3.

67. *GUB* 2/506/1296 (Weddingen); 3/81/1304 (Jerstedt?); 3/245/1311 (Astfeld); 3/259/1311 (Langelsheim); 3/264/1311 (Jerstedt); Schmidt, ed., *Urkundenbuch der Hochstifts Halberstadt*, 2/37/1311 (Quenstedt, Werstedt, and other places); *GUB* 3/551/1321 (Astfeld); 3/714/1325 (Kirch-Nauen, Bodenstein); 4/158/1341 (Jerstedt); Conrad von Schmidt-Phiseldeck, ed., *Die Urkunden des Klosters Stötterlingenburg* (Halle: Waisenhaus, 1974), 193/1424 (Bühne).

68. *Erfexe* is derived from *erve*: see Schiller and Lübben, *Mittelniederdeutsches Wörterbuch*, pp. 740–41. The *erfexe* has to be understood as a freeholder with unrestricted rights concerning the common.

69. "quasdam terrulas ad usum pecorum dudum habitas ad agriculturam propter sui ubertatem redegissent et quosdam alios agros minus fertiles in compensationem dictarum terrularum ad eundem usum communiter reliquissent": *GUB* 3/ 264.

70. Schwineköper, "Die mittelalterliche Dorfgemeinde," p. 145, emphasizes the close relation between the parish and the community.

71. "ea de causa, ut ipsi proprium habentes plebanum sacramenta ecclesiastica commodius possent recipere divinisque ministeriis liberius interesse": *GUB* 4/538/ 1356.

72. Zobel, *Das Heimatbuch*, p. 124 ff.

73. "quod . . . in ecclesia Maldem sacerdos in sacrae missae actione a quodam perditionis filio instigatione dyabolica fuerit interfectus, ipsique parochiani in eadem ecclesia pro audiendo divino officio congregati malefactorem hujusmodi nec prohibuerint nec post perpetratum scelus ipsum curaverint detineri": *GUB* 3/886/ 1331.

74. *GUB* 3/172–3/1307; the previous history can be found in *GUB* 2/322, 324/ 1285.

75. A chapel is mentioned in Mahlum beginning only in 1360: *GUB* 4/692/ 1360; 857/1365.

Gerhard Jaritz

8. The Material Culture of the Peasantry in the Late Middle Ages: "Image" and "Reality"

Research into the life of medieval peasants may, in many respects, still be placed "at the stage of specialized . . . probing."[1] Generalized conclusions can be drawn only for some, often small areas. As at the beginning of the 1980s, we are still walking on "pathways,"[2] looking for "traces"[3] of peasant culture and life in the Middle Ages. Although some new results have been achieved during the last decade,[4] we are still confronted with a large number of unsolved problems.

In dealing with medieval peasants, and particularly with the material culture and the everyday life of peasants in the Middle Ages, we regularly encounter difficulties in analysis that are not present to the same extent when we investigate the nobility or townspeople. The reasons are clear. The chief problem is the "reality" of the representations of the peasants in the sources. Since one of our aims is to reconstruct the "realities" of the past in the most accurate way, we must consider that our view of matters may be very different from that of medieval people. Sources that emphasize the truth and reality of their contents may show medieval "truth" and medieval "reality," but not necessarily anything that we would understand as "our" truth or reality.

Most aspects of medieval material culture—as we all know—were normally not considered sufficiently important to be written down, or told to other people, or passed on to other generations. This was true especially for those spheres that were part of everyday life. When it did seem worthwhile to record material culture, it was for economic reasons or because what was being described was unusual, special, new, deviant, or subject to criticism. And that holds true even more strongly for the material culture and life of the peasants, who, in many respects or even generally, were not considered worthy of mention by members of the higher social ranks. Moreover, when peasants were the subject of certain treatises, chronicles, medieval literary works, or images, they were not often the readers, the audience, or the beholders.

Our sources about peasants were, in most cases, written or painted or drawn for other members of society, but in general those others were not directly interested in what the peasants actually were doing. The exceptions occurred (1) if peasant life was important for economic reasons; (2) if it seemed to be deviating from normality, from the order of the system, and thus might threaten the position of others with respect to their food, their dress, their social exclusiveness, and so on; (3) if it could serve as a positive or negative model for those who had deviated from certain rules of society, or could prevent others from doing this; (4) if it might amuse the members of higher social classes and in that way motivate them in different directions.

All the written and pictorial sources dealing with peasant life should be investigated in light of these considerations, which admittedly restrict the possibilities of reconstructing "realities." This holds true for legal sources as well as for literary sources with a didactic intention, chronicles, religious and secular pictures, and so on. An emphasis on negative and positive images of peasants and a resulting strong polarity are evident. Medieval attempts to describe "reality"—or better, what we might understand as reality—are rare and must be treated very carefully.

The exception to the above-mentioned patterns are the archaeological sources. These can certainly provide accurate information on houses, farming tools, peasant diet, vessels and pots, the structure of rural communities, and so on.[5] Quite often they supply material that can only to a limited degree also be traced in the written and pictorial evidence. On the other hand, archaeological sources may permit us to use comparative methods for only a limited number of aspects in a small region.

To come closer to our aim of reconstructing "reality" we must, from the beginning, use all the available sources, which means not only written sources, but also artistic representations and archaeological evidence. Studies that are based on such sources used separately and not comparatively often reach very disparate conclusions that are not always the result of regional or local differences, but rather the consequence of the fact that the sources have different intentions that have not been taken into consideration. Trying to reconstruct peasant dress, for example, by using only literary sources is simply impossible. Adding the evidence from inventories, chronicles, sumptuary laws, plays, sermons, pictures, and archaeological objects may be helpful by making contradictions and identities clearer.

Nevertheless, we often will be unable to reconstruct "realities," but may discover only the attitudes of others toward the peasants: we may see the function of the peasants as being useful or threatening to certain other groups in society.

Literary texts of a didactic character, sermons, pictures—most of them dealing with peasants but not having peasants as their audience—were supposed to influence others, to have an emotional effect. The regular use of peasants for this purpose can be demonstrated. Peasants were the lowest class of society. For that reason they could be the "ideal victim" and function as good or as bad examples.[6] They serve as examples of good by their simple and hard way of life; they are a model for people who illegitimately try to change their status in the hierarchical order of society or want to lead a pampered life not equivalent to their rank. A large number of examples could be given.

The peasant is used as a bad example particularly to show the results of unjustifiably leaving one's given place in society. Clumsy and awkward peasants begin to wear clothes reserved for the nobility, to hold feasts like knights, and to eat like rulers. Historians of the nineteenth century and well into this century interpreted such descriptions in the sources as real, without considering that the representation of reality and truth was not the aim of such sources. Stories were never told from a pure interest in what the masses of society were doing; rather, they were didactic and satiric in order to motivate others to live a better life. And it did not appear contradictory for the same author to describe peasant life as the positive ideal for human existence and, at the same time, the negative model, as, for example, Hans Rosenplüt of Nuremberg, writing in about 1450[7]:

Ich sprich, es ist in dreißig Jorn
Rechter Paurn nit vil geporn.
Das ist wol an irer Hoffart Schein,
Sie wölln all Herren sein.

[I tell you that for the last thirty years
no right peasant has been born.
This is due to their arrogance,
for all of them would like to be masters.]

Ich will loben den edeln frumen Paur.
Wann warumb? Es wirt im oft saur.

Wenn er mit seinem Pflug fert,
Damit er alle Werlt ernert.
. . .
Gott grüss dich, du edler Ackerman!
Wann dein niemant enpern kann.

[I would like to praise the noble and pious peasant.
When and why? For he often turns sour,
Plowing his field,
To nourish all the world.
. . .
May God be with you, you noble countryman!
For nobody can do without you.]

Similar tendencies have been noted by Keith Moxey, who analyzed the images of peasants in sixteenth-century German woodcuts.[8] He showed, on the one hand, that the depiction of peasant holidays or peasant marriages could not be interpreted as real images of joyful peasants who had obtained more self-confidence during the Peasant Wars.[9] Rather, they must be seen as didactic elements intended for other members of medieval society—particularly in the towns—as examples of the complete disorder represented by peasants. On the other hand, the same artists could produce a positive image of the peasant, depicting him as the common man to whom the message of the Reformation was directed.[10]

In medieval religious plays peasants and shepherds often were used to arouse emotion in the audience. They took the stage at a point when humorous details seemed to be needed in advance of a more relevant and interesting part of the play.[11] This could have either negative or positive connotations. Saint Joseph is portrayed as a peasant in the Christmas plays of the late Middle Ages, a convention that also was taken over in the pictorial arts—old, a bit stupid, clumsy, made fun of by maidservants, only useful to prepare pap for the baby Jesus: he certainly was a caricature rather than a real peasant.[12] He bore enough elements of a real peasant to be recognizable to the audience, but many other aspects were exaggerated in order to move, to arouse emotion, to evoke laughter. Would these plays have been insulting to a peasant audience? Were they regularly, or ever, seen or heard by peasants?

If we want to extract from the sources as many elements of "real" medieval peasants as possible as well as the attitudes of other groups of society toward the peasants, we *must* use a comparative approach. One decisive aspect is the already mentioned role of the audience, of the public, and of the public nature of the media that are available as our sources.

We may deal with the (few) sources that were meant to be received and used by the peasants themselves: first, directly—for example, religious pictures in village churches, normative sources and by-laws, sermons to some extent, and archaeological objects, particularly those derived from excavations of deserted villages. The second type of source is indirect—for example, rural economic sources such as account books of landholders, inventories, wills, and so on. Such sources generally had a different function from those non-economic, mostly literary or pictorial sources meant to have a didactic effect on people of a higher social rank. That does not mean, though, that the latter are of minor importance for our analysis.

Recognizing that the methods of reconstructing the reality of medieval peasant life applied in the nineteenth century and later evidently led to a number of wrong conclusions,[13] historians have had to start from the beginning in many fields of analysis. Earlier, uncritical interpretations had the "advantage" that many aspects seemed rather clear and obvious. For these writers, what was said about the peasants, for example, in the literary sources of the Middle Ages, reflected true and "real" tendencies in the peasant life of the period. Having become aware that this certainly is incorrect, we now are confronted with a number of problems and questions that seemed several decades ago to have been solved. We must admit that a number of efforts to reconstruct "realities" have failed and that attempts to investigate attitudes and mentalities must still be considered vague and uncertain. The lack of sources, moreover, is a problem that often cannot be overcome.

An important point for our analysis is the fact that peasants, even when depicted satirically or with strong positive or negative connotations, still had to be recognizable as peasants. This means that they always had to wear or possess certain signs or characteristics that clearly made them "peasants." These became the basis for every exaggeration, change, or falsification. Such characteristic signs could be personal attributes, some significant type of labor, or the surroundings in which an event was placed. A typical example of the satirical use of the Labors of the Months (Figure 21) made the peasant recognizable only by his work and not through his apparel, which can be considered "anti-peasant." His tight doublet and hose, his long, curled hair usually signify young noblemen or their servants, and never peasants. On the other hand the young man is certainly

Figure 21. Satirical image of the "noble" peasant. Detail, Labors of the Months, August. Vienna, Österreichische Nationalbibliothek Cod. 3085, south German, 1475, fol. 74. Photo: Institut für Realienkunde, Krems, Austria.

not a nobleman or a servant of one, not only because of the work he is performing (binding sheaves) but also because of his gesture, his posture, and the roughness of his hair. We see here satirical elements aimed at arousing laughter. Every medieval spectator was supposed to recognize and understand this. We today may often lack this knowledge, which may lead to crucial misinterpretations. In a line of people on their way to Hell on the

Figure 22. To recognize a peasant woman one sign (the straw hat) might be enough. Detail, panel painting, Last Judgment, Austrian, c. 1490. Heidenreichstein, Kinsky collection. Photo: Institut für Realienkunde, Krems, Austria.

occasion of the Last Judgment, the peasant woman (Figure 22) is recognizable only, but still very clearly, by her straw hat. This was one of the most relevant signs for recognizing peasants, particularly when they were not portrayed in connection with rural work. Similarly, it might have been very evident from a single sign that the representation showed a "false" peasant. A woman working in the fields (Figure 23) who possessed virtually every

Figure 23. Satirical image of peasant woman with pointed shoes. Detail, Labors of the Months, August. Vienna, Österreichische Nationalbibliothek Cod. 3085, south German, 1475, fol. 74. Photo: Institut für Realienkunde, Krems, Austria.

legitimate attribute of her peasant rank must have still become a caricature when she was portrayed wearing pointed shoes. And even if everything was meant to be correct—when, for example, the "Good Regime" was visualized by showing peasants laboring (Figure 24)—the picture must be reanalyzed with the knowledge that an ideal was being presented with all its possible attributes, from cleanliness to a general view of perfection sym-

Figure 24. Idealized peasants as part of the "Good Regime." Detail, fresco, Labors of the Months, August. Trento (Upper Italy), Castello di Buonconsiglio, Torre d'Aquila, beginning of the fifteenth century, Lombardian or Bohemian. Photo: Institut für Realienkunde, Krems, Austria.

bolized by the absence of any sign of disorder. Again, "reality" might be lacking.

Some further examples may illustrate these points more clearly. One of the problems of research into the material culture of medieval peasants is their dress. Although written and pictorial sources sometimes are quite rich, archaeological evidence has often been lacking. We are, on the one hand, confronted with detailed literary descriptions of the clothing of certain peasants who were trying to escape their social class. This is particularly true for the period from the thirteenth century onward. Such descriptions of peasant dress cannot be seen in any way as real, but rather must be interpreted as a pattern of a general warning to others not to abandon the system of society; or as an effort to attack political adversaries by comparing them with fictionally disobedient peasants.[14] Those descriptions of dress, therefore, may sometimes be useful for reconstructing the clothing of knights, the attitudes against indecency in connection with dress, or the general importance of clothes in medieval society for classifying and recognizing people. But they can never be used as evidence for a general image of "real" peasant dress, even if we assume that peasants— like members of any other social group in the Middle Ages—tried to draw themselves closer to the higher social classes, particularly by using certain characteristics of dress.

Similar problems may occur when we use regulations governing dress or sumptuary laws concerning peasants. They show the law, the normative side of life. Although peasants may have been the direct recipients, the interpretation of these laws still involves many uncertainties. We usually cannot determine the extent to which the law was obeyed or disobeyed. What we are able to prove, though, are sometimes significant changes in the laws that evidently corresponded to changes in reality. Some Austrian and south German examples may be given. The so-called "Seifried Helbling," a poet of the thirteenth century, referred to a (fictitious?) dress regulation of Duke Leopold V of Austria, who ordered peasants to use only coarse gray woollen cloth for their garments.[15] Only on Sundays and other feast days were they permitted to wear coarse blue woollen cloth. The first part of this regulation obviously describes simple peasant work clothing, reaching to the knees, which can be seen in hundreds of pictures and illuminations of the Middle Ages. The color "gray" meant a natural, undyed color. The blue of the Sunday clothing, although dyed, was a cheap color. The so-called *Kaiserchronik* of the mid-twelfth century refers to a fictitious law of Charlemagne ordering peasants to wear gray or black garments and to use not more than seven ells of cloth for a shirt and breeches, the latter to be made of coarse linen cloth.[16] Again, a minimum is specified. In the Bavarian *Landfrieden* of 1244, we find a regulation that peasants must wear only cheap gray clothing.[17] Only peasants who had an official position were allowed to wear a better type of garment.

Evidence from almost three hundred years later offers a different image. In 1518, regulations were published for Austria that contain rather detailed remarks on dress.[18] Peasants again were placed at the low end of the hierarchy, but they were not restricted to the cheapest cloth nor to cheap colors. Now the maximum price for cloth was one guilder per ell. Gold, pearls, velvet, and silk were explicitly prohibited. No regulation of the thirteenth century would have found it necessary to deal with gold, pearls, velvet, or silk with regard to peasants. Such references are found only in satirical literature. To treat them factually would have been unimaginable. But at the beginning of the sixteenth century they apparently had to be mentioned. Moreover, something very important became actual—at least normative—reality: peasant women were allowed half an ell of silk to use for the hem of their dresses. And everyone was permitted to use London cloth ("lindisches Tuch"), which meant that peasants could as well. This was not cloth produced in and imported from London, which it might have been three hundred years before, but rather something that was produced locally, in the style of "London cloth." Still, we may suppose that it was rather special for the lower classes of the social hierarchy, who were no longer restricted to the cheapest coarse woollen cloth.

Changes had taken place. A rather general development may be assumed, at least from the evidence of the normative sources. More accurate regional trends, though, are still very difficult to grasp. Some other elements have to be added to our interpretation. We particularly must consider the difference between work days and Sundays or feast days. The work clothes that peasants wore do not seem to have changed significantly over many centuries.[19] The coarse (woollen) robe reaching to the knees, either undyed or a cheap color, can be accepted as reality for the whole of the Middle Ages. Many illustrations of the Labors of the Months show such a garment.[20] The garments worn on Sundays and feast days, however, were different. There we may find, or at least might suppose, rather important changes through the centuries. The permitted use of silk, even if only for hems, is a very relevant change. Problems with regard to public rivalry with people of higher social rank might have arisen. Such dress might have threatened other groups in society.

Sources that bear a higher degree of "reality" support the occurrence of such a development in a rather impressive way. Appliqués, buttons, and belt buckles made of copper or sometimes even silver were used on the clothing of Burgundian peasants as early as the fourteenth century,[21] and were found in inventories of Hungarian peasants and in archaeological investigations of Hungarian villages in the fifteenth and sixteenth centuries.[22] Hungarian inventories show that peasants possessed one or two Sunday suits of clothes, often made of more expensive cloth, in different colors. They also wore furs (particularly sheepskin and fox). These Sunday clothes were not much different from the clothes worn by members of higher social classes. Moreover, the discovery, in some archaeological finds, of silver threads in what are apparently peasants' clothes leads us into a milieu that seems neither exclusively rural nor peasant-like.

The possibility of using a large quantity of sources—as Françoise Piponnier was able to do with Burgundian inventories[23]—is very rare in the rural environment. An investigation of 150 inventories of peasants' possessions in the fourteenth century—which evidently did not include rich peasants—shows a more modest situation than the one known for Hungary at the end of the fifteenth and the beginning of the sixteenth century. Simple clothes predominate, comparable to those in illustrations of working peasants. The same is true for their color, which we already know from normative sources to be blue. But again, generalizations are dangerous.

The inventories of apparently impoverished members of the lower nobility, for example, from Franconia at the beginning of the sixteenth century,[24] show that at least in some cases their possessions, including clothing but excepting a few typical and commonly used luxury objects, could not have been too far from peasant culture or, rather, from the stereotype of the peasant situation. But we are still confronted with results from very different types of sources. General comparisons on this level still seem to be problematic.

Let us summarize our discussion of peasant dress and ask whether we can find similar situations and tendencies in other areas of rural life. First, the stereotype of the simple dress of medieval peasants seems to be more or less true for work clothes, which apparently did not change significantly during the centuries of the Middle Ages. Second, generalizations about peasants forgetting or ignoring their social class by wearing clothing that would characterize them as members of the aristocracy are literary fictions. We cannot conclude that such remarks in literary sources prove a particular or stronger tendency of the peasants in certain periods to try to escape their low position in the hierarchy. Comparisons with sources pertaining to other social groups in medieval society again and again show similar efforts to use dress to imitate or even to attain a higher rank in the social hierarchy. It is by no means typical of peasants. Third, in connection with that phenomenon, peasants can be seen as substitutes for other persons or groups in society that are the targets of criticism. The peasant is to some extent being used in their place. Fourth, all this does not mean that the tendency of peasants to aspire to higher social classes, at least by adopting their dress, is completely fictitious or generally less than in other social groups. Particularly with regard to the clothing worn on Sundays and feast days we may assume such a development, one that can be supported particularly by the change in regulations concerning dress. Finally, in regard to the sources, we cannot directly compare or commingle the contents of literary sources about the luxurious peasant and some singular remarks in other sources showing "realistic" situations (inventories etc.) with respect to elaborate rural Sunday clothes. Both can exist side by side without necessarily being connected to each other. Such connections have been emphasized too much in some historical research of the past, and attempts have been made to connect "images" with "realities" too forcefully, so that both—"image" and "reality"—illegitimately lost their relevance and independence.

Tendencies similar to those mentioned above can be seen if we consider diet. One cliché about peasants in the Middle Ages is that they ate only grain products, dark bread, and vegetables—particularly cabbage, turnips, beans, and peas—and drank water. Again, such a generalized stereotype occurs not in more or less reliable contemporary medieval descriptions of rural life, but rather in sources dealing with exceptional situations. We find it in literary descriptions concerning the question of what would or should be adequate for a peasant's sustenance, often accompanied by the criticism that this standard had been seriously ignored. On the other hand, we see it also when members of the upper classes were confronted by hunger. Famine forced even people in the higher ranks to eat dark bread, grain pulps, and vegetables, and to drink water. That kind of diet, as is sometimes explicitly emphasized, was supposed to be peasant food and drink and on one occasion was considered fit only for the enemy.[25] Peasants and the stereotypes of their material culture again are used to make an extraordinary situation understandable, to exaggerate, and to arouse emotion. Peasants are used to represent a certain pattern in situations that might not actually have touched them in any way.

Once again, particularly in literary sources, peasants are described as trying to flee from their modest milieu, desiring to ignore the (food) regulations for their class. Again we find the stereotypical image of peasants—especially at festivities—being extravagant, eating and drinking heavily, and so on.[26] The normative sources do not emphasize such a trend. They do sometimes regulate, as for other groups of society, the number of guests and courses to be served at special occasions, particularly at wedding or baptism banquets.[27]

In literary sources, it is assumed that meat is not often served at peasant tables. Archaeological investigations have shown a very small number of animal bones at some excavated medieval (deserted) village sites, but, of course, the situation varied locally.[28] This does not mean, however, that peasants never ate meat. Regulations concerning villeinage or labor services for the lord sometimes state that peasants and rural workers received meat regularly from their lords, though apparently in relatively small amounts, for example a piece of meat in boiled cabbage.[29] Cabbage with meat, though, was not typically or exclusively peasant food but could also be found at upper-class tables. At an eleven-course dinner given for a Venetian legation by the archbishop of Salzburg in 1492, cabbage with pork seems to have occupied a suitable place as the sixth course of that very exclusive meal.[30]

A "real" and, moreover, general situation again seems difficult to reconstruct even though some results have been obtained from archaeological investigations. Systematic studies of the rich sources for Italy have been conducted and may furnish results that are lacking for other regions in Europe.[31] Massimo Montanari, for example, was able to prove that the diet of Italian peasants was in fact dominated by grain (bread) and vegetables, but he also strongly warned against generalizations. There were villages where meat was eaten regularly, and these cannot be seen as exceptional, but rather as different cases resulting from different economic situations and developments.[32] Nevertheless, the stereotypical image of the peasant living predominantly on bread, grain, and vegetables in many cases can be shown to be correct.

On the other hand, if we read a fifteenth-century text concerning the inhabitants of the region of Styria (in Austria) that says:

Sy haben ze essen ain gutten prein,
und dar auff trinken ain guten wein,
fleysch, ayr und prots genug . . . [33]

[They have enough millet-gruel to eat,
they have enough wine to drink
and enough meat, eggs and bread . . .]

we are confronted with something that was meant to characterize a rural population and at the same time to show an ideal image—mainly by the mention of wine and meat and by the word "enough." This contradicts the images of peasant food mentioned earlier, which were used in situations where comparisons with other social classes were being made or where poverty was being emphasized. We also know for certain that the "ideal" of drinking water is generally fictitious and that every effort must have been made to change to wine, cider, or beer. A number of sources show that, particularly on Sunday, peasants regularly visited inns, often run by the parish priest himself.[34] Peasants drank wine in the parsonage after the Sunday service,[35] a habit that was prohibited in Styria, in 1445, by the later emperor Frederick III.[36] Sermons lamented the habit of drinking wine before the service,[37] and a satirical poem written sometime after 1400 stated:

"Als pald daz selb ein end hat
sy eilend aus der chirchen drat,
zu dem wein stet in der sinn
da habent sew ein pesseren gewinn . . ."[38]

[After the end (of the Sunday service)
they (the peasants) hurry out of church,
they are in the mood for drinking wine
by that they think to have better profit . . .]

These texts lead us again to one of the main differences that must be considered in our investigations of "image" and "reality": working days and Sundays or feast days or other special days (for example, when the peasants worked in their lords' fields). As we have already emphasized when dealing with dress, we must stress once more that work days and Sundays can mean two very different situations. In talking about topoi, stereotypes, "images," and "reality," we must never forget this characteristic polarity.

The role of feast days in peasant life is characterized by one type of picture that is very significant for rural areas in virtually all of Europe, particularly in the fourteenth and fifteenth centuries: the representation of "Christ of the Trades."[39] Mainly seen in frescoes situated on the outer walls of village churches, they often show a very large number of tools and other objects signifying the different types of agricultural work, or they show the work itself (Figure 25). Designed to motivate their beholders to abstain from work on Sundays and feast days, these pictures offer a very large array of recognizable objects that were known and used by peasants. In depicting so many objects, these representations certainly do not reflect "reality" as a whole, but rather a large number of singular "realities" that are combined in order to motivate not only by the mere appearance of the well-known objects but also by their quantity. Nevertheless, these images contain comparatively dense information on the material culture of peasant work in the late Middle Ages. Dealing with single objects out of such a spectrum requires careful analysis of context dependencies, the aim of the images, the role of the beholders, and certain local characteristics.

A typical example of these kinds of problems may be given from another context. The harvesting of grain was regularly done with a sickle. Use of the medieval type of scythe meant an immense loss of grain. In the so-called *Kaiserchronik* of the twelfth century, the use of a scythe for mowing grain is called an act of madness.[40] Therefore, a picture like that shown in Figure 26, from the mid-fourteenth century, must make us suspicious. We cannot be satisfied by just describing it or even stating that the image represents one of the earliest examples of the use of the scythe for mowing grain. We have to ask what the context is. And here the context solves many problems. This picture's "reality" had a very special aim: to show a negative and reprehensible deed and, in that way, to arouse feeling. The depicted scene is the destruction of the fields of Israel by the Midianites (Judges 6: 1–4). And how could this destruction be made clear? The horses are eating the grain; the enemies—the Midianites—are using scythes to mow it. In this way, the loss of grain is shown: an act that would certainly have been understood by a medieval audience.[41] Everyone would see at first glance that something was wrong. Image and reality, therefore, could and still can be understood only if one is equipped with certain previous knowledge.

This also holds true for quite a number of other tools and objects related to agricultural work. As we have already emphasized, the archaeo-

Figure 25. Detail, fresco, Christ of the Trades. Outer wall of parish church, Saak, Carinthia, 1465. Photo: İnstitut für Realienkunde, Krems, Austria.

Figure 26. The Midianites destroy the fields of Israel. Detail, Lilienfeld (Lower Austria), Stiftsbibliothek Cod. 151, Concordantiae Caritatis, c. 1355, fol. 26v. Photo: Institut für Realienkunde, Krems, Austria.

logical evidence may offer much more accurate and "realistic" evidence than any other source. Nevertheless, to avoid erroneous conclusions we must be very careful and must concentrate, for example, on omissions in the archaeological evidence. The phenomenon that iron objects have seldom been found in excavations of late medieval deserted villages may serve as an example. Knowing the high cost of iron in many areas of Europe, we might conclude that few iron objects were used. We also must consider—and this seems more relevant—that people who left their village, for whatever reason, undoubtedly tried to take with them everything that was expensive or seemed to be of value.[42] Iron objects were expensive items. In using archaeological evidence, therefore, we must bear in mind that the excavated objects were of course used by medieval people, but that such evidence cannot lead to a complete and general image. Very significant parts of life cannot be traced or related to other parts. The frequency of use of certain objects is a particularly difficult question, and our knowledge must remain fragmentary.

A final example based on pictorial sources points up some additional problems of analysis. Descriptions of houses and medieval villages in the written sources occur much less frequently than remarks or treatises about peasant dress or food or festivities. Again, archaeological sources are of major importance. If we use additional pictorial evidence,[43] we must consider the possibility that it was not necessarily a real village that might have been important, but just the type "village."[44] In a scene of the legend of Saint Catherine we find, in the distant background, the depiction of a late medieval village (Figure 27). Everyone looking at the picture, in the Middle Ages as well as today, would have recognized it as a village. The question is to what extent such a picture can be used as local evidence, in this case for late medieval Lower Austria, where the picture was painted for the Benedictine monastery of Melk in the 1480s. The problem is solved when we come upon an engraving by Martin Schongauer, who worked in Alsace, titled "On the Way to the Market" (Figure 28). There, in the background, we find our village. It was not a real Lower Austrian village that was the model for the Melk picture, but an engraving of Schongauer's. Moreover, in the *Weltchronik* of Hartmann Schedel, printed in Nuremberg in 1493, when we look at pictures showing Poland, Lithuania, and northern Italy ("Welschland"), we notice first that all three depictions are identical, and then we recognize our Schongauer village (Figure 29). It was not the local evidence (which was lacking) that was important, but only the type.

We have tried to show that our comprehension of "reality" often may not be the right one for analyzing peasant life. Rather, we need another, much more open and broader "image," or let us say a "medieval reality" designed to enable those who experienced it to understand, recognize, become informed, be moved emotionally, be influenced. Of major importance was the "type"—of a village, of a good or bad peasant, of peasant dress, or of peasant food. We are confronted with a number of different or equivalent "images"; we also are confronted with a number of different or equivalent "realities," with a number of different or equivalent attitudes toward peasants. All of them depend on *context*. To attain a kind of context sensitivity, to try to understand those contexts and in that way "images," "realities," and attitudes, we must increase our knowledge—a

Figure 27. A "village" type. Detail, Hans Egkel, Execution of St. Catherine, panel painting, c. 1490. Melk (Lower Austria), Stiftsgalerie. Photo: Institut für Realienkunde, Krems, Austria.

Figure 28. Model of the "village." Detail, Martin Schongauer (d.1491), On the Way to the Market, engraving. Marianne Bernhard, ed., *Martin Schongauer und sein Kreis: Druckgraphik, Handzeichnungen* (Munich: Südwest, 1980), p. 115.

knowledge comparable to that of medieval people, particularly a knowledge of signs. As we gain success in some aspects, we should then be better able to comprehend "images," "realities," and attitudes. If they contradict one another, we should not wonder but rather ask why such a contradiction might have been necessary or useful to medieval people. To admire the peasants and at the same time to condemn them, instead of being strictly

Figure 29. Another copy of the "village." Detail, Hartmann Schedel, *Buch der Croniken und geschichten mit figuren und pildnussen von anbeginn der welt bis auf dise unsere zeit* (Nuremberg, 1494), fol. 263r (Poland), 278r (Lithuania), 285v ("Welschland"). (Reprint Grünwald: Konrad Kölbl, 1975).

contradictory, seems, as I have tried to emphasize, to have been rather normal.

Sources on material culture and everyday life are one of the most important types of evidence for showing such phenomena regularly and clearly. Many aspects still may seem strange, even to historians. In many ways we cannot rid ourselves of our cultural filters. Particularly with regard to "image" and "reality" we must resist the fact that we are accustomed to know, or, better, accustomed to believe that we know, so much about truth and about what is real and what is false. Being confronted with the more or less fuzzy sets of sources on the material culture of medieval peasantry, such attitudes may hinder our work.

Notes

1. J. Ambrose Raftis, "Introduction," in *Pathways to Medieval Peasants*, ed. Raftis (Toronto: Pontifical Institute of Mediaeval Studies, 1981), p. vii.

2. See the title of Raftis's book (note 1 above).

3. See Vito Fumagalli and Gabriella Rossetti, eds., *Medioevo rurale: Sulle tracce della civiltà contadina* (Bologna: Il Mulino, 1980).

4. See, for example, the following general overviews: Werner Rösener, *Peasants in the Middle Ages*, trans. Alexander Stützer (Urbana: University of Illinois Press, 1992); Karl Brunner and Gerhard Jaritz, *Landherr, Bauer, Ackerknecht: Der Bauer im Mittelalter—Klischee und Wirklichkeit* (Vienna: Böhlau, 1985); *Bäuerliche Sachkultur des Spätmittelalters*, Veröffentlichungen des Instituts für mittelalterliche Realienkunde österreichs 7, Sitzungsberichte der österreichischen Akademie der Wissenschaften, phil.-hist. Kl. 439 (Vienna: Verlag der österreichischen Akademie der Wissenschaften, 1984).

5. See, especially, the very stimulating book by Jean Chapelot and Robert Fossier, *The Village and House in the Middle Ages*, trans. Henry Cleere (London: Batsford, 1985).

6. See, for example, Helga Schüppert, "Der Bauer in der deutschen Literatur des Spätmittelalters," in *Bäuerliche Sachkultur*, pp. 128–29; Walter Achilles, "Bemerkungen zum sozialen Ansehen des Bauernstandes in vorindustrieller Zeit," *Zeitschrift für Agrargeschichte und Agrarsoziologie* 34(1) (1986), 1–30; Reginaldo Grégoire, "Il contributo dell'agiografia alla conoscenza della realtà rurale: Tipologia delle fonti agiografiche anteriori al XIII secolo," in *Medioevo rurale*, ed. Fumagalli and Rossetti, pp. 343–47.

7. Günther Franz, ed., *Quellen zur Geschichte des Bauernstandes im Mittelalter*, Freiherr vom Stein-Gedächtnisausgabe 31 (Darmstadt: Wissenschaftliche Buchgesellschaft, 1974), pp. 548–52. See also Gerhard Jaritz, *Zwischen Augenblick und Ewigkeit. Einführung in die Alltagsgeschichte des Mittelalters* (Vienna: Böhlau, 1989), pp. 111–12.

8. Keith P. F. Moxey, *Peasants, Warriors and Wives: Popular Imagery in the Reformation* (Chicago: University of Chicago Press, 1989), pp. 35–66 and passim. See also Hans-Joachim Raupp, *Bauernsatiren: Entstehung und Entwicklung des bäuerlichen Genres in der deutschen und niederländischen Kunst ca. 1470–1570* (Niederzier: Lukassen, 1986).

9. Moxey, *Peasants, Warriors and Wives*, pp. 35–39, 140–41.

10. Ibid., pp. 57–58.

11. Wolfgang Greisenegger, *Die Realität im religiösen Theater des Mittelalters: Ein Beitrag zur Rezeptionsforschung*, Wiener Forschungen zur Theater- und Medienwissenschaft 1 (Vienna: Braumüller, 1978), passim; Greisenegger, "Bauer und Hirt im szenischen Spiel des Mittelalters," in *Bäuerliche Sachkultur*, pp. 177–91.

12. Leopold Schmidt, "Sankt Joseph kocht ein Müselein: Zur Kindlbreiszene in der Weihnachtskunst des Mittelalters," in *Europäische Sachkultur des Mittelalters*, Veröffentlichungen des Instituts für mittelalterliche Realienkunde Österreichs 4, Sitzungsberichte der österreichischen Akademie der Wissenschaften, phil.-hist. Kl. 374 (Vienna: Verlag der österreichischen Akademie der Wissenschaften, 1980), pp. 143–66.

13. See, for example, Alfred Hagelstange, *Süddeutsches Bauernleben im Mittelalter* (Leipzig: Duncker und Humblot, 1898).

14. Ursula Liebertz-Grün, *Seifried Helbling: Satiren kontra Habsburg* (Munich: Beck, 1981), esp. pp. 67–69; Hans-Dieter Mück, "Oswald von Wolkensteins Liedpropaganda gegen die hoffärtigen Bauern," *Der Schlern* 60 (1986): 330–43; Gerhard Jaritz, "Realienkunde der bäuerlichen Welt des Spätmittelalters: Zum Aussagewert von Bildquellen, Schriftzeugnissen und Ergebnissen der Wüstungsarchäologie," in *Mittelalterliche Wüstungen in Niederösterreich*, Studien und Forschungen aus dem Niederösterreichischen Institut für Landeskunde 6 (Vienna: Niederösterreichisches Institut für Landeskunde, 1983): 180–82. For some descriptions of such extravagant peasant dress in literary sources, see Sieghilde Benatzky, "Österreichische Kultur- und Gesellschaftsbilder des 13. Jahrhunderts auf Grund zeitgebundener Dichtungen (Seifried Helbling und Meier Helmbrecht)," unpublished phil. dissertation, Vienna, 1963, pp. 134–51 (though with no critical interpretation); Ulrike Lehmann-Langholz, *Kleiderkritik in mittelalterlicher Dichtung: Der Arme Hartmann, Heinrich "von Melk," Neidhart, Wernher der Gartenaere und ein Ausblick auf die Stellungnahmen spätmittelalterlicher Dichter*, Europäische Hochschulschriften ser 1, vol. 885 (Frankfurt and New York: Peter Lang, 1985), passim; Gabriele Raudszus, *Die Zeichensprache der Kleidung: Untersuchungen zur Symbolik des Gewandes in der deutschen Epik des Mittelalters*, Ordo: Studien zur Literatur und Gesellschaft des Mittelalters und der frühen Neuzeit 1 (Hildesheim and New York: Olms, 1985), esp. pp. 158–70 and passim.

15. Joseph Seemüller, ed., *Seifried Helbling* (Halle: Waisenhaus, 1886; reprint Hildesheim and New York: Olms, 1987), p. 69, II, vv. 70–75.

16. Franz, ed., *Quellen*, p. 220 n. 82.

17. Ibid., pp. 326–27 n. 122.

18. Hartmann J. Zeibig, "Der Ausschuss-Landtag der gesammten österreichischen Erblande zu Innsbruck 1518," *Archiv für österreichische Geschichte* 13 (1854): 254.

19. For some changes, see Perrine Mane, "Émergence du vêtement de travail à travers l'iconographie médiévale," in *Le Vêtement: Histoire, archéologie et symbolique vestimentaires au moyen âge*, ed. Michel Pastoureau (Paris: Cahiers du Léopard d'Or, 1989), pp. 93–122.

20. Wilhelm Hansen, *Kalenderminiaturen der Stundenbücher: Mittelalterliches Leben im Jahreslauf* (Munich: Georg D. W. Callwey, 1984).

21. Françoise Piponnier, "La qualité de la vie en milieu rural: exemples bourguignons," in *Bäuerliche Sachkultur*, pp. 277–90.

22. András Kubinyi, "Bäuerlicher Alltag im spätmittelalterlichen Ungarn," in *Bäuerliche Sachkultur*, pp. 245–50.

23. Piponnier, "La qualité," p. 289; Monique Closson, Perrine Mane, and Françoise Piponnier, "Le costume paysan au moyen âge: Sources et méthodes,"

Ethnographie (1984): 302–7. Concerning Normandy, see Jean Glénisson and John Day, eds., *Textes et documents d'histoire du moyen âge, XIVe–XVe siècles* 2: *Les Structures agraires* (Paris: Société d'Édition d'Enseignement Supérieur, 1977), pp. 264–66.

24. Werner Endres, "Adelige Lebensformen in Franken im Spätmittelalter," in *Adelige Sachkultur des Spätmittelalters*, Veröffentlichungen des Instituts für mittelalterliche Realienkunde Österreichs 5, Sitzungsberichte der österreichischen Akademie der Wissenschaften, phil.-hist. Kl. 400 (Vienna: Verlag der österreichischen Akademie der Wissenschaften, 1982), p. 88.

25. Theodor von Karajan, ed., *Michael Beheims Buch von den Wienern, 1462–1465* (Vienna: P. Rohrmann, 1843), pp. 128–29, in which the young Maximilian, the later emperor, is said to have complained about the lack of food during the siege of Vienna in 1462: Maximilian had to be content with barley and peas, and did not get any meat. This was not adequate to his rank, and such food should rather have been given to the enemies. See Gerhard Jaritz, "Der Einfluß der politischen Verhältnisse auf die Entwicklung der Alltagskultur im spätmittelalterlichen Österreich," in *Bericht über den sechzehnten österreichischen Historikertag*, Veröffentlichungen des Verbandes österreichischer Geschichtsvereine 24 (Vienna: Verband österreichischer Geschichtsvereine, 1985), p. 529.

26. Benatzky, "Österreichische Kultur- und Gesellschaftsbilder," pp. 116–25.

27. See, for example, Gerhard Jaritz, "Zur materiellen Kultur der Steiermark im Zeitalter der Gotik," in *Gotik in der Steiermark, Ausstellungskatalog* (Graz: Kulturreferat der Steiermärkischen Landesregierung, 1978), p. 42.

28. See, for example, Vladimír Nekuda, "Mährische Wüstungen als Quelle spätmittelalterlichen Dorflebens," in *Bäuerliche Sachkultur*, p. 211.

29. See, for example, Jaritz, "Zur materiellen Kultur," p. 41.

30. Helmut Hundsbichler, "Stadtbegriff, Stadtbild und Stadtleben des 15. Jahrhunderts nach ausländischen Berichterstattern über Österreich," in *Das Leben in der Stadt des Spätmittelalters*, Veröffentlichungen des Instituts für mittelalterliche Realienkunde Österreichs 2, Sitzungsberichte der österreichischen Akademie der Wissenschaften, phil.-hist. Kl. 325 (Vienna: Verlag der österreichischen Akademie der Wissenschaften, 1977): 129.

31. Massimo Montanari, *L'alimentazione contadina nell'alto Medioevo*, Nuovo Medioevo 11 (Naples: Liguori, 1979); Anna Maria Nada Patrone, *Il cibo del ricco ed il cibo del povero: Contributo alla storia qualitativa dell'alimentazione—l'area pedemontana negli ultimi secoli del Medioevo*, Biblioteca di Studi Piemontesi 10 (Torino: Centro Studi Piemontesi, 1981).

32. Massimo Montanari, "Rural Food in Late Medieval Italy," in *Bäuerliche Sachkultur*, pp. 316–20. Also see Montanari, *The Culture of Food*, trans. Carl Ipsen (Oxford: Basil Blackwell, 1994).

33. Anton Schönbach, "Steirisches Scheltgedicht wider die Baiern," *Vierteljahrschrift für Litteraturgeschichte* 2 (1889): 326–27.

34. For this and the following, see Jaritz, "Zur materiellen Kultur," pp. 41–42.

35. Angeli Rumpleri, abbatis Formbacensis . . . , "Historia inclyti monasterii," in Bernardus Pez, *Thesaurus anecdotorum novissimus I/III* (Augsburg and Graz, 1721), col. 479 (for the year 1502).

36. Burkhard Seuffert and Gottfriede Kogler, eds., *Die ältesten steirischen*

188   Gerhard Jaritz

Landtagsakten 1396–1519 1, Quellen zur Verfassungs- und Verwaltungsgeschichte der Steiermark 3 (Graz: Stiasny, 1953): 107.

37. Anton Schönbach, "Miszellen aus Grazer Handschriften, 12: Der Prediger von St. Lambrecht," *Beiträge zur Kunde steiermärkischer Geschichtsquellen* 33, N.F. 1 (1903): 88.

38. Graz, University Library, Cod. 1017, fol. 108v.

39. Edgar Breitenbach and Thea Hillmann, "Das Gebot der Feiertagsheiligung, ein spätmittelalterliches Bildthema im Dienste volkstümlicher Pfarrpraxis," *Anzeiger für schweizerische Altertumskunde* N.F. 39 (1937): 23–36; Robert Wildhaber, "Der 'Feiertagschristus' als ikonographischer Ausdruck der Sonntagsheiligung," *Zeitschrift für schweizerische Archäologie und Kunstgeschichte* 16 (1956): 1–34; Wildhaber, "Feiertagschristus," *Reallexikon zur deutschen Kunstgeschichte* 7 (Munich: Alfred Druckenmüller, 1981), cols. 1002–10.

40. Edward Schröder, ed., *Die Kaiserchronik eines Regensburger Geistlichen,* MGH Deutsche Chroniken 1 (Hannover: Hahn, 1895; reprint Dublin: Weidmann, 1969), p. 282, v. 10978; see also Ulrich Bentzien, *Bauernarbeit im Feudalismus: Landwirtschaftliche Arbeitsgeräte und -verfahren in Deutschland von der Mitte des ersten Jahrtausends u.Z. bis um 1800,* Veröffentlichungen zur Volkskunde und Kulturgeschichte 67 (Berlin: Akademie Verlag, 1980): 74.

41. Jaritz, "Realienkunde," pp. 174–76.

42. See, for example, Vladimír Nekuda, *Pfaffenschlag: Zaniklá středověká ves u Slavonic—Příspěvek k dějinám středověké vesnice* [Pfaffenschlag: A Medieval Deserted Village near Slavonice—A Contribution to the History of the Village in the Middle Ages] (Brno: Moravske Muzeum, 1975), pp. 134–35 and 257.

43. Sarah M. McKinnon, "The Peasant House: The Evidence of Manuscript Illumination," in *Pathways to Medieval Peasants,* ed. Raftis, pp. 301–9.

44. Concerning this problem of the use of types in medieval and early modern pictorial sources, see, for example, Elisabeth Rücker, *Die Schedelsche Weltchronik: Das größte Buchunternehmen der Dürer-Zeit* (Munich: Prestel, 1973), esp. pp. 48–77. For the following, see Jaritz, "Realienkunde," pp. 158–64; Carsten-Peter Warncke, *Sprechende Bilder, sichtbare Worte: Das Bildverständnis in der frühen Neuzeit,* Wolfenbütteler Forschungen 33 (Wiesbaden: Otto Harrassowitz, 1987), pp. 64–80.

PART III

Literary Representations

9. "A thing most brutish": The Image of the Rustic in Old French Literature

For a long time now scholars have been investigating the vernacular literatures in order to gain information about medieval society. It is worthwhile
taking all the evidence that can be gleaned into account since, apart from
their primary purpose of instruction or entertainment, literary texts may
provide us with insights into the society in which they came into being. To
look for "realism" in literature is today no longer fashionable or considered
advisable, and we recognize that the writer's representation of reality is
often, whether purposely or accidentally, distorted and deficient. Even the
so-called "realistic" genre of the fabliaux, which is one of our main sources
here, cannot be taken at face value, since its authors did not in the first place
aim to depict their contemporary reality, but rather intended to criticize,
satirize, or amuse.

In particular with regard to the fabliaux, short tales written in the
thirteenth and fourteenth centuries, detail is scant and information about
society is necessarily incomplete: such functional data as are mentioned
exist solely to develop the action. Other snippets of "real life" as can be
formed are purposely misinterpreted or magnified; they may nonetheless
be valuable as evidence of certain mentalities. Indeed, beggars can't be
choosers. We cannot afford to neglect any documentary evidence, however
seemingly negligible; in submitting texts to comparative investigation, in
sorting out what is inherent to the genre and what we may see as proceed-
ing from the representation of reality, we may be able to gain some insight
into the way contemporaries perceived their society.[1]

Some data may be considered as trustworthy because they occur frequently—or, better still, because of their gratuitous nature. Most of them
are not surprising: one expects to learn about peasant children guarding
flocks, about the kind of utensils laborers need to acquire, about techniques of manuring fields.[2] Of course one can find mention of markets
where farmers sell or buy livestock and produce; sometimes prices are men-

tioned.[3] Rich as well as poor peasants are very much in evidence: the former enjoying their well-stocked cellars, proud owners of several ploughs, oxen, and horses (the last, proof of real wealth[4]), the latter possessing nothing but a cow or a miserable nag, which they see themselves obliged to sell for lack of fodder.[5] Hired laborers are also mentioned, as are their main tasks and their pay.[6] All this is not surprising, although one must point out that the frequent occurrence of rich villeins does not necessarily reflect contemporary reality.[7]

Food is not always the banal item it may seem. The texts give us some information about the staple diet, such as bread—its composition, varieties, and consumers.[8] Among garden produce green peas feature prominently: their presence on the table—next to milk and cheese—was so customary that the epithet "peas-eaters" became an insult leveled at rural dwellers.[9] Meat, reserved for Sundays and feast days, deserves mention for its very rarity.[10] Even at a wedding feast celebrating the union of Robinet with the lord's daughter, guests have to be content with pastry, porridge, and milk.[11] In this case such frugality may have an ironic undertone: the occasion is after all one of those mixed marriages that will be discussed later. All information about food may have special connotations, to be appreciated by an audience that is likely to understand the contrast between rustic cooking and select cuisine. The menu of another peasant wedding tends to prove this: noble meats seasoned with precious spices are aped in a burlesque way by dressing a sow with wild berries.[12] In fact, the specific features of all rural fare—heavy, stodgy, and frugal—alert us to the social distribution of food.[13]

The same probably holds true in respect to clothing, another issue, it would seem, of purely documentary interest. The corroborating evidence of manuscript illuminations shows that the literary descriptions tend to be accurate:[14] in other words, they are not meant to be caricatures. Since clothes, however, reflected social status, it was easy for townsfolk to distinguish countryfolk from burghers. Even if their details are based on reality, some medieval writers may have tried to take advantage of the rustic aspect of peasant clothing by implying that it was not proper dress.

Other aspects of the portrayal of the peasant may be ascribed to a deliberately subjective view. Some observations are favorable: for instance, a few authors mention the abstinence of country dwellers who are satisfied with drinking water instead of wine; another speaks highly of their diligence in saving money.[15] Their frankness is readily stressed, as is their plain speaking, which at times induces them to confront other people with unpleasant truths.[16] The wisdom of this group is a recurrent theme in literature: they are credited with some innate common sense—once more, nature has the upper hand on culture—a common sense that is illustrated by the many sayings and proverbs ascribed to the *vilain*.[17]

As far as the rural dwellers' occupations and daily life are concerned, we meet completely conflicting observations. At times we are regaled with an almost Theocritean country idyll populated with charming shepherds and shepherdesses,[18] whose sole preoccupation is with their love affairs, their food, and their childish games. At other times authors show some understanding of the realities of life: they acknowledge that country people do not live well and have to scratch a bare living from an unyielding soil, struggling against the hardships nature inflicts on them. The *Conte des Vilains de Verson*, for instance, depicts the statute labor and heavy dues that overwhelm them all year round; every month brings its own burden, and the author addresses his complaints to God in the name of the villagers.[19] According to Étienne de Fougères, in the *Livre des Manières*, farmers suffer great ills and are imposed on by their betters.[20] Étienne stresses the countryman's industriousness. So do quite a few fabliau authors;[21] farmers are proper workaholics, very early risers who toil till evensong, who are never deterred from their labors even by a thunderstorm nor inclined to go home at midday; they have their first meal at six in the morning or even earlier and the last one after sunset. Farmers even work on Sundays. Étienne again emphasizes how the peasant has to sow, harrow, and reap without respite. He is not rewarded for all this: "Never will he eat any good bread; the better grain will be kept for us, and he will be left with the bran" (vv. 689–92).

Even such apologetic awareness is sometimes subverted, as in one author's grudging tribute to the people whose labor provides his food: "Indeed, it is to please the scholars and the others, that farmers produce wine and bread. Thanks to them we are provided with everything, while they put up with the cold and the heat. However, they don't suffer for our sake, but for the love of money."[22] Many authors are convinced that country people enjoy a comfortable existence—a prosperity often associated with a blatant avarice.[23]

And this is only the first shortcoming of an impressive list. Thus many proverbs still current at the time of Rabelais depict the villein's ingratitude: "anyone who gives him anything is the loser";[24] "scratch a villein's testicles and he will defecate in your hand";[25] "be soft with him and he will sting you—sting him and he will be soft."[26] Étienne de Fougères regrets that the hardworking country dwellers lose any credibility they might have by re-volting against their condition: they refuse to be patient and takes issue with God Himself. They even try to trick the Almighty by cheating Him out of the tithes. As Étienne puts it, they shake down their sheaves of corn on the threshing floor before calculating their dues (vv. 709–56). The *Besant de Dieu* also criticizes the peasants' grumbling, their envy, and their refusal to accept their poverty.[27]

Even if they seem more superficial, other reproaches are nonetheless significant. The peasant's natural coarseness is pointed out, for example, when one of the shepherds in *Robin et Marion* proposes to break wind for fun (v. 485), a feat that Trubert, a fabliau hero, accomplishes when sitting at a duke's table. Of course, Trubert is too uncouth to appreciate the nobleman's way of life: when offered a bed worthy of a king, he cannot fall asleep, since he is unaccustomed to such luxury.[28] Another character, when hearing the beautiful song of a bird, thinks at once about putting it in a cage or in a casserole.[29]

The much-vaunted peasant wisdom is inconsistent with their credu-lity, clumsiness, sometimes almost animal stupidity,[30] a tradition that goes a long way back, as Jacques Le Goff reminds us when remarking that medieval Latin *rusticus* already meant "ignorant."[31] In the fabliau, deceit is recurrent: wives cuckold their husbands, servants cheat their employers. And a character called Brifaut is the victim of a thief who takes from him the roll of cloth he was carrying on his shoulders, then makes fun of him by showing it to him without it being recognized and by blaming him for his carelessness in not tying his wares to his neck, as he himself did. The Vilain de Bailluel is a witness to his wife's adulterous activities, after being persuaded that he has died.[32]

One of the most significant items concerns the peasant's lack of cour-age.[33] The *Oustillement au Vilain* advises him in wartime to stay at home: it is better to let his rusty sword hang at the foot of the bed. "Keep clear anyway," he is told, "of the first blows: one never loses anything by being the last one."[34]

When we strike a balance of all the scenes where peasants are mocked, or even beaten,[35] we realize that these situations are inspired by one par-ticular motivation: peasants are victimized because they are despised. In the *Renart*, even the term "serf" is used, a terrible insult, meaning that one is without any dignity and not worthy of any respect.[36]

One also finds hatred. Sixty-three of the 106 lines of the prologue to

Le Vilain qui n'ert pas de son osté sire deal with an action brought against the villein: He is so wicked and so ill-natured that nobody will ever say anything positive about him, declares this writer, who wishes he were king of the villeins so that he could make them suffer all the more.[37] Another text contains a prayer beseeching the Lord to inflict great harm upon them, sickness of heart and head, of mouth and teeth, as well as several kinds of gout.[38] The Lord hates them, says the *Despit au Vilain*; that is why He sent them great suffering. The poet Rutebeuf even denies that they are human and in his *Pet au Vilain* proclaims that none of them will ever be admitted to Heaven; Hell itself will not receive them, as their soul leaves the body through another orifice than usual.[39]

The same kind of rejection is found in the analysis of the physical image. The peasant seldom has a pleasing appearance: he is badly shaven if at all, his hair is unkempt; he generally looks dirty.[40] One Rigaudin is hirsute, his cheeks are dark and tanned, for they have not been washed for seven months; the only water ever to have wetted them is rainwater.[41] Some characters are shown living in dirt and dung, like the hero of *La Crote*, who sits beside the fire and scratches himself; a scene in the *Roman de Renart* shows us a peasant immersed in mud—an indication of what would appear to be his only natural habitat.[42] The villein cannot smell incense, says one poet. We are reminded of the story of the *Vilain Asnier*, the donkeyman who, in a street where spices and aromatic herbs are sold, falls down in a faint when inhaling these strange fragrances, but regains consciousness when someone places some familiar dung beneath his nostrils.[43]

The emphasis on the peasantry's filthiness results in too unpleasant a picture to agree with reality; but it reveals all the more a mental attitude. In those cumulative descriptions so often found in medieval rhetoric, dirtiness very frequently goes together with ugliness and deformity. Constant du Hamel is dirty, badly shaven, and shaggy; he also has a bald brow, bloodshot eyes, red teeth, and a big face; he has a dark countenance and looks like a scarecrow. "One never saw or met a more hideous creature," adds one scribe.[44] Bérangier the cowherd is one-eyed; he squints and he limps. His colleague Raoul is not better off: not only is his face dirty and ugly, but his hands are repulsive, their palms hard and itchy. The peasant in *Le Vilain de Bailluel* is hideous: even his wife does not like him.[45] One remembers, of course, the famous and terrifying cowherd described in detail in *Aucassin et Nicolette*, rather like a medieval Caliban: he has a big shock of hair, his face—the author calls it a snout—is blacker than coal dust, his cheeks are very large, his nose is flat with flaring nostrils, his lips are redder than broiled meat, his teeth are yellow, the palm of your hand would fit between his eyes (XXIV, 12–24).

The word "snout" is also used for Rigaudin, whose eyes are again very widely spaced, thus accentuating his bestiality (v. 8817). Indeed, peasants are readily associated with animals: one poet thinks they look as frightful as wolves or leopards, while another tells of a character who makes faces that remind people of a grinning ape. Similarly, the *XXIII. Manières* describe villeins with the nature of a pig, a dog, a donkey, and a baboon.[46] The grotesque bull-guardian in Chrétien de Troyes' *Chevalier au Lion* is extraordinarily hideous: "his head was larger than that of a packhorse, he had the eyes of an owl, the nose of a cat; his mouth was split like a wolf's, his teeth were sharp and red like a wild boar's, his ears were large like an elephant's."[47] Bestial in appearance, the villein is also like an animal in manners and behavior, as the poet of the *Despit au Vilain* told his public: "Instead of coming to the table, these people ought to go and graze the fields together with the cattle, and walk naked, on all fours."[48]

Obviously, the peasant's ugly appearance is in the first place a matter of literary convention, as is borne out by its mention in Matthieu de Vendôme's rhetorical treatise[49] as well as by the hyperbolical manner of all such descriptions: all features are extreme and the ugliest ever seen—a foil, perhaps, for the no less conventional embellishment and idealization of characters in courtly romance. Of course, besides being a cliché, the portrait may have a comical function, but we should take into account a third possibility. In medieval literature, physical appearance reflects the moral personality: thus the ugliness of the peasant takes on a profound significance.

This seems to be the case not only on the psychological but also on the social level. In accordance with medieval esthetics, authors do not depict individuals but types:[50] all peasants are thus represented as being ugly—their ugliness having become distinctive evidence of the lower class and its inferiority. A literary character comparable to the pagan in the chanson de geste, who is likewise described as belonging to a different order, the peasant becomes the representative of a class that is rejected as a whole.[51] Stressing the marginal in comparison with the norm leads to the marginalization of this group in society and, perhaps in some cases, to exclusion from humankind.[52]

When confronted with the muteness of a fellow-being whom he likens to an animal, the knight in Chrétien's romance wonders whether he is dealing with a creature endowed with speech and reason; the bull-herd gapes dumbly at him, which elicits the question: "Just tell me what you are." The answer is plain and pathetic: "I am a man" (v. 328). As if, indeed, the knight had not recognized his interlocutor as such or had not wished to do so.[53]

One more element ought to be added to the moral and psychological image: the love life of the peasant. This subject may have been difficult to handle in an impartial manner in an era dominated—at least in literature—by courtly ideology: while his lack of courage makes the villein unfit for chivalry, it is of course his uncourtliness that accounts for his ineptness at loving. The pastourelle provides an excellent opportunity for social encounters: in its setting many a naïve shepherdess allows herself to be seduced by a soft-spoken passing knight.[54] In the fabliau, peasants are frequently cuckolded—but this doesn't mean that the part of the lover is generally entrusted to the knight; in fact, the lover is usually a cleric.[55]

More important, there seems to be a general feeling about the villein's incapacity to love: several authors declare him utterly incapable of experiencing love. It comes as no surprise when Andreas Capellanus, in his well-known *De Amore*, written at the behest of Marie de Champagne, compares the rustic's passion with that of a mule or a horse. The author of a thirteenth-century French treatise reminds us of their bestiality: "Passion changes them into animals, they cannot love in a courtly manner, as theirs is not love but a kind of fury." [56] The rustic again serves here as a foil both to courtly ideals and to society, the point being that love is possible only between equals.[57]

There must indeed have been a social confrontation, not only in literature but also in real life, as is evident from the theme of mixed marriages in the fabliaux. Even in fiction, people do not always marry within the same class: unions between villeins and noble daughters do exist. Such couples, however, always seem to disturb the storyteller: time and again, trouble or misfortune will occur. Thus, in *Aloul* and in *le Vilain Mire*, the bridegroom quickly becomes very jealous. Moreover, the hero of the latter tale confides to us that he did not want a noble bride, considering that whenever he would be ploughing, the chaplain would be visiting his conjugal bed.[58]

To stress the point that such matches are ill matched, one poet makes use of the ugliness motive,[59] perhaps as a way to vindicate the lady's unfaithfulness. Other shortcomings are quoted in the same context, such as the husband's avariciousness, which is fundamentally an uncourtly trait to the same extent as jealousy. The husband arouses even more complaints due to his wickedness and his savagery—the *Vilain Mire* decides to batter his wife every morning; she will cry all day and this will cool the ardor of any suitor—or else by his inexperience in sexual matters.[60]

But if these unions make noble young women unhappy, why do they take place? A possible explanation is that the bride's father happens to be up to his ears in debt and sees himself compelled to accept a proposal of marriage from a well-off villein: "His friends went and saw the knight; they asked for the hand of his daughter on behalf of the peasant who had so much gold and silver, so much wheat and so much cloth. And the girl, who was well-bred, did not dare cross her father."[61] The true state of affairs is not concealed either in the *Fol Vilain*, where a rich peasant couple finds a beautiful noble wife for their feeble-minded son: "He got her for his inheritance, not for his wit," explains the author, who then goes on to tell how the young lady manages to take her revenge.[62]

In the *Chastelaine de Saint Gilles*, the groom exults: "Money delivers the chatelain's daughter to the villein." Everybody tries to discourage the girl from getting married, but her father insists. And of course, just married, she elopes with a friend and tells her husband: "You bought me as if I were an animal, you will not have me." Very philosophically, he then returns to his labors, to earn more money, and concludes it was a foolish thing to address himself to so high a station.[63]

One of our authors compares social mixing in a marriage to a pear that has been grafted on a cauliflower or a turnip. Nothing, he declares, will ever change a peasant's condition or mind; after their marriage to a lady, these so-called "grafted villeins" will go on ploughing. Even after being dubbed a knight, one of them still prefers to make hay instead of wielding shield and lance: "Moreover the 'knight' wouldn't give two pieces of garlic for fame, glory, or chivalry. What he liked was tarts and baked custards."[64]

Clearly, if these poets pretend not to believe in a possible assimilation of knights and peasants, it is because the two worlds look too different to them. Indeed, they warn their audience against those who use its financial problems in order to penetrate a class that is not theirs by right: this will inevitably lead not to integration but to degradation—degradation of the lineages. "They marry below themselves for money and ought to be greatly ashamed at the harm they do to themselves. From this sort of people who love gold and silver more than chivalry, you get the base, wretched, cowardly sort of knights and that is how nobility perishes."[65]

Any change of social status for the villein, by marriage or otherwise, seems to be suspect and deprecated. "Nul ne se doit desnaturer"—nobody should change his nature, says the author of the *Vilain Asnier* (v. 51). Indeed, as another poet states, nobody can: "Let a villein possess all the treasures in the world, yet he is and will always remain a villein." Or, as a third writer puts it: "They will become gentlemen when dogs sell meat."[66] This is a common saying that is illustrated by Trubert's promotion: as the knight Hautdecuer, he becomes everybody's laughing stock. His armor is ever so splendid, his horse impetuous, his involuntary charges impressive. But everything goes wrong because he does not know how to lace his helmet and puts the holes for the eyes at the back of his head (vv. 1843–46). This inverted helmet sums up symbolically the entire situation: a peasant can only ape a knight, he will never be one.

Robert de Blois, in his *Enseignement des Princes*, warns the reader: it is unnatural to exalt those Nature decided to humble. "Serfs they are because they have to serve." And the author of a fabliau reminds his audience that God established three orders: the ownership of the land he gave to the knights; the alms and the tithes he bestowed upon the clergy; and he assigned the work to be done to the workers: "Asena les laborages as laboranz, por laborer."[67]

It is quite clear that these authors think it intolerable that a peasant might disturb the divine order; for them even the slightest breach of custom might look like the start of a revolution. This explains the many misadventures they attribute to social climbers: upstarts have to be kept in their place.

We can recognize here the medieval aversion to any change, the dislike of anything new, which makes moralists and satirists alike preach moderation and resignation. Their hierarchical attitude confronts us with a static conception of the social order.[68] But all the same, things do change. Newly rich peasants, who live off the work of others, seek to move up the social ladder. No wonder they become the chosen targets of all kinds of writings that, by ridiculing the expectations of the newcomers, try to deny them any hope of bettering themselves.

As Le Goff puts it, literature is a mirror that also reveals something about the makers of the mirror.[69] Through the portrayal of the peasants and their classification as a kind of pariah, we perceive indeed the views and mental attitudes of the target audience. The writers deny the possibility of social dynamism and thus show concern for the peace of mind of their upper class patrons; at the same time, they accidentally reveal the latter's expectations, doubts, and fears. In so doing, they acknowledge that change is on the way, through conflicts, tensions, and interplay. Thus literature submits a commentary on both the new and the ancient order. Even if its vision is sometimes crude and gross, it wonders and makes us wonder in turn about society, its roots, and its future.

Notes

1. For the debate about the problem of "reality," particularly in the fabliau, one should compare the remarks of Per Nykrog, *Les Fabliaux: Étude d'histoire littéraire et de stylistique médiévales* (Copenhagen, 1957; nouv. ed. Geneva: Droz, 1973), pp. 105–7, and Philippe Ménard, *Les Fabliaux: Contes à rire du moyen âge* (Paris: Presses Universitaires de France, 1983), pp. 47 and 105–7.

2. Most textual references to the fabliaux are taken from Willem Noomen and Nico van den Boogaard, eds., *Nouveau recueil complet des fabliaux* (Assen: Van Gorcum, 1983–) (hereafter *NRCF*), and Anatole de Montaiglon and Gaston Raynaud, eds., *Recueil général et complet des fabliaux des XIIIe et XIVe siècles*, 6 vols. (Paris: Librairie des Bibliophiles, 1872–90; reprint New York: B. Franklin, 1964) (hereafter *RGF*). The reference is to *Jouglet*, vv. 1–5 (*NRCF* 2); see Oskar Reich, *Beiträge zur Kenntnis des Bauernlebens im alten Frankreich auf Grund der zeitgenössischen Literatur* (Göttingen: Hänsch, 1909), pp. 27–28; *De l'Oustillement au Villain* (*RGF* 2); *Du Vilain Asnier* (*RGF* 5).

3. *Brifaut* (*NRCF* 6); *Les Deus Chevaus* (*NRCF* 5); *Boivin* (*NRCF* 2). For some prices, see Wilhelm Blankenburg, *Der Vilain in der Schilderung der altfranzösischen Fabliaux* (Greifswald: J. Abel, 1902), p. 64.

4. Tibout, in *Le Segretain Moine* (*RGF* 5); also the instances quoted by Blankenburg, *Vilain in der Schilderung*, pp. 60–61.

5. See Blankenburg, *Vilain in der Schilderung*, p. 61, and Reich, *Beiträge*, p. 83.

6. See Reich, *Beiträge*, pp. 41–42.

7. Caution is advisable when it comes to the Old French *vilain*, from vulg. Lat. *villanus*, which originally and literally meant "countryman, farmer." Its meaning widened in that it came to indicate all those who are "non-aristocratic," that is to say rural dwellers and townsfolk alike. It then underwent a pejorative development, its most important connotation being that of "uncourtly." See K. J. Hollyman, *Le Développement du vocabulaire féodal en France pendant le haut moyen âge: Étude sémantique* (Geneva: Droz, 1957), pp. 72–78, 162 ff., and Glyn S. Burgess, *Contribution à l'étude du vocabulaire précourtois* (Geneva: Droz, 1970), pp. 35–42 (ch. 2, "Vilain"). This study, however, tends to concentrate on texts that clearly evoke the inhabitants of the countryside.

8. See among others, *Le Vilain qui donna son âme au deable*, vv. 36–37 (*RGF* 6) and the data given by Reich, *Beiträge*, p. 101.

9. They are called "loucheour de pois" in *Le Vilain au Buffet*, v. 107 (*NRCF* 5); for cheese etc., see Reich, *Beiträge*, p. 101. Peas—and cheese—were often associated with simplemindedness or madness, as is the case in Adam de la Halle's two plays (see J. Dufournet, "Du 'Jeu de Robin et Marion' au 'Jeu de la Feuillée,'" in *Études de langue et de litteratures du moyen âge offertes à Félix Lecoy* [Paris: Champion,

1973], pp. 73–94, esp. 75–76) and elsewhere: see esp. Philippe Ménard, "Les fous dans la société médiévale," *Romania* 98 (1977): 441–42.

10. See Reich, *Beiträge*, p. 104, and Nykrog, *Fabliaux*, p. 91.

11. *Jouglet*, vv. 101–5.

12. *Le Fol Vilain*, vv. 230–34, in *Le Jongleur Gautier Le Leu*, ed. Charles H. Livingston (Cambridge, Mass.: Harvard University Press, 1951), p. 154; see Nykrog, *Fabliaux*, p. 92.

13. For another use of the food code, see Herman Braet, "'Cucullus non facit monachum': Of Beasts and Monks in the Old French 'Renart' Romance," in *Monks, Nuns and Friars in Medieval Society*, ed. Edward B. King, Jacqueline T. Schaefer, and William B. Wadley (Sewanee, Tenn.: Press of the University of the South, 1989), p. 165. See also the following works by Massimo Montanari: *L'Alimentazione contadina nell'alto Medioevo* (Naples: Liguori, 1979), *Campagne medievali: strutture produttive, rapporti di lavoro, sistemi alimentari* (Turin, 1984), and *Alimentazione e cultura nel Medioevo* (Rome: Laterza, 1988), esp. ch. 2: "Il linguaggio del cibo," pp. 23–34.

14. Ménard, *Fabliaux*, p. 52, refers to various studies on the history of costume.

15. See Blankenburg, *Vilain in der Schilderung*, p. 54, and Reich, *Beiträge*, p. 106.

16. For example, *Le Vilain qui conquist paradis par plait* (*NRCF* 5) and *Aucassin et Nicolette*, ed. Mario Roques, 2nd ed. (Paris: Champion, 1935), XXIV, 43–44.

17. Besides the collection of *Li Proverbe au Vilain*, ed. Adolf Tobler (Leipzig: S. Hirzel, 1895), one often comes across sayings in texts of different genres introduced by the formula "ce dist li vilains / li v. dist." See Elisabeth Schulze-Busacker, *Proverbes et expressions proverbiales dans la littérature narrative du moyen âge français* (Paris: Champion, 1985), and Eckhard Rattunde, *Li Proverbes au Vilain* (Heidelberg: C. Winter, 1966).

18. Especially in the so-called "bergeries," of which the second part of the *Jeu de Robin et Marion* by Adam de la Halle, ed. K. Varty (London: Harrap, 1960) is an illustrious dramatization, as well as in a number of pastourelles, for example, those of Guillaume le Vinier (*Les Poésies de Guillaume le Vinier*, ed. Philippe Ménard, 2nd ed. [Geneva: Droz, 1984]).

19. Quoted by Marie-Thérèse Lorcin, *La France au XIIIe siècle: Économie et société* (Paris: Nathan, 1975), p. 86.

20. *Le Livre des Manières*, ed. R. Anthony Lodge (Geneva: Droz, 1979), p. 84, vv. 681–92.

21. A number of texts are quoted by Blankenburg, *Vilain in der Schilderung*, p. 62, and Reich, *Beiträge*, pp. 113–15.

22. Prologue of *Des Vilains*, ed. Arthur Långfors, *Romania* 64 (1938): 254, vv. 8–17.

23. *Roman de Carité*, ed. A.-G. van Hamel (Paris, 1885), CXCVIII, 1–4; *Dit des Planètes*, ed. Achille Jubinal, in *Nouveau recueil des contes, dits, fabliaux, et autres pièces inédites des XIIIe, XIVe et XVe siècles*, 2 vols. (Paris: E. Pannier, 1839–42) 1: 379. See also Blankenburg, *Vilain in der Schilderung*, p. 50.

24. Douin de Lavesne, *Trubert*, ed. G. Raynaud de Lage (Geneva: Droz, 1974), v. 551.

25. Joseph Morawski, ed., *Proverbes français antérieurs au XVe siècle* (Paris: Champion, 1925), no. 834, p. 30.

26. François Rabelais, *Gargantua*, ed. Gérard Defaux (Paris: Librairie Générale Française, 1994), ch. 22. Cp. this twelfth-century epigram by Evrart de Béthune: "Quando mulcetur, villanus peior habetur. Pungas [*for* Ungas] villanum, polluet ille manum. Ungentem pungit, pungentem rusticus ungit," in *Carmina Medii Aevi*, ed. Francesco Novati (Florence: Dante, 1883; reprint Turin: Bottega d'Erasmo, 1961), p. 26.

27. Guillaume le Clerc de Normandie, *Le Besant de Dieu*, ed. Pierre Ruelle (Brussels: Éditions de l'Université de Bruxelles, 1973), vv. 115–20 and 1146–62.

28. Adam de la Halle, *Robin et Marion*, vv. 524–25 and 572–79.

29. *Le Donnei des Amanz*, ed. G. Paris, *Romania* 25 (1896): 517, vv. 1000–1001; see also the *Lai de l'Oiselet*, ed. R. Weeks, in *Medieval Studies in Memory of Gertrude Schoepperle Loomis* (New York: Columbia University Press, 1927), p. 341, vv. 204 ff.

30. Peasant characters are readily introduced by fabliau authors as being "sot," "fol," or "nice": *Jouglet*, vv. 6–7; *Trubert*, v. 38; *La Sorisete des Estopes*, v. 1 (*RGF* 4). See also *Des Vilains ou Des .xxiii. Manières de Vilains*, ed. Edmond Faral, *Romania* 48 (1922): 256, vv. 31–34 and 49–51; *Le Despit au Vilain*, in *Jongleurs et trouvères*, ed. Achille Jubinal (Paris: Merklein, 1835; reprint Geneva: Slatkine, 1977), p. 107, vv. 60 and 65. Gautier de Coincy tells about the slowness of the peasant's intellectual processes in *D'un Vilain*, in *Les Miracles de Nostre Dame* 4, ed. V. Frederic Koenig (Geneva: Droz, 1970): 173–74, vv. 502–4.

31. Jacques Le Goff, "Peasants and the Rural World in the Literature of the Early Middle Ages (Fifth and Sixth Centuries)," in Le Goff, *Time, Work, and Culture in the Middle Ages*, trans. Arthur Goldhammer (Chicago: University of Chicago Press, 1980), p. 96.

32. *Brifaut* (*NRCF* 6); *Le Vilain de Bailluel* (*NRCF* 5).

33. See Robin's cowardliness when faced with the knight, in *Robin et Marion*, and the instances quoted by Philippe Ménard, *Le Rire et le sourire dans le roman courtois en France au moyen âge (1150–1250)* (Geneva: Droz, 1969), p. 170.

34. *L'Oustillement au Vilain*, in *Poèmes français sur les biens d'un ménage*, ed. U. Nyström (Helsinki: Imprimerie de la Société de la Litterature Finnoise, 1940). The farmer Lienart in the *Roman de Renart*, ed. J. Dufournet and A. Méline (Paris, 1985), br. IX, v. 1871, also has a sword covered with rust.

35. *Le Vilain Mire* (*NRCF* 2); *La Male Honte* (*NRCF* 5); *Le Meunier et les .II. Clers*, vv. 317–18 (*NRCF* 7); *Robin et Marion*, vv. 314–17.

36. *Roman de Renart*, br. IX, vv. 1301 and 1352.

37. *Le Vilain qui n'ert pas de son osté sire*, ed. L. F. Flutre, in "Un manuscrit inconnu de la bibliothèque de Lyon," *Romania* 62 (1936): 1–16.

38. *Des Vilains ou Des .xxiii. Manières*, pp. 256–60, vv. 1–28.

39. *Despit*, vv. 57–59; *Pet*, vv. 10–30 and 64–65 (*NRCF* 5); cf. *Vilain qui conquist paradis par plait*.

40. *Boivin*, vv. 12–14; *Constant du Hamel*, vv. 57–59 (*NRCF* 1); *Vilain au Buffet*, v. 97; *Fol Vilain*, v. 10.

41. *Garin le Loherain*, ed. Josephine Elvira Vallerie (Ann Arbor, 1947), vv. 8819–21.

42. *La Crote*, vv. 8–9 (*NRCF* 6); *Renart*, br. Ib, vv. 2593–94.

43. *Vilain Asnier*, in *Fabliaux français du moyen âge* 1, ed. Philippe Ménard (Geneva: Droz, 1979) and *NRCF* 8.

44. *Constant du Hamel*, var. ms D, vv. 746–47.

45. *Aloul*, vv. 700–703 (*NRCF* 3); *Vilain au Buffet*, vv. 96–99 and 173–75; *Vilain de Bailluel*, vv. 8–11.

46. *Du Prestre et du Chevalier*, vv. 109–12 (*RGF* 2); *Sorisete des Estopes*, vv. 124–27. Cf. the *Despit*, v. 60: "villeins are like asses," and v. 65: "he who created the villein, created the wolf."

47. Chrétien de Troyes, *Chevalier au Lion*, ed. Mario Roques (Paris: Champion, 1960), vv. 293–302.

48. *Despit*, vv. 39–42; cp. vv. 63–64: "they ought to live in the woods." Jean Le Fèvre, author of the fourteenth-century *Les Lamentations de Matheolus*, ed. A.-G. van Hamel (Paris: E. Bouillon, 1892; reprint Geneva: Slatkine, 1983) 1: 289, Bk. IV, 687, declares that peasants live like animals. There is also the phrase coined by a Latin poet writing around 1100: "rustici qui pecudes possunt appelari," in *Das Streitgedicht in der lateinischen Literatur des Mittelalters*, ed. Hans Walther (Munich: Beck, 1920; reprint New York: G. Olms, 1984), p. 141.

49. *Ars versificatoria*, in *Les Arts poétiques du XIIe et du XIIIe siècle*, ed. Edmond Faral (Paris: Champion, 1924), pp. 130–31, vv. 6–18.

50. See Edgar de Bruyne, *Études d'esthétique médiévale* 2: *L'Époque romane* (Bruges: De Tempel, 1946; reprint Geneva: Slatkine, 1975), pp. 107–8.

51. For the comparison of the portrayal of villeins and pagans, explicit in Chrétien's "vilein qui resanbloit Mor" (*Chevalier au Lion*, v. 286), see Josef F. Falk, *Étude sociale sur les chansons de geste* (Nyköping: Imprimerie de la Société du Sodermanlands lans tidning, 1899); S. L. Galpin, "Cortois and Vilain," unpublished Ph.D. dissertation, Yale University, 1905.

52. The reasons for this exclusion can be found not only in the fact that the rustic lives geographically on the margins of human habitation but also in his mode of existence. His appearance and his behavior are unfit for civilized society. Even his diet consists in part of foods associated with madness. Of course the peasantry is not the only group seen as marginal in society; it shares this fate, and some of the reasons for it, with pagans and infidels, giants and wild men, fools and, naturally, animals. See, among others, Paul Bancourt, *Les Musulmans dans les chansons de geste du cycle du roi* (Aix-en-Provence: Université de Provence, 1982); W. W. Comfort, "The Literary Rôle of the Saracens in the French Epic," *Publications of the Modern Language Association* 55 (1940): 628–59; Richard Bernheimer, *Wild Men in the Middle Ages: A Study in Art, Sentiment, and Demonology* (Cambridge, Mass.: Harvard University Press, 1952); Joel Lefèbvre, *Les Fols et la folie: Étude sur les genres du comique et la création littéraire en Allemagne pendant la Renaissance* (Paris: C. Klincksieck, 1968), esp. pp. 51–54 about the link between peasants, fools, and wild men.

53. The intense impact of this tradition of seeing the *vilain* as someone of subhuman nature is borne out by its tenacity: in the seventeenth century one still

finds its echoes in the writings of La Bruyère: "One sees certain wild animals, both male and female, scattered over the countryside, swarthy, livid, completely burned by the sun, anchored to the soil which they till and turn over with an invincible stubbornness; they possess the semblance of articulate voice and when they stand upright, they show a human face, and in fact they are men." *Les Caractères ou les moeurs de ce siècle*, "De l'homme," 128, in *Oeuvres de La Bruyère*, ed. G. Servois (Paris, 1922), III 1, p. 61.

54. See Michel Zink, *La Pastourelle: poésie et folklore au moyen âge* (Paris: Bordas, 1972).

55. Nykrog, *Fabliaux*, p. 116.

56. "For a farmer hard labor and the uninterrupted solaces of plough and mattock are sufficient," adds the chapter "de amore rusticorum": Andreas Capellanus, *The Art of Courtly Love*, trans. John Jay Parry (New York: Ungar, 1941; reprint New York: Columbia University Press, 1990), p. 149; "L'Arbre d'amours," ed. Arthur Långfors, *Romania* 56 (1930): 367. The distinction between love and fury is made by another thirteenth-century writer, Jean de Thuin, *Li Hystore de Julius Cesar*, ed. Franz Settegast (Halle: Max Niemeyer, 1881), p. 170.

57. In the *Roman de la Rose* the Lover is invited to kiss the mouth of the God of Love—"a villein or swineherd was never permitted to do this," ed. Félix Lecoy, 1 (Paris: Champion, 1968), vv. 1935–38.

58. *Vilain Mire*, vv. 42–52.

59. *Vilain de Bailluel*, vv. 8–11.

60. *Jouglet*, vv. 128–35 and 276–90; *Sorisete des Estopes*, vv. 2–5.

61. *Vilain Mire*, ms A, vv. 25–33; cp. *Jouglet*, vv. 15–39, and *Berengier au lonc Cul*, vv. 14–23 (NRCF 4).

62. *Fol Vilain*, vv. 186–87; cp. *Vilain Mire*, vv. 15–19.

63. *La Chastelaine de saint Gilles*, vv. 70, 207–15 and 257–62 (RGF 1).

64. *Des .xxiii. Manières* 2: 51–53; *Berengier*, vv. 43–46.

65. *Berengier*, ms D, vv. 27–34.

66. *Despit*, vv. 66–68; *Chanson satirique*, in *Poésies inédites du moyen âge*, ed. Edélestand du Méril (Paris: Franck, 1854; reprint Bologna: Forni Editore, 1969), p. 340, v. 24.

67. Robert de Blois, *Enseignement des Princes*, in *Robert de Blois: Sämmtliche Werke*, 3, ed. Jacob Ulrich (Berlin, 1895; reprint Geneva: Slatkine, 1978), vv. 1141–42 and 1149; *Les Putains et les Lecheors*, vv. 1–11 (NRCF 6). See also *La "Bible" au seigneur de Berzé*, ed. Félix Lecoy (Paris: Droz, 1938), vv. 180–86.

68. "God does not want the poor to become full of pride," explains the *Proverbe au vilain*, vv. 193–96. Another proverb states: "He shames God who elevates a villein to a higher station," in Morawski, ed., *Proverbes français*, no. 870, p. 32. On the folly of changing one's estate, see Ruth Mohl, *The Three Estates in Medieval and Renaissance Literature* (New York: Columbia University Press, 1933), pp. 332–40.

69. Le Goff, "Peasants and the Rural World," p. 88.

Jane B. Dozer-Rabedeau

10. *Rusticus*: Folk-Hero of Thirteenth-Century Picard Drama

The vivid portrayals of *rusticus* in the tavern scenes of Jehan Bodel's *Jeu de Saint Nicolas*, the anonymous *Courtois d'Arras*, and Adam de la Halle's *Jeu de la Feuillée* and *Jeu de Robin et Marion* appear at first reading to condemn him to eternal damnation. In the earlier works especially, he is shown as a thief, profligate, gambler, or drunk, and his female counterpart as a scheming prostitute. One need only consider, for example, the ill-fated thieves of the *Nicolas* who must return the royal treasure they have stolen from a pagan king; or the Courtois d'Arras who is tricked out of his inheritance by two prostitutes; or any one of the many commoners of the *Feuillée*: Dame Douche, the drunken Henri, or even the protagonist, Adam, who is brought to inaction by the dramatic performance itself. In the *Robin et Marion* as well, we observe common people who tempt the Wheel of Fortune and break with earlier dramatic tradition at the beginning of the thirteenth century to experiment with and define by their movement a new concept of dramatic space—comic space—peopled not by saints and religious figures but by *rusticus*, a new breed of comic folk-hero, who is the only one in a rigidly controlled Christian hierarchy to enjoy the freedom to laugh and to cause laughter. For the development of medieval drama as performable art, this clearly suggests that we owe the discovery and affirmation of comic space to *rusticus*.[1]

The Larousse *Dictionnaire de l'ancien français*[2] provides four meanings for *ruiste, ruste* adj. (déb. XIIe s. . . . ; lat. *rusticus*) of which two apply to this study:

> 1° Fort, vigoureux [strong, vigorous]. 2° Rude, violent, terrible [base, violent, terrible].

For the same listing, there are also three meanings for the related noun *rustie* n.f. (XIIIe s. . . .):

1° Grossièreté [crude language and behavior]. 2° Violence [violence]. 3°
Tapage [unruliness].

For *tapage* Larousse continues:

> *Mener, faire rustie*, faire un grand vacarme, en se battant, en buvant, etc. [to
> create disorder by brawling, drinking, etc.].

On the basis of these entries, this study of *rusticus* will not focus on socio-
economic questions pertaining to medieval peasantry per se but rather on
ribald tavern scenes characterized by gross speech, violence, and unruly
drinking behavior. For an urban audience such as that which supposedly
viewed these plays in thirteenth-century Arras,[3] such uncivilized, risible
behavior might have been attributed to an extra-urban group such as an
uncultivated peasantry, but the conjecture must remain beyond the scope
of this paper.

Le Jeu de Saint Nicolas

In the *Jeu de Saint Nicolas*,[4] for example, composed by Jehan Bodel of Arras
(ca. 1200), a wandering Christian *preudom*[5] is captured during a crusade
and is taken to a pagan king, who questions him about his religious beliefs. When the prisoner claims that he is protected by Saint Nicolas, whose
miraculous power will protect and multiply whatever is entrusted to his
care, the king decides to test the power of the saint. He proposes a wager
that stakes his royal treasure against the *preudom*'s life. The unguarded treasure will be exposed to public view, protected only by a statue of Saint
Nicolas. If anyone steals the treasure, the king will require the life of the
preudom as repayment. As the plot unfolds, the treasure is stolen, and the
king orders the death of the Christian. But Saint Nicolas himself inter-
cedes; the treasure is restored; and the king converts to Christianity.

This briefly is the plot of the first extant miracle play written entirely in
the vernacular, a mixture of Francien and Picard. It is a text of 1,540 lines, of
which the first 114 lines represent a narrative prologue that introduces and
summarizes the action of the subsequent drama.[6] The paradoxical relationship between the prologue and drama furnishes a challenging case study
of inter/intra-textuality and leads to a heightened appreciation of Bodel's
comic tavern scenes. According to Tony Hunt, Bodel

provides an abstract of the plot, . . . a prescribed device for soliciting the audience's attention, but he holds the details of his originality in reserve, withholding all references to his audacious introduction of low-life realism, in order to secure the maximum effect later in the play.[7]

While earlier critics were less generous in their assessment of the tavern scenes,[8] later scholars (including Alfred Jeanroy, F. J. Warne, Albert Henry, Patrick R. Vincent, and F. W. Marshall) recognized the fundamental performability of the *Nicolas* and provided us with various guides for dramatic performance.[9] More recently, however, Henri Rey-Flaud, in his important study *Pour une dramaturgie du moyen âge* (1980), attempts to reorient *Nicolas* criticism both thematically and structurally in terms of the psycho-social experience created by a medieval theater in the round:

> This theater is a universe closed in on itself, turning its back on the world and on reality, recreating another fundamental reality, one of phantasm. The inhabitants of an entire city are here, packed in against one another, in the circle of wooden scaffolding rising floor upon floor, the image of the social hierarchy. But around them there is nothing more than a little sand, a desperately empty stage without sets. And here in this derisive circle, civilization all across Europe played out its entire existence for nearly five centuries.[10]

I would like to build on Rey-Flaud's analysis by proposing an interpretation of dramatic space in the *Nicolas* based on the premise that beyond questions of sociology, we are dealing first and foremost with a dramatic composition intended for performance.

Let us begin the study of the *Nicolas* tavern with a close examination of dramatic space as it is defined by the movement of *rusticus*, for I disagree with Rey-Flaud who argues that the audience creates space: "In a theater of the Middle Ages the spectator is active. It is [the spectator] who creates space and time at each moment of the action."[11] There is after all a predetermined text that, if it does not define specific movement in the acting space, at least fixes the spoken word that the actors must verbalize. Gilbert Dahan, for example, sees in the *incipit* a new awareness of the audience, which, he claims, had to be rediscovered by medieval stagecraft.[12] He adds that the medieval prologue commonly begins with an invitation to the audience, which is generally expressed by the verb *ouïr* [to hear].[13] In the *Nicolas* we find:

Oiiés, oiiés, seigneur et dames,
Que Diex vous soit garans as ames (vv. 1–2)

[Hear ye, hear ye, lords and ladies,
May God be with your souls.]

The imperative seems to be more than an attention-getting device; rather it equates the speech act with movement in dramatic space.

The dramatic roles associated with the *Nicolas* tavern can be divided into three groups. First we can identify the *Tavreniers* (innkeeper), the only non-noble employer of the work, who provides food and wine to his clients, establishes credit, settles disputes, and employs both a crier (Raoulet) and a tavernboy (Caignet). In a second group are the employees: Auberon, Connart, Raoulet, and Caignet. Auberon is courier to the pagan king; Connart and Raoulet are both public criers; and Caignet works in the tavern. In a third category are the thieves (Cliquet, Rasoir, and Pincedés), upon whom will focus nearly half of the dramatic action as they move from tavern to royal treasury and back.[14] As we shall see, these examples of *rustici* are the actors whose movement and actions will define the tavern as the *locus* for comedy. Moreover, this *locus* will receive the greatest spatial emphasis of the *jeu* (49 percent), with 696 lines devoted to tavern scenes (vv. 251–314, 595–998, 1023–1190 and 1281–1340).[15]

The first tavern scene (vv. 251–314)[16] includes sufficient textual evidence to identify its location. No rubric is needed to substantiate the *Tavreniers'* claim:

Chaiens fait bon disner, chaiens!
Chi a caut pain et caus herens,
Et vin d'Aucheurre a plain tonnel. (vv. 251–53)

[Herein is a good dinner, herein!
Here is warm bread and fresh herring
And Auxerre wine by the barrel.]

Soon he invites Auberon to sit "en ceste achinte" (v. 261), which Henry interprets to mean an open area in front of the tavern.[17] Auberon is engaged by Cliquet in a dice game ("petit gieu" [low score])[18] to decide which of the two will pay for Auberon's drink. The scene early equates the tavern with games of chance and plunges the audience unquestionably into the human sphere, where chance, not God, controls the outcome of events.

The second and third tavern scenes (vv. 595–998 and 1023–1190), which precede and immediately follow the theft, are the longest and most fully developed segments of the drama and are essentially a demonstration of four dice games much like those found in other game books of approximately the same period.[19] Here the thieves again play "petit gieu," followed by "plus poins" [high score], "pour les des" [for the roll], and "hasart" [chance].[20] However, since cheating appears to be the game of preference, the text includes many caveats: The dice may not be perfectly cubed (vv. 832–34); they may not be tossed directly in front of the player's hand (v. 854); a sleeve may cover the throw (v. 907); the playing surface may be moved during play (v. 1079); the surface may not be level (vv. 1086–87); or the room may be too dark to see (vv. 1093–96). In addition, the scenes are punctuated by numerous examples of colorful gambling terms: "hocherons as crois" (v. 807), "en wanquetinois" (v. 907), "preng je ches nois" (v. 908), "mailles de musse" (v. 1067), and "bele couvee" (v. 1145), to name but a few.[21] The text also points out the advantage of registering dice with municipal authorities (v. 851), suggesting that chance is a powerful but capricious force that can be manipulated by properly instructed human interference. Clearly the *Nicolas* tavern delineates an identifiable playing space with games, rules, and language of its own, which is separate from, though it may mirror, the pagan-Christian conflict that underlies the dramatic work as a whole.[22]

In the fourth tavern scene (vv. 1281–1340), Saint Nicolas appears in person and frightens the thieves into returning the royal treasure,[23] after which they decide to go their separate ways. Rasoir plans to steal a bridal trousseau; Cliquet will try to rob the mayor of Fraisne; and Pincedés will steal laundry that he has seen nearby (vv.1368–81). That the royal theft is unsuccessful, however, thwarted by Christianity, demonstrates that in the *Nicolas* the element of chance and indeterminacy is still checked. Certainly every roll of the dice is prescribed by the text; every movement of *rusticus* is controlled by the written word regardless of the details of performance. In short, the tavern *locus* does not yet permit the full expression of dramatic freedom that will be found later, for example, in the plays of Adam de la Halle, in which there will be no conversion or *Te Deum* to confirm Christian space.[24]

Le Courtois d'Arras

The *Courtois d'Arras* is an anonymous early thirteenth-century piece of only 664 lines that elaborates on the New Testament parable of the Prodigal Son (Luke 15: 11–32) in which a youth spends his inheritance on "riotous living."[25] Unlike the parable, however, which emphasizes the father's joy at the return of his repentant son, the *Courtois* focuses two-thirds of the dramatic interest on the son's experiences away from home: first in a tavern and then as a swineherd.

The spatial requirements of the *Courtois* are simple and straightforward and can be adapted without difficulty to either a linear or circular arrangement of space. There must be three dramatic *loci*: one to indicate Courtois's home for the performance of 176 lines (27 percent) of the dramatic work (vv. 1–90 and 580–664);[26] one to represent the tavern for the performance of 328 lines (49 percent) of the drama (vv. 103–430); and one for the hog farm for the performance of 114 lines (17 percent) of the drama (vv. 451–564). There also must be an undefined area for Courtois's wanderings, which account for 7 percent of the drama (vv. 91–102, 431–50 and 565–79).[27]

Courtois (whose name should be construed ironically since he is merely a naive, if not lazy, young peasant) spends his time drinking and gambling.[28] Unlike the *Nicolas* thieves, whose corrupt behavior may be construed as an urban phenomenon, Courtois's self-indulgence in the tavern can be blamed on his isolated, rural background. Although he is an example of *rusticus* (viz. "country bumpkin") by virtue of his non-urban roots, one can imagine a hypothetical biography according to which the "real" scenes of his gambling transpired prior to the performance in his native rural environment, where he may have exhibited more typical *ruste* behavior. Yet it would appear that he comes from a relatively comfortable farming background since his father possesses work animals, cultivates vegetables for domestic consumption, and even has ready cash to give Courtois, who demands his inheritance before leaving home.

Since the youth has no trade or other means of support, he hopes to increase his inheritance by gambling and thus to make his fortune (vv. 74–81). In the course of the play, however, Courtois is never shown gambling and uses gambling vocabulary merely to establish the credibility of his claim to independent adulthood. Nor in his wanderings does he ever flaunt his inheritance or express any interest in "riotous living" but rather is concerned with finding food, drink, and a place to rest:

Qui eüst un cambon salé
et plain pot de bon vin sor lie,
sor un petit de raverdie
se fesist ja trop bon mucier! (vv. 98–101)

[Whoever has ham
And a full pitcher of good, clear wine
On a small knoll
Has found a very good spot!]

When he arrives at a tavern, he orders wine and praises God for leading him to such material abundance:

Hé! Dieus, aorés soies tu,
qui m'as mené en tel contree
ou jou ai tel plantet trovee! (vv. 114–16)

[Oh! God, may you be praised,
For you have led me to such a place
Where I have found great abundance!]

He is especially pleased to learn that credit is readily available (vv. 120–22), which means that, theoretically, wine and food will be obtainable even if his funds are depleted. Not so for poor Courtois!

Not knowing what lies ahead, Courtois asserts that the tangible luxuries of the tavern are more welcome than the intangible spiritual benefits of the Church (v. 125),[29] for clearly the tavern does not represent a priori a gambling spot for him but rather a *locus amoenus* (garden of earthly delights)[30] which, according to the *Ostes* (innkeeper), offers boundless material delights:

Çaiens sont tuit li grant delit,
cambres pointes et soef lit
haut de blanc fuerre et mot de plume,
fait a le françoise coustume,
covertures bieles et netes
et orelliers de violetes . . . (vv. 133–38)

[Herein are all great delights,
Attractive rooms and soft beds,
Full of white straw and many feathers,
Prepared according to French custom,
Beautiful, clean covers
And violet (-scented) pillows . . .]

Soon Courtois meets Manchevaire and Pourette,[31] who easily seduce him
and, with the help of the *Ostes*, dupe him out of his inheritance. Thus he
learns from experience that the only rule of the tavern is dishonesty, for he
is left drunk and penniless, unable to pay even his own tavern bill. Unlike
the *Nicolas* thieves, the prostitutes never return Courtois's inheritance but
simply abandon the youth. Moreover, no saint appears to intercede on his
behalf.

Here the tavern represents not just a departure from the material conditions of economic oppression that Courtois rejected at home but also a
radical opposition to them. Hard work is replaced by frivolity and leisure,
material privation by self-indulgent amusement, and Christian wisdom by
human folly. According to L. C. Porter:

> The first concern of the author of the *Courtois d'Arras* was surely not to
> amuse the audience. For him, amusement must have been a secondary ele-
> ment, serving to emphasize, by contrast, the moral lesson he was making. If
> we divide the play in three sections (revolt, flight and penitence), it is clear
> that the first and third have no comic intent. On the contrary, the tavern scene
> should at least have made the listeners laugh.[32]

Certainly in a broader context, the events of the *Courtois* are dictated by
the Biblical source and, since they are predetermined, were not subject to
alteration by the unknown medieval author. Only the non-Christian *mansion* of the tavern is free to house the previously unexplored possibilities of
comic theater.

Moreover, Courtois's return home, which is interrupted by a brief
period of employment as a swineherd (vv. 451–564),[33] is motivated by fear
of starvation—hence material and non-religious—and is not the result of
any spiritual rebirth. Although he has been promised a place to sleep, bread
each day, and four sous, he soon finds that the bread is not to his liking:

Je deuisse mangier, je quic;
mais mes pains resanble bescuit
plains est de mesture et de drave. (vv. 498–99)

[I must eat, I believe;
But my bread resembles toast
Full of wheat and rye.]

Later he laments that it contains pieces of straw:

Ha! Dieus, com cis pains me dehaite!
Je cuic k'i soit d'avaine u d'orge:
ja m'aront trenchie la gorge
les pailles et li festu lonc. (vv. 508–11)

[Oh! God, how this bread makes me sick!
I believe it may be of oats or barley:
Already has my throat been scratched
By the straw and long husks.]

Convinced he will starve on this meager diet of *pute blee* (v. 525), he decides
to leave in search of food and ultimately returns home. Although he claims
to repent (vv. 506, 605),[34] his universe is centered even now around human,
material existence, not spiritual needs. Only the final *Te Deum*, invoked by
Courtois's father, will reclaim the dramatic *locus* for Christianity.[35]

Le Jeu de la Feuillée

Of particular interest to our study of *rusticus* is the *Jeu de la Feuillée*, written
by Adam de la Halle (ca. 1276).[36] Interpretations of the work are numerous,
and many seem contradictory. Is it a biographical work, a religious satire,
a parody of the traditional *congé*, or a new genre?[37] Certainly, for many,
the apparent fragmentation of the plot is disconcerting. Yet it is my belief,
both as critic of the work and as stage director on two occasions,[38] that the
entire dramatic action, as fragmented as it is, finds unity in the very process
of dramatic performance, that is, by the spatial constraints imposed by the
medium itself.

In contrast to the *Nicolas* and the *Courtois d'Arras*, the action begins
in an undefined *locus* and will remain abstract for 565 lines until a table is
mentioned. In this first scene (vv. 1–181),[39] Adam attracts the audience's attention with a *congé*, or farewell, to his wife and friends as he prepares to
return to Paris to continue his studies.

Seigneur, savés pour coi j'ai men abit cangiét?
J'ai esté avoec feme, or revois au clergiét;
Si avertirai chou ke j'ai piech'a songiét,
Mais je voeil a vous tous avant prendre congiét. (vv.1–4)

[Lords, do you know why I have changed my cape?
I have been with my wife; now I return to the clergy.
Thus I will undertake what I have been dreaming of for a long time.
But first I want to bid you all farewell.][40]

In the second scene (vv. 182–245), Adam's father claims to be ill and con-
sults a *Fisisciens* who declares that his only "illness" is that of avarice. In the
third scene (vv. 246–321), the *Fisisciens* also examines Dame Douche, who
is pregnant. She accuses one of Adam's friends, Rikier, of being the father
of her child. In the fourth scene (vv. 322–589), a Monk appears with a relic
of Saint Acaire and claims to be able to perform miracles for a fee. After
he fails to cure two fools (Walet and a *Dervés*), the Monk joins Adam and
his friends to wait for the arrival of Dame Morgue and her fairy companions. Their visit constitutes scene 5 (vv. 590–875), which takes place around
a table indicated by Rikier:

Dame Morgue et se compaignie
Fust ore assise a cheste tavle. (vv. 564–65)

[Dame Morgue and her companions
Were seated at this table before.]

No rubric is necessary to define the space, for it has been done from within
the text by *rusticus*.

Scene 6 (vv. 876–1099) formally introduces the tavern as the dramatic
locus for the performance of the remainder of the play, and it too is indicated by the presence of a table signaled this time by Hane, another of
Adam's friends:

Alons ent dont ains ke li gent
Aient le taverne pourprise.
Eswardés, li tavle est ja mise . . . (vv. 899–901)

[Let's go in before (other) people
Have filled the tavern.
Look, the table is already set . . .]

Although it is impossible to ascertain from the text alone whether or not the tables of lines 565 and 901 are one and the same, there is nothing to preclude the use of a single table since the two scenes are not acted simultaneously or juxtaposed physically. Instead there is sufficient indication of movement *away* from the space occupied by the fairies and *to* the tavern that we can assume a circuitous path returning to the same *locus* now (re)defined by the *Ostes*[41]:

Sire, bien soiés vous venus! . . .
Vous voeil je fester, par saint Gile! (vv. 907–8)

[Lords, welcome! . . .
I want to toast you, by Saint Gilles!]

Rikier replies:

Or me prestés donques un voirre,
Par amour, et si seons bas.
Et che sera chi li rebas
Seur coi nous meterons le pot. (vv. 915–18)

[So give me a glass,
By God, and then let's sit down.
This sill will be a (good) spot
To put the pitcher on.]

This mention of *li rebas* (the sill) is the only other textual indication of scenery found in the Feuillée and serves as yet another example of how *rusticus* defines the acting space.

In short, the spatial distinction between the important tavern scene and all previous action is confused by the absence of textual information locating the earlier scenes. Nor can we tell initially what kind of a tavern this is. Should we expect an aggressive Saint Nicolas or clever prostitutes to appear, as in the *Nicolas* or *Courtois d'Arras*? On the contrary, the tavern of the *Feuillée* is depicted as the convivial meeting place of a largely hetero-geneous group of non-noble urban dwellers who derive pleasure not from stealing, gambling, and brawling but rather from tricking the Monk into believing he has lost an imaginary dice game and must pay their tavern bill. Here, as in the *Courtois*, the game of chance takes on significance pre-cisely because it is *not* played; but unlike the *Nicolas* and *Courtois*, it is the innkeeper who suggests the ruse:

> Metons li ja sus k'il doit tout
> Et ke Hane a pour lui jué. (vv. 965–66)

> [Let's all agree that he owes everything
> And that Hane rolled for him.]

At this point, the play undergoes visible thematic fragmentation with the recapitulation of the earlier comic scenes and the reappearance of the main characters.[42] Although there is no evidence of structural cohesion, it is the physical presence of the tavern table of line 901 that spatially unifies the scene, the *unité de lieu* provided by the dramatic *locus* itself. Moreover, although only 20 percent of the *Feuillée* is specifically housed in the tavern (224 out of 1,099 lines), this figure ignores the possibility that the entire performance may have been acted in the round[43] or in front of a single *mansion* representing a tavern or even a real, historic tavern in Arras, as Walton has suggested.[44]

Finally Gillot, Rikier, and the others clear the table and prepare to leave the tavern to light a candle at the shrine of Notre-Dame (vv. 1077–80).[45] Only the Monk remains to witness the departure of the entire com-pany. Alone in the acting space, he closes the play, inviting the audience to follow him not into a traditional Hell Mouth but into the arms of the Church:[46]

> N'il n'i a mais fors baisseletes,
> Enfans et garchonaille. Or fai,
> S'en irons; a saint Nicolai
> Commence a sonner des cloketes. (vv. 1096–99)

> [There remain only girls,
> babies, and boys. So be it.

Let's go. At Saint Nicolas
They are beginning to ring the bells.]

The noise and chaos that must have accompanied their exodus marks a
striking contrast to the silence surrounding the emptiness that is left behind, a tangible reminder of the ephemeral quality of dramatic space.

In sum, the dizzy fragmentation of the dramatic action at the conclusion of the *Feuillée* is such that only abstract space is left after the departure
of *rusticus*. It has finally been given its complete autonomy to function as
a seemingly spontaneous comic *locus* and continues to exist spatially even
in the absence of the actors. Although one has the impression of complete
spontaneity, the pandemonium is nonetheless prescribed by the text and
controlled by the medium itself. I would argue, moreover, that it is *rusticus*, not the spectator (as Rey-Flaud maintains[47]), whose movement and
laughter define space and breathe life into the performance.

Le Jeu de Robin et Marion

The *Jeu de Robin et Marion*, also by Adam de la Halle (ca. 1300), is an "espèce
de bergerie" (shepherds' play) of 780 lines that will conclude our study of
rusticus.[48] Generally speaking, criticism of the *Robin et Marion* centers on
the question of genre.[49] Ernest Langlois, for example, affirms in passing
that it is "a 'pastourelle' written in a new framework."[50] He and others (including Claude-Alain Chevallier and Grace Frank) have pointed out that
if the narrative portions of the lyric "pastourelle" were eliminated, the remaining dialogue would form the dramatic basis of the *Robin et Marion*.[51]
According to them, the work marks the genesis of a new dramatic genre,
the "pastourale," based on the conventional thirteenth-century lyric form.

Critics have paid little attention, however, to the important role of
rusticus within the work especially in terms of the use of dramatic space.
Marion, for instance, who is supposedly a simple shepherdess, is generally
treated only in relation to Robin or to the knight who abducts her. Her
ability at first to pun with the knight, to say nothing of her single-handed
escape from his aggressive advances, is often ignored or misconstrued as
an indication of naive peasant ignorance rather than a clever display of her
contempt for the knight and her faithfulness to Robin. As I have argued
elsewhere, Marion, who is the strongest character of the play, functions as
an on-stage director (*meneur de jeu*) for the performance.[52]

In his modern French edition of the *Robin et Marion*, Chevallier divides the action into nine scenes, but, in general, editors regard the work as a spatially unified composition based on the use of simultaneous scenes (*décor simultané*).[53] According to Langlois, for example, the overall acting space must necessarily include the following areas: a small field, bordered by a hedge behind which the peasants hide and watch the abduction of Marion; a rock or small abutment to be used as a throne during the second game; the house (or houses) of Gautier and Baudoin; and a large pasture in the distance for Peronnelle's sheep.[54] I would argue, however, that only a single space is needed for the entire performance since only 44 lines (6 percent) of the play take place outside Marion's pasture. Instead, the dramatic *locus* is constantly changing, abstractly redefined by each dance, game, and song. In other words, the overall dramatic space is conceptually altered when it becomes a pasture, dance floor, village, and finally a space for playing games. In this, I believe, lies the dramatic genius of Adam de la Halle, who seems to have understood in both the *Feuillée* and *Robin et Marion* that drama can (through the spoken word, gesture, song, and game) transform empty space into something else.

In scene 1 Marion opens the action with a song already familiar to the medieval audience:

Robins m'aime, Robins m'a;
Robins m'a demandee, si m'ara. (vv. 1–2)

[Robin loves me. Robin has me.
Robin asked for me. So he will have me.]

This is an excellent attention-getting device to quiet the audience and leads gradually into the dramatic form. Soon she is joined by a knight returning on horseback from a tournament. He recognizes her song and asks her to become his *amie*. After some punning in which Marion shows that she is more clever than the knight, he rides off and ends the scene. This represents the first instance of Marion's strength of character over the knight and establishes her pivotal role in the drama. As the play progresses, she is joined by her shepherd friend Robin, who sings and dances with her. After he leaves and Marion is alone once again, the knight returns and forcibly abducts her. She manages to free herself, although Robin and several friends have been watching the struggle from behind some trees.[55]

After the departure of the knight, a celebration is in order. They play various games, sing and dance, and sit down to a hearty peasant banquet. Here we have fine examples of *rustici* at their best. There is none of the unsavoriness of the *Nicolas* thieves or the cunning manipulation of the *Courtois* prostitutes; these *rustici* are a new breed of "rustics" at play.[56]

The first game played is called "Saint Coisne" (vv. 445–78). All the players must offer a present to the one chosen to be the saint, who, in turn, attempts to make them laugh. If that person is successful, roles are reversed and a new saint is chosen:

. . . Quiconques rira
Quant il ira au saint offrir,
Ou lieu saint Coisne doit seïr. (vv. 449–51)

[. . . Whoever laughs
When he goes to make an offering to the saint
Must take the place of Saint Coisne.]

When Marion finds the game to be *trop lais* [too ugly] (v. 479), Gautier suggests another: "Faisons un pet" [Let's fart] (v. 485). However, the "game" is rejected as being offensive to the women. In the second game, "au Rois et as Roïnes" (vv. 495–596), the "king" may command anything or ask any question, and his will must be obeyed in all things. Baudoin becomes "king" and is properly crowned with Peronnelle's straw hat (*capel de festus*, v. 518), with the dramatic *locus* now transformed into a mock court. The game soon turns to questions about sex but is interrupted by the news that one of Marion's sheep has been caught by a wolf (v. 597–98). Robin rescues the sheep after which there is much joking about sex, marriage, and food. Robin now sings two brief songs (vv. 675–80 and 684–88) and leaves to find some musicians. While he is gone, the women prepare a picnic spot (vv. 699–716) with Peronnelle's apron serving as an analogue for the tavern table of the earlier plays. By their gesture, the women redefine the dramatic *locus*, preparing the space for the singing, dancing, and eating of the final scene (vv. 717–80). At the conclusion of the *jeu*, Robin leads everyone in a *farandole* and gradually dances them away from the audience's view in much the same way that the *Feuillée* concludes with the affirmation of autonomous dramatic space. Moreover, there is no *Te Deum* to close the play on a solemn note but rather a lively dance entirely consistent with the general tone of the work. Not only do these peasants move exclusively in the human sphere but also their activities, games, and songs repeat recognizable forms of entertainment probably known to the audience.

Conclusion

The focus of the *Robin et Marion* is clearly on the terrestrial preoccupations of *homo ludens*. Like the somewhat earlier *Feuillée*, the *Robin et Marion* is filled with jubilant enthusiasm and laughter of a sort that was non-existent one hundred years earlier in the *Nicolas* and *Courtois*. Perhaps this change of attitude can be explained on the basis of the improved economic situation of Arrageois society as the thirteenth century progressed.[57] Or perhaps changing theology engendered more favorable attitudes toward human existence.[58] Clearly these explanations may be related. It is my view, however, that beyond questions of sociology and theology the real contribution of thirteenth-century Picard drama lies in its unprecedented experimentation with movement in dramatic space.

In the *Nicolas* tavern we found that the Christian world was displaced by chance and that *rusticus* claimed this *mansion* for the depiction of comic, human space. In the *Courtois* the tavern served as a false *locus amoenus* that betrayed *rusticus* and sent him wandering again through non-specific, human space. In the *Feuillée* the tavern was again juxtaposed to a spatial void, which was then filled with a series of comic scenes performed by various incarnations of *rusticus*: a student, his wife, father, and friends, a physician, a few peasants, and a variety of fools. Finally the *Robin et Marion* expanded the abstract, human space of the *Feuillée* and affirmed the autonomy of the dramatic *locus* reserved now exclusively for *rustici* at play. In the four works examined here, the tavern scenes represent the antithesis of Christian dogma: Human space is valid, and play can be enjoyed without guilt. Moreover, the tavern is the logical space for comedy to be tolerated in an essentially Church-dominated society, for it is *rusticus* who, at the beginning of the thirteenth century, breaks with traditional dramatic forms in Latin to explore the possibilities of comic space in vernacular drama. In the tavern of the *Nicolas, Courtois*, and *Feuillée* and its analogue in the *Robin et Marion*, we observe many examples of *rustici* who all share one important characteristic: They thrive in non-Christian space, in which chance, not God, controls their individual and collective destinies.

What can be said of the other *dramatis personae* who people the tavern *locus* yet are not examples of *rusticus*? In the *Nicolas*, for example, the saint appears briefly in the tavern (but in very unsaintly fashion) and terrorizes the thieves into returning the royal treasure. He is an outsider; the tavern is not his space. In the *Feuillée* too, the fairies depart before the tavern is actually introduced, and the Monk falls asleep, missing both the fairies' visit and most of the tavern scene. Nor does the knight of the *Robin et Marion* witness the peasant games. Clearly it is to *rusticus* alone that we owe the discovery and affirmation of comic space.

But is *rusticus* a folk hero? Certainly the *Nicolas* thieves bring about the monumental conversion of the pagan world; the *Courtois* prostitutes cause not only the youth's ruin but also his ultimate repentence; the tavern *habitués* of the *Feuillée* cause Adam to postpone his departure, thereby witnessing their chaotic *speculum mundi*; and the peasants of the *Robin et Marion* play familiar games that divert the audience and tease their sensibilities. In each case, *rusticus* creates a spatial unit—a tavern or its analogue—into which he introduces laughter. Thanks to his new-found freedom of movement and action, he is able to claim this space for the salvation of suffering humanity through joy rather than tears, through comedy rather than the tragedy of Christ's Passion. In conclusion, I would agree with Rey-Flaud: "In medieval society theatrical play is, therefore, the only form of play which has no limitation."[59] But I would add that, in these four dramatic works, true freedom of action is reserved exclusively for the various incarnations of *rusticus*.

Notes

1. Although I recognize that there are numerous similarities to recent works by Henri Rey-Flaud (*Pour une dramaturgie du moyen âge* [Paris: Presses Universitaires de France, 1980]); David Raybin ("The Court and the Tavern: Bourgeois Discourse in 'Li jeus de Saint Nicolai,'" *Viator* 19 [1988]: 177–92); and especially Jean-Claude Aubailly ("Réflexions sur le 'Jeu de Saint Nicolas': pour une dramatologie," *Moyen Âge* 95[3] [1989]: 419–37), I refer the reader to my Ph.D. dissertation (Jane B. Dozer, "The Tavern: Discovery of the Secular Locus in Medieval French Drama," unpublished Ph.D. dissertation, UCLA 1980 [University Microfilms, Ann Arbor, Mich., 1980, AAC8023298]). This paper is an outgrowth of that research and is dedicated to the memory of my mentor, Professor Lora S. Weinroth (UCLA).

Among the manuscripts I have studied, thanks to a generous grant from the French government, the following pertain specifically to this article: B. Arsenal 3203 (*Le Jeu de la Feuillée* [=facs. of Vat. MS. Reg. 1480]); B.N. MSS. fr. 837 [in facs. 4° 422] (*Les Congiez Jehan Bodel; Le Courtois d'Arras*; XIIIe); 1553 (*Le Courtois d'Arras*; XIIIe); 1569 (*Li Jeus du bergier et de la bergiere*; XIIIe–XIVe); 19152 (*Le*

Courtois d'Arras; XIIIe); 25566 (. . . *Li jus de s. nicholai*; . . . *Li jus des esquies*; . . . *Li congie jehan bodel*; XIIIe); MS. lat. 11331 (. . . *Ludus super Iconia Sancti Nicolai*; XIIe); B. Méjanes, MS. 572 (*Mariage et Robin et de Marote*; XIVe); and B.M.L. Add. 22414 (*Le Jeu de saint Nicolas* [fragment]; XIe). See also below, note 19.

2. A. J. Greimas, *Dictionnaire de l'ancien français* (Paris: Larousse, 1980), pp. 574–75.

3. Although the *Jeu de Robin et Marion* may have been performed initially at the court of Robert II, Count of Artois, in Italy, our interest here is limited to performances in Arras. See Ernest Langlois, ed., *Adam le Bossu, trouvère artésien du XIIIe siècle, "Le Jeu de Robin et Marion," suivi du "Jeu du Pèlerin"* (Paris: Champion [CFMA], 1968), pp. iv–v; and Kenneth Varty, "Le mariage, la courtoisie et l'ironie comique dans le *Jeu de Robin et de Marion,*" *Marche romane* 30(3–4) (1980): 287–92. For a discussion of the moneyed aristocracy of thirteenth-century Arras, see Henri Guy, *Essai sur la vie et les oeuvres littéraires du trouvère Adan de le Hale* (Paris, 1898; Geneva: Slatkine, 1970), pp. xi–lviii; Marie Ungureanu, *La Bourgeoisie naissante: Société et littérature bourgeoises d'Arras aux XIIe et XIIIe siècles*, Memoires de la Commission des Monuments historiques du Pas-de-Calais, 8th ed. (Arras: CNRS, 1955), pp. 23–62; also Raybin, "Court and Tavern," pp. 189–92.

4. All citations and line references are from Alfred Jeanroy, ed., *Jean Bodel, trouvère artésien du XIIIe siècle: "Le Jeu de Saint Nicolas"* (Paris: Champion [CFMA], 1974).

5. F. J. Warne, ed., *Jean Bodel: "Le Jeu de Saint Nicolas"* (Oxford: Blackwell, 1968), characterizes the *preudom* as a "worthy and valiant man" (p. 107).

6. See David H. Carnahan, "The Prologue in the Old French and Provençal Mystery," Ph.D. dissertation, Yale, 1904 (published, New Haven: Tuttle, 1905); Gilbert Dahan, "Note sur les prologues des drames religieux (XIe–XIIIe siècles)," *Romania* 97 (1976): 306–26; and Bethany A. Schroeder, "The Function of the Prologue in 'Le Jeu de Saint Nicolas,'" *Romance Notes* 10 (1968): 168–73. For specific discussion of the questionable paternity of the prologue, see Albert Henry, ed., *Le Jeu de Saint Nicolas de Jehan Bodel*, Travaux de la Faculté de Philosophie et Lettres 21, 2nd ed. (Brussels: Presses Universitaires, 1965): 9–16; Gaston Raynaud, "Les Congés de Jean Bodel," *Romania* 9 (1880): 216–47; Tony Hunt, "The Authenticity of the Prologue of Bodel's 'Le Jeu de Saint Nicolas,'" *Romania* 97 (1976): 252–67; Rey-Flaud, *Pour une dramaturgie*, p. 57; and Aubailly, "Réflexions," p. 425.

7. Hunt, "Authenticity," p. 259.

8. Pierre J.-B. Legrand d'Aussy, for example, wrote: "Comme cette pièce n'est en grande partie que le miracle du prologue un peu étendu; qu'elle est très longue et encore plus ennuyeuse, je crois suffisant d'en donner un court extrait" (*Fabliaux ou contes, fables et romans du XIIe et du XIIIe siècle*, 3rd ed. (Paris: Renouard, 1829) 1: 341). Cf. Gaston Paris, *La Littérature française au moyen âge (XIe–XIVe siècles)*, 3rd ed. (Paris: Hachette, 1905), p. 265.

9. In addition to Jeanroy, ed., *Jean Bodel*, Warne, ed., *Jean Bodel*, and Henry, ed., *Jeu de Saint Nicolas*, see Patrick R. Vincent, *The "Jeu de Saint Nicolas" of Jean Bodel of Arras: A Literary Analysis*, Johns Hopkins Studies in Romance Literatures and Languages 49 (Baltimore: Johns Hopkins University Press, 1954); F. W. Marshall,

"The Staging of the 'Jeu de Saint Nicolas': An Analysis of Movement," *Australian Journal of French Studies* 2 (1965): 9–38.

10. "Ce théâtre est celui d'un univers clos, fermé sur lui-même, tournant le dos au monde et à la réalité, recréant une autre réalité fondamentale celle-ci, celle du fantasme. Les habitants de toute une cité sont là, serrés les uns contre les autres, dans le cercle de bois des échafauds, superposés les uns au-dessus des autres, à l'image de la hiérarchie sociale. Mais autour d'eux il n'y a rien qu'un peu de sable, une aire de jeu désespérément vide, sans décors. Et c'est là, dans ce cercle dérisoire, qu'une civilisation à travers toute l'Europe, et cela pendant près de cinq siècles, a joué toute son existence." Rey-Flaud, *Pour une dramaturgie*, p. 166.

11. "Dans un théâtre du Moyen Age le spectateur est actif: c'est lui qui crée l'espace et le temps à chaque moment de l'action." Ibid., p. 170.

12. Dahan, "Note sur les prologues," p. 308.

13. Ibid., pp. 321–22.

14. I disagree with Raybin, who, referring to the "mimetic mirror" of the tavern, writes of the thieves: "They are like us; they are of us" ("Court and Tavern," p. 183). Perhaps this value judgment could be made about the Crusaders or the *Tavreniers*, but I suspect *not* of the thieves. It is precisely because they are *not* like us, *not* of us, that we can laugh at them. Cf. Sigmund Freud, *Jokes and their Relation to the Unconscious*, trans. James Strachey (London: Routledge, 1960); also Rey-Flaud, *Pour une dramaturgie*, pp. 92–101.

15. See Dozer, "The Tavern," p. 154; cf. Aubailly, who places 48.5 percent of the dramatization in the "abitacle" ("Réflexions, pp. 426, 428).

16. Here the term "scene" is used to denote a spatially unified action; cf. Vincent, "*Jeu de Saint Nicolas*," p. 95, and Aubailly, "Réflexions," p. 429.

17. Henry, ed., *Jeu de Saint Nicolas*, p. 197.

18. For the names and details of various medieval dice games, see Franz Semrau, *Würfel und Würfelspiel im alten Frankreich*, Beihefte zur Zeitschrift für romanische Philologie 23 (Halle: Niemeyer, 1910): 35–60, who discusses the origins of dicing, the manufacture of dice, dicing customs, and especially ways of playing and cheating. See also C. E. Cousins, "Tavern Bills in the 'Jeu de Saint Nicolas,'" *Zeitschrift für romanische Philologie* 56 (1936): 85–93; Dozer, "The Tavern," pp. 155–61; Alfred L. Foulet and Charles Foulon, "Les scènes de taverne et les comptes du Tavernier dans le 'Jeu de Saint Nicolas' de Jean Bodel,'" *Romania* 68 (1944): 422–43; Charles Foulon, *L'Oeuvre de Jehan Bodel*, Travaux de la Faculté des Lettres et Sciences Humaines de Rennes, ser. 1, vol. 2 (Paris: Presses Universitaires de France, 1958), pp. 688–91; Grace Frank, "Wine Reckonings in Bodel's 'Jeu de Saint Nicolas,'" *Modern Language Notes* 50 (1935): 9–13; and Raybin, "Court and Tavern," pp. 185–86.

19. I have examined B.N. MSS. fr. 1173 (*Le Gieu des eskies* [Nicholes de S. Nicholai], which also contains *Les Partures des tables* and *Le Jeu des merelles*; XIIIe); 1688 (*Les Horoscopes du Jeu des dez*; XVe); 1999 (*Livret de divers jeux du tablier*; XVe); 14776 (*Livre de fortune du jeu des dés*; n.d.); and 25566 (*Li jus des esquies* [Engrebans d'Arras]; XIIIe).

20. See C. E. Cousins, "Deux parties de dés dans le 'Jeu de Saint Nicolas,'"

Romania 57 (1931): 436–37; Charles A. Knudson, "'Hasard' et les autres jeux de dés dans le 'Jeu de Saint Nicolas,'" *Romania* 63 (1937): 248–53; and Willem Noomen, "Encore une fois la partie de 'hasard' dans le 'Jeu de saint Nicolas,'" *Romania* 81 (1960): 139–41. See also above, note 18.

21. For additional examples see Henry, ed., *Jeu de Saint Nicolas*, pp. 398–403.

22. For a discussion of *agôn*, *alea*, *mimicry*, and *ilinx*, see Roger Caillois, *Les Jeux et les hommes: Le masque et le vertige*, 7th ed. (Paris: Gallimard, 1958), p. 75; also Johan Huizinga, *Homo Ludens: A Study of the Play Element in Culture* (1938; reprinted Boston: Beacon, 1970), p. 10: "Into an imperfect world of confusion, game brings a temporary, a limited perfection."

23. I take issue with Vincent regarding the supposed nobility of Saint Nicolas's character (*"Jeu de Saint Nicholas,"* p. 99). In three speeches (vv. 1281–85, 1288–91, and 1294–1306), the saint clearly presents himself to the thieves as a terrifying figure, not sublimely noble. I also disagree with Rey-Flaud, who describes the saint as a "terrible, mais bon enfant" (*Pour une dramaturgie*, p. 134). Although Jeanroy fails to list Saint Nicolas among the *dramatis personae*, he has an active, speaking role of 22 lines.

24. Omer Jodogne discusses the importance of the *Te Deum* to medieval drama in "Le Théâtre français du moyen âge: Recherches sur l'aspect dramatique des textes," in *The Medieval Drama*, ed. Sandro Sticca (Albany: SUNY Press, 1972), pp. 1–21, esp. pp. 5–8.

25. All citations and line references are from Edmond Faral, ed., *"Courtois d'Arras": Jeu du XIIIe siècle* (Paris: Champion [CFMA], 1967); see p. iv regarding the date of authorship.

26. I have not included 36 lines following v. 90, which present Courtois's sister and which appear only in MS. B (B.N. 837); see Faral, ed., *"Courtois d'Arras,"* pp. 24–25. The inclusion of these lines would alter not only the total number of lines in the *Courtois* (700 not 664) but also the percentages: home (30 percent), tavern (47 percent), farm (16 percent), and wanderings (7 percent).

27. Courtois's wanderings should end at v. 579 rather than at v. 599 as Faral indicates by his arbitrary division between scenes X and XI (20–21). Although Courtois does not address his father until v. 600, it is clear from v. 583 that he is close enough to the *mansion* to see his father inside.

28. On "tremeriel" (v. 26), "hasart," and "plus poins," (v. 74) see above, note 18.

29. If one accepts Bakhtin's analysis of the comic implications of a system of imagery based on eating, digestion, urination, and sex, one might argue that Courtois's odyssey represents a regenerative quest based on non-Christian ritual. Mikhail Bakhtin, *L'Oeuvre de François Rabelais et la culture populaire du moyen âge et sous la Renaissance*, trans. Andrée Robel (Paris: Gallimard, 1970).

30. On the *pays de Cocagne* see Legrand d'Aussy (*Fabliaux*, 1: 302). Joseph Bédier refers to the *pays* as "une sorte de vallée de Tempé bourgeoise" (a kind of bourgeois Vale of Tempe) in "Les Commencements du théâtre comique en France," *Revue des Deux Mondes*, 3rd ser. 99 (1890): 871–97.

31. For the etymology of the names "Manchevaire" and "Pourette" see Dozer, "The Tavern," pp. 227–28.

32. L. C. Porter, "Le Rire au moyen âge," *L'Esprit Créateur* 16 (1976): 5–15, esp. p. 8.

33. According to C. Pfeiffer and E. Harrison, ed., *The Wycliffe Bible Commentary* (Chicago: Moody Press, 1962), the feeding of swine was considered to be "the lowest possible humiliation for a Jew" (p. 1054).

34. Courtois admits that his sin is one of pride (*orgueil*, v. 506).

35. See above, note 24.

36. All citations and line references are from Langlois, ed., *Le Jeu de la Feuillée*. Regarding the scribal notation in MS. P, which gives the actual title as "li dis adan," see Jane B. Dozer, "Mimesis and 'li jeus de le fuellie,'" *Tréteaux*, Bulletin de la Société Internationale pour l'Étude du Théâtre Médiéval 3 (1982): 80–89.

37. See Dozer, "The Tavern," pp. 235–78.

38. Middlebury College (July 1972) and UCLA (May 1977).

39. For purposes of consistency I shall use the term "scene," although I do not equate the term here with spatial division. See above, note 16; also Guy, *Essai*, pp. 355–56.

40. In spite of Joseph Dane's excellent analysis of the *Feuillée* as a parody of the traditional *congé* ("Parody and Satire in the Literature of Thirteenth-Century Arras," *Studies in Philology* 81(1–2) (1984): 1–27), I cannot accept his summary of the plot (pp. 13–14). Adam does *not* "discuss [his departure] with his father . . . , a fool and *his* father, a prostitute, a monk, and a physician." His interaction is limited to his friends and ends after his second long speech (vv. 81–174), at which time his role is reduced to that of witness for most of the remaining action. Nor do I understand Dane's reading of vv. 876–1099: "The [fairies'] fête breaks up; the Arras men return to the tavern where they have left the monk with their bill." (p. 14). In fact, after the fête, the Monk and Hane proceed to the tavern (vv. 876–904), where they are joined by Rikier, Gillot, Adam, and Maître Henri. Soon the monk falls asleep (vv. 912–976), and a trick is played on him at that time.

41. My reading of this spatial transition is in perfect agreement with Rey-Flaud's theory of theater in the round although it is not necessarily limited to the staging techniques he suggests.

42. According to Jean Dufournet (*Sur le "Jeu de la Feuillée": études complémentaires* [Paris: SEDES, 1977], pp. 319–29), Adam is no longer needed *in propria persona* since the Monk now functions as his *porte-parole* and since the *Dervés* serves as his double in a significant number of ways (pp. 335–40). Cf. Edelgard Dubruck, "The 'marvelous' madman of the 'Jeu de la Feuillee,'" *Neophilologus* 58 (1974): 180–86; also Normand R. Cartier, *Le Bossu désenchanté: Étude sur le "Jeu de la Feuillee"* (Geneva: Droz, 1971), p. 151, who argues that Adam's spirit is already in Paris.

43. Henri Rey-Flaud, *Le Cercle magique: Essai sur le théâtre en rond à la fin du moyen âge* (Paris: Gallimard, 1973).

44. Thomas Walton, "Staging 'Le Jeu de la Feuillée,'" *Modern Language Review* 34 (1941): 344–50.

45. Adolphe Guesnon ("Adam de la Halle et 'Le Jeu de la Feuillée,'" *Moyen Age*, 2nd ser. 19 [1915–16]: 173–233) adds that the shrine was displayed in a public *place* known as *follye* (p. 34, n. 2); cf. Cartier, *Le Bossu désenchanté*, p. 159, and Dufournet, *Sur le "Jeu de la Feuillée*," p. 219.

46. One is reminded of Archibald's concluding remarks on the "architecture de vide et de mots" in Jean Genêt's *Les Nègres*; cf. Rey-Flaud's discussion of "le degré zero de la place" [the nothingness of the acting space] in *Pour une dramaturgie*, pp. 39–40.

47. See above, note 11.

48. Regarding the date see Grace Frank, *The Medieval French Drama* (Oxford: Clarendon, 1954), p. 209; Claude-Alain Chevallier, ed., *Théâtre comique du moyen âge*. Ser. 10/18: 752 (Paris: Union Générale d'Éditions, 1982), p. 147; and Dufournet, *Sur le "Jeu de la Feuillée,"* p. 63, who believes the work predates the *Feuillée*. Langlois gives no date. The description "espèce de bergerie" is a post-medieval addition found on fol. 1 of MS A (B. Méjanes 572). All citations and line references are from Langlois, ed., *Jeu de Robin et Marion*.

49. Kevin Brownlee, "Transformations of the Couple: Genre and Language in the 'Jeu de Robin et Marion,'" *French Forum* 14, supp. 1 (1989): 419–33. A notable exception is Varty, "Le mariage, la courtoisie et l'ironie comique."

50. "Une pastourelle écrite dans un cadre nouveau," Langlois, ed., *Jeu de Robin et Marion*, pp. vi–vii.

51. Ibid., p. vi; Chevallier, *Théâtre comique*, pp. 141–42; Frank, *Medieval Drama*, p. 231.

52. Jane B. Dozer, "Medieval Woman as Stage Director: Adam de la Halle's 'Jeu de Robin et Marion,'" conference paper presented to the Philological Association of the Pacific Coast, University of British Columbia, Vancouver, Nov. 9–11, 1984.

53. Cf. Rey-Flaud's argument against the term "décor simultané" in *Pour une dramaturgie*, p. 167.

54. Langlois, ed., *Jeu de Robin et Marion*, pp. vii–ix.

55. Langlois's synopsis of the plot is inaccurate on this point. He claims (p. vii) that the knight abducts Marion during Robin's absence and that she is alone to defend herself ("Marion est seule à se défendre contre ses tentatives"). On the contrary, Robin watches the abduction but fails to stop the knight (vv. 349–56). Later he and friends observe Marion's struggle from behind some bushes (vv. 372–74), but they do not come to her aid.

56. Unlike the *Nicolas* where the audience does not identify with the thieves (see above, note 14), here it is arguable that an audience composed primarily of Arrageois patricians would have liked to identify with these peasants, like Marie-Antoinette, who escaped real-world concerns by pretending to be a shepherdess.

57. See Ungureanu, *Bourgeoisie naissante*, pp. 23–62; also Raybin, "Court and Tavern," pp. 189–92.

58. Rey-Flaud, *Pour une dramaturgie*, pp. 15–22, 129–48.

59. "Dans la société médiévale, le jeu théâtral est donc la seule forme du jeu qui ne reçoive pas de limitation." Ibid., p. 81.

Robert Worth Frank, Jr.

11. The "Hungry Gap," Crop Failure, and Famine: The Fourteenth-Century Agricultural Crisis and *Piers Plowman*

Famines gather history around them. In the Blantyre District of Southern Malawi many people remember the famine which occurred in 1949–50, and many more have been told about it. There are stories and songs and recollections of the famine: accounts of foraging and migration; of "famine disease" and death; of both the cohesion and the disintegration of communities and families; of selflessness and the extremes of individualism. Survivors of the famine can give close accounts of the events of that year. They begin with the abnormal weather conditions—what the clouds looked like and what this meant; how high the crops grew before they died, and which crops survived; the names of the children who starved and the husbands who left; the wild foods eaten and their methods of preparation; the minute details of the famine relief system—the coarseness and the colour of the grain distributed, the size of the tins which acted as measures, the behaviour of the queuing people who waited to receive it. Older people can tell stories of famines that went before—the 1922 famine and the famine of 1903 which drove people into the area from Mozambique. They say that nothing comparable to 1949 has happened since. People calculate their ages by reference to it, and women consciously keep the communal memory of the event alive when they sing the pounding songs they composed then.[1]

There is a half-line early in the ploughing scene in *Piers Plowman* whose significance a modern reader may easily overlook. Piers has just volunteered to guide the company of "pilgrims" to Truth, but first he must perform the task assigned his estate in the social order: he must plough the land and provide all society with food. This he promises to do. As long as I live, he says, "I shall provide them with the necessities for life—unless the land fails" ("but if the londe faille" B.6.17).[2]

To us in the twentieth century, unless we are from a Third World country, the line seems innocuous enough. We are vaguely aware that in some years the wheat or corn crop is less than average. There is upset at the Chicago Board of Trade. We hear of farm foreclosures. Congress may pass an emergency agricultural aid bill. Piers's half-line may sound at most to us like a hedging against grain futures.

But for Langland and many in his audience the half-line, found in all three texts (A.7.16; B.6.17; C.8.15), would be charged with dramatic meaning. Economic historians are now pretty much agreed that the fourteenth and fifteenth centuries were for western Europe, including England, a period of agricultural crisis and depression. Though questions of cause and effect are still under debate, this crisis involved a number of factors: the exhaustion of marginal lands and overworked soils; three centuries in which population had tripled; changing social relations and dislocations as feudal societies moved increasingly into a market economy; the end of the warming period from 750–800 to 1150–1200 known as "the little optimum" and the coming of colder weather; and an increasing number of famines.[3]

Piers's half-line reflects a harsh truth: In England the land *had* failed, more than once, in the fourteenth century. In 1315–17 there had been a terrible famine in which 10 percent or more of the population had died.[4] And the land had failed in a number of years in Langland's lifetime, the failure reaching famine intensity on at least two, possibly three, occasions, as we shall see.

For many in England in the fourteenth century, hunger and starvation were familiar and menacing facts of life. Though the Black Death in 1348–49 and the plague's subsequent visitations are, in their recorded horror and destructiveness, so dramatic that they dominate *our* image of the age of Langland, famine and the threat of famine were a powerful competitor in terror and suffering in those years. Indeed, a modern historian has speculated that, except in the Third World, the "lack of awareness of famine as a possible force in our lives is one of the things that most critically divides us off from our own past."[5] As a recurrent possibility, famine haunted imaginations with each cycle of the seasons. It haunted Langland's imagination, certainly, and hunger and famine are haunting presences in his poem.

We can distinguish among three sources of hunger in the fourteenth century, three sources and degrees of inadequacy of the food supply: first, famine proper; second, "bad years," that is, harvests of markedly lower than usual yield (of cereals—wheat, rye, barley, and oats); and third, summer scarcity, what specialists refer to as "the hungry gap."

The Hungry Gap

Let us begin with the most pervasive but possibly least familiar of these in-adequacies, the "hungry gap." This is an almost universal phenomenon of peasant societies.[6] Commonly, the year's stocks of cereals begin to dwindle and run out by early or mid-summer, weeks or months before the new harvest. Langland tells us the rich and the powerful were not affected by this "hungry gap," but that the poorer levels of society were: landless workers, beggars and the unemployed, working peasants with small holdings, village artisans, laborers in towns. The cost of food rose and the kind of food accessible was less attractive and often less nutritious—and rations were leaner. Patience prays for poor people, who suffer much grief and who seldom are fully fed in the summer (B.14.173, 174, 177). Earlier, Patience had observed that the rich lived happily, with respect to food and clothing, all summer long, but "around Midsummer, beggars dined with no bread" (that is, with no food; see below, note 9) (B.14.156, 158–59). For many, hunger was an annual, recurring experience.

The scene with Hunger that follows the ploughing of the half-acre is primarily—though not exclusively—a dramatization of the yearly summer scarcity, the "hungry gap."[7] The details do not suggest famine. There is grain to be threshed (B.6.183–84; A.7.171–72; C.8.179–80), probably sheaves left over from the previous year's harvest. There is farm work to do. More important, for those who will work there *is* food, though not the most palatable or nourishing: barley bread, pease-loaf (peas, beans and bran baked together), horse-bread (B.6.193, 214), hound's bread, and beans (B.6.214–15). (We must remember that bread then was literally "the staff of life." The English peasant in this period consumed on the average between four and five pounds of bread a day. Bread was essentially the peasants' only article of diet, certainly the principal article.[8]) Such coarse fare is the alternative to languishing or dying from hunger that Piers offers the able-bodied.

Hunger's demand to be fed before he departs takes us deeper into summer scarcity. Though Piers is a hard worker, his cupboard is almost bare. He does have a few curds and cream, a cake of oats, two loaves of beans and bran baked (for his children), parsley, leeks, and garden greens: these must be his family's food till Lammastide (B.6.280–89). And we see the diet of even poorer people during the hungry gap in what they bring Hunger: pea pods, beans and baked apples, spring onions, chervil, ripe cherries, green leeks, peas, and, in desperation, green cabbage and peas

(B.6.292–98). No bread, notice. When harvest comes, Hunger's menu of course changes radically for the better.

Crop Failure

The second source of widespread hunger was the "bad year," when the grain harvest was markedly below normal. The crisis in the food supply would be intensified. The rich, of course, rode comfortably through the hard times; Patience notes that scarcity does not harm them, nor does drought nor excessive rain, identifying the two main weather patterns leading to bad harvests. The rich were immune: whatever they wished for or wanted they got. But in such bad years God should comfort the poor, who suffer greatly, because of drought, all their lives (B.14.170–75).

There had been a number of bad years in Langland's early life. A study by J. Z. Titow of the account rolls for the bishopric of Winchester indicates that there were "outstandingly bad" harvests—that is, yields 15 percent or more below average—in 1339, 1343, 1346, 1349, and 1350. In a thirteen-year period, only two years, 1338 and 1344, were "outstandingly good," with yields 15 percent or more above average.[9] The next twenty some years, critical years in the history of Langland's poem, were no better for either weather or harvests, as a second study of these same records reveals.[10] The brutal economic consequences stand out starkly in the figures for the price of wheat from 1350 to 1375: in only three years was the price low or average (*moyen*): 1354, 1362, 1367. For twenty-one of the twenty-six years the price was high (*élevé*), and in two years, 1352 and 1370, it was very high.[11] The high price of wheat in this period is attested to by other economic historians.[12]

The chronicles, notably the *Brut*, which was of London provenance, give us the same picture of these years in somewhat more dramatic language. In 1353 was the great scarcity of food called the "dere somer" (*Brut*, 304).[13] In the following year, 1354, a drought lasted from March to July: no rain fell; fruit, grain, and vegetables were for the most part lost. As a result, England, which had been a land of plenty, now, because of disease among people and cattle and failure of crops, had to import food from other countries (*Brut*, 304). In 1362 there was a new drought that led to scarcity of grain, fruits, and hay (*Brut*, 313). Capgrave reported that as a result, in that year, a quarter of wheat sold for 15 shillings (174). That same year there was a destructive rain at haying and harvest time (*Brut*, 315). In 1365 a severe frost lasted from November 30 to April 14, so that spring ploughing, sowing, and other field work were delayed, because the earth was frozen and hard (*Brut*, 315). In 1366 rain at haying time destroyed grain and hay (*Brut*, 316). And heavy rains in 1369 led to the loss of so much grain that the next year a bushel of wheat sold for 40 pence (*Brut*, 321). 1353, 1354, 1362, 1363, 1365, 1366, 1369, 1370—a plethora of bad years that Langland and his world had seen.[14]

The shock of all these bad years can be observed, unexpectedly, in the scene with Anima, long after the ploughing scene. Inveighing against the present failure of charity and belief among the people, Anima suddenly bursts forth with a lament on the breakdown of weather prophecy. Both sailors and shepherds, who have practical experience, Anima says, and learned clerks and astronomers, who once could foretell the weather by reading the heavens, can do so no longer. They have lost confidence in what the skies can reveal. Formerly, those who ploughed and worked the land, by examining the seed, could tell their masters how much of the crop they could sell and how much to keep for food and for the next year's sowing. Now workers on land and sea—shepherds, ploughmen, sailors—cannot tell what to do. And astronomers especially are at their wits' end. What happens is just the opposite of what they predict (B.15.355–69).

The weatherlore based on folklore, tradition, and experience, as well as the "science" of astronomers, has failed. Among other concerns, Langland is referring to the terrible weather of recent decades, with special reference to crop failures and bad harvests (B.15.362–67). The reworking of this passage in the C text reveals Langland still preoccupied some years later with the weather and still trying to explain it. Not surprisingly, the ultimate explanation is that it is a punishment for humankind's sinfulness (B.15.352–55; C.17.85–87). That explanation holds out some promise of control. But both texts convey a sense of bewilderment: to blame human failure seems not completely satisfying. Indeed, in C, Langland feels obliged to deny other possible explanations, as though they had occurred to him. Where's the fault? he asks. Not God's: he is good. So is the land. The sea and the seed, the sun and the moon, do their duty, day and night. If we did ours, there would be perpetual peace and plenty (C.17.88–93). And note that the discussion of the weather ends in puzzlement in C as in B, with astronomers in despair and farm laborers ("follwares") and shipmen unable to choose one course rather than another (C.17.103–6).

This is not irritation over gaffes in the daily weather report or the long-range predictions of Accu-Weather. Here is fear and uncertainty about weather patterns that have already brought great suffering and threaten survival itself. For the bad weather had been responsible not only for "bad years," but for the third source of hunger, that ultimate disaster, famine itself.

Famine

If hunger is a disease, David Arnold remarks, famine is an epidemic.[15] There had been the terrible famine of 1316–17,[16] in which mortality was so high "that, but for the Black Death itself, these years might well have their mark in historical records and popular memory as the years of highest mortality in the Middle Ages."[17] And in Langland's lifetime, three of the "bad years" probably qualify as famine years. There was 1353, the "dere somer" (*Brut*, 304). (Famine years were often fixed in popular memory by names.)[18] There was 1362 (*Brut*, 314). And there was 1369–70 (*Brut*, 321), which Haukyn alludes to so sardonically: "I believe all London likes my cakes [bread]," he says, "and looks sullen when they can't get them. It's not so long ago that the people were troubled when no bread carts came to London from Stratford. Beggars wept, and laborers were a bit aghast. This will be long remembered: it was on a dry day in April, the year of our Lord 1370, when Chichester was mayor, that my cakes were in short supply" (B.13.263–70).

One thing seems certain: Langland was familiar with chronic hunger and near-starvation, quite possibly intimately familiar. In the final passus of the A-text, the narrator refers to a time of great hunger in his youth when he met Hunger in person, who had set out, he says, to kill Life—a chilling mission. Faint from lack of food himself, the narrator is given coarse bread, eats it too ravenously, and suffers great pain and a swelling of the gut (A.12.59–76). And in the Half-Acre sequence, the violent scene where Hunger attacks Wastour and the Breton (A.7.159–70; B.6.174–80; C.8.171–78) is a vivid expression of the suffering starvation causes, the caution that must be observed in feeding a victim of starvation (A.7.168–69 and Z.7.165–67 only), and the imminence of death in time of famine and dearth. He knows too about hunger edema ("the bollynge of hire wombe" [B.6.179–80; cf. A.7.200–201; C.8.226]). He knew well the terror and horror of hunger and starvation. This horror and terror are, in fact, the ultimate weapon by which he seeks to energize the agricultural economy of his time so that the constant threat of hunger and starvation hanging over his generation can be warded off and humankind can survive. The agricultural laborer is, in his view, the key to the problem: "I warn you workers," he says as the Ploughing Scene draws to a close, "earn your wages while you can, for Hunger is hurrying here. Heavy rains will rouse him to punish wastrels. Famine will be here within five years, floods and foul weather will cause crops to fail" (B.6.320–24). And his last grim words in the scene are these: "Death will stand aside, and Famine will sit in judgment. And Dawe the ditchdigger will die of hunger—unless God in his goodness grants us a truce" (B.6.328–30).[19] The great fear is that "derthe" (famine), crueller than "deeth" (death), will replace it as God's stern justice. The wordplay "deeth-derthe" makes the two one and the same: scarcity, famine, *is* death itself.

But what is famine? We can give a dictionary definition, of course, but that is experientially empty and lifeless. We need to work toward a sense of the grim reality of famine, its ramifications, its impact on people as it runs its course, and, also important, its aftermath. In recent years, partly because of famine in the Third World, demographers and historians have begun to study famine in depth. Famine is a "historical" event that can, sad to say, be as it were replayed and observed. While no two famines are alike in every respect, they do possess a sufficient number of common elements for some generalizations to be made with confidence. Readers of *Piers Plowman* would do well to take advantage of such studies.

What emerges beyond doubt is the enormity of famine's impact. A country-wide famine is a devastating event, a terrifying social catastrophe. Says David Arnold:

> The horror of famine was compounded of many things—having to subsist by scavenging and begging, being reduced to eating "unclean things" [note Hunger's heartless suggestion for Wastour and the rebellious Bretoner, "Lat hem ete with hogges" (B.6.181)], being abandoned by family and friends, dying alone, away from home and uncomforted, being devoured unburied by vultures and wolves. The conventions, the constraints, the securities of everyday life one by one dropped away or were brutally inverted. Men behaved like animals; dogs gnawed at human corpses. The ultimate descent, difficult though it is for us to comprehend it or perhaps even read about without incredulity, was to cannibalism.[20]

We should read some of the horrifying accounts of famines—such as the 1846 Potato Famine in Ireland, or the 1942 Honan famine in China, in which an estimated two to three million people died and as many more abandoned their homes in search of food. Until we do so, I do not believe we can understand Langland's world.[21]

The extended report on the 1315–17 famine in England in the *Annales Johannis de Trokelowe* will, however, serve well as an introduction.[22] It corroborates many of the generalizations formulated concerning famine today. And since famines are long remembered among the people, it gives a glimpse of the nightmare that the famines in Langland's own day revived or realized.

Trokelowe begins by reporting the scarcity of meat, eggs, dairy products, and fowl and the sharp increase in the price of wheat, malt, and oats. As the crisis deepened because rain prevented the crops from ripening, the bread deteriorated in nutritional value so that even large quantities left one hungry. The powerful families (*magnates*) and the monastic communities began to cut down on their household staffs and withdrew their customary charitable support of the poor (*eleemosynes* [sic] *subtrahebant*). People began to turn to thievery. Jeremiah's portrait of famine in his land (14: 18) seemed to be coming true for England. People weakened by hunger came into the city. The poor and the sick, overcome by famine, could be seen lying neglected and dead in the villages and along the roads.

The famine became more severe in 1316 and the death rate was even higher; so many died there were scarcely enough living to bury them. There was dysentery, caused by spoiled food; it was accompanied by a severe rash and high fever (typhus? smallpox?). Cattle and swine sickened from rotted hay and dropped dead.[23] No one could have foreseen such dearth and famine or such mortality as followed. Physicians could find no fitting remedy for the sickness in this pestilence as they had been accustomed to do in the past, for medicinal herbs, because of the bad weather, instead of providing relief to the sick, brought poison rather than strength.

Four pennies' worth of coarse bread would not suffice for an ordinary person. Customary meats were very scarce, but horse meat was a delicacy to those who were stealing fat dogs. Many claimed that in many places both husbands and wives secretly ate their own and others' children. Prisoners devoured newly arriving thieves at once, half-alive (92–98).

With a fuller awareness of the eventful history of hunger in Langland's day, we can try to judge the impact of this history on *Piers Plowman*. One result, obviously, is the scene with Hunger. The experiential significance of the scene, its overtones and associations, should be clearer now. Though Piers summons Hunger to drive the rebellious idlers to work, the summons is an unnecessary fiction. Hunger would have come uncalled for, did come regularly, had come often and stayed much too long. Piers's summons is only a reminder of the omnipresent fact of Hunger.

And the scene tells of real hunger, violent, gut-gripping hunger that leaves sunken cheeks and rheumy eyes and brings the sufferer close to death (B.6.171–80). It inspires a reaction bordering on terror and creates a frenzy of action among faking cripples, hermits, malingerers, beggars, and—note—plain poor men (B.6.183–96). The lean rations, coarse breads, and scavenged greens of the "hungry gap" (B.6.278–98) are familiar and real. And over the scene hangs the threat that one of these days Dawe the Dykere may die "for hunger" (B.6.329). The very word "hunger," let along its personified presence, would be frightening in Langland's world. "Even the words 'hunger,' 'drought,' or 'famine' might themselves be enough decades after the event, to spark a train of bitter recall."[24] One senses this in the words "hunger and derthe" in the poem of complaint "Whii werre" (*The Simonie*), written at least ten years after the famine of 1315–17.[25] For these reasons I cannot accept R. E. Kaske's learned argument that "a hunger for righteousness" is the issue in the Hunger scene. The terrible reality evoked by the word and the personification would leave no room for any other, much less such a rarefied meaning.[26]

The Hunger scene is also an eating scene, and is not the only important moment where eating is a central action. There is also the scene with Patience and the Doctor of Divinity (B.13.21–175). The two scenes suggest, I believe, the impact of hunger on Langland's imagination. There is nothing carnivalesque, nothing Rabelaisian or Bakhtinian, about the eating in either. There is rather a joyless voraciousness, the feeding of the half-starved or the greedy, not the happy. Langland is almost uneasy about eating. There is also a surprising volume of references to eating in the poem and a rich imagery based on eating, hunger, and food.[27] I suggest it needs separate study. It is found in some of Langland's most striking comments: Not everything's good for the soul that one's gut asks for (B.1.36); some chaplains are chaste, but charity they chew up and complain they need more (B.1.193); and, he reports, Lechery, repentant, vowed to drink only with the duck and dine but once a day (B.5.74). The robust figures of Haukyn in B and Activa Vita in C are wafer sellers, and though they embrace many other forms of involvement with the working world, we see them most clearly as bringers of food to the streets of the city. Masticating humanity is never long outside the range of Langland's imaginative vision.

More important is the relation of hunger and famine to several social/ moral issues in the poem. With the growth of towns in the thirteenth and fourteenth centuries a new population came on the scene: the urban poor. Throughout this period, pauperism was on the rise.[28] Ironically, famine made a sizable contribution in England to the population of wandering, rootless, unemployed poor that Langland deplores, fears, anguishes over. This is an inevitable consequence of famine. When there is no food in the home territory, people emigrate to towns, cities, seaports, wherever they hope to find food.[29] Trokelowe has told us how in 1315–16 large households, baronial and monastic, tightened their belts by cutting their staffs. As crops failed, agricultural laborers were let go and joined the force of vagrants. Artisans and workers in towns were released as the economy shrank. And the poorer peasants and their families, their food stores exhausted, took to the road as well. The disruptions of the plagues swelled their ranks, of course, but the local famines in the 1350s and 1360s added a constant stream. Begging and stealing became ways of life, of survival. Capgrave, writing of 1315–16, says some died of hunger; those that were dismissed from the manor house went out into the countryside and robbed the poor (141). This was happening in the famine years in Langland's day as well.

For Langland these poor were a deeply disturbing presence. They were telling evidence that all was not right with his world, morally and economically. He and his contemporaries rarely managed to disentangle economics from morality. We can see the entanglement in his term "wastours." Langland was all too aware of the uneasy balance of the food supply, constantly teetering on the edge of extended scarcity and famine. The unemployed poor were a drain, needing to be fed but contributing nothing. "You are wasters," Piers says, "you waste what men win by hard work and suffering" (B.6.130, 133). Langland's answer is to put them to work, using hunger as the spur. He is indignant at the demands for higher wages made by laborers after the plague, and he confuses this group with those dislocated by the plagues and by famine—the displaced peasants, broken families, released demesne workers.

The crisis, however, was of a magnitude and complexity beyond his— or anyone's—ability to understand. Recent studies have come to see famines as more than "accidents," more than events that occur and then vanish. A famine is both "event" and "structure." Famine "casts a hard and clear light on the nature and problems of the societies it afflicts and of the world in which they exist. It has been and remains at the centre of the great turning points in world history, in social and economic structures and in human relations," writes R. I. Moore.[30] And discussing famine as "structure," David Arnold asserts that famine is rarely "a purely chance misfortune." "More commonly . . . famine acts as a revealing commentary upon a society's deeper and more enduring difficulties." It "bring[s] to the fore a society's inner contradictions and inherent weaknesses."[31] Langland's world was changing; the process was painful and was prolonged over several centuries. His society was indeed plagued with difficulties. The famines of the 1350s and 1360s helped reveal some of these. Langland saw them without understanding them, except as the working out of a baffling divine plan. But it is to his credit that he did not blink at them.

One thing he saw was that the term "wastours" would not cover all the poor and dislocated. Derek Pearsall has analyzed with great sensitivity Langland's struggle to discriminate between the deserving and the undeserving poor and his continuing "scrupulous concern" for the undeserving poor.[32] But the excessive suffering of the guiltless poor was inescapable and an even greater moral burden to him. Though plague is relatively sparing of the rich and privileged, it is democratic in the swinging of its scythe compared with famine, which displays a ravenous appetite for the poor and underprivileged.[33] "Starvation," a modern economist observes, "is the characteristic of some people not *having* enough to eat. It is not the characteristic of there *being* not enough to eat."[34] "The critical issue is the way in which food is distributed rather than the volume of food itself."[35] Langland views the moral implications of famine in similar terms, reacting strongly against the immunity of the rich and the vulnerability of the poor to suffering in time of scarcity. A great injustice of his society was revealed to him by famine, as was the failure of charity to cope adequately. This injustice and this failure fired his conscience.

It did more. Conservative though he was, Langland, in the light cast by hunger, was led to half-question certain orthodoxies of his time. Though he never quite makes the comparison, he is aware that the clergy contribute no more than the "wastours" to the food supply and yet must be fed. Piers reviews this arrangement as though it is open to question and does not agree to feed all who come under this title, "clergy." He will help those attached to cloisters or churches or those whom bishops have licensed to preach. But he will not support wandering, unlicensed preachers—"Robert Renabout" and "postles." Most interesting, he will also help anchorites and hermits who eat only once each day and no more till the next day—but the language is grudging (B.6.145–50). And we see, when the threat of hunger activates the community idlers, a "heep of heremytes" grab spades, cut their copes short so their arms can move freely, and set to work digging frantically (B.6.187–90). Piers is hedging ever so slightly in his view of one of the estates.

Hunger leads him to hedge slightly with another. The knightly class does not shine. It produces nothing, but Piers attempts to get some useful work from it. He suggests that some of the fine ladies who can only wield the needle should sew sacks for holding wheat—a likely prospect! And the knight, who would help but does not know how to plough, is requested to put his talents for hunting to good use ridding the land of the animals that eat crops or poultry or break down Piers's hedges (B.6.9, 21–32). When the laborers on the half-acre go on strike, Piers asks the knight to get them working again, but the knight is shown to be a marvel of ineptitude, a Bertie Wooster aristocrat, and the laborers defy him rudely and with impunity (B.6.159–69). There is no direct criticism, and Langland may not quite know what he has done. But a social myth comes out looking poorly.

Finally, it is difficult to believe that Langland's world of hunger and famine had nothing to do with larger matters, with his doctrine of work, his doctrine of patient poverty, even his selection of a ploughman as his society's salvation. I would not presume to say that this fourteenth-century world of hunger and famine explains these matters completely, nor would I wish to ignore the biblical texts cited as sources.[36] On the other hand, I find it impossible to ignore the daily lessons of his time, or to believe that Langland ignored them. Food was scarce for the lower levels of society. The only known way to remedy this was to work the land—and to endure patiently. If one did not do the first, one would, many would, die. If one did not do the second, one despaired and was damned.

And the constant threat of scarcity and famine must have made unmistakable to Langland the absolutely essential role for society to survive: that of the despised, mocked, abused, exploited worker in the muck and dung of the field, the ploughman. In other words, it was a time when diurnal event and sempiternal text coalesced.

After the Half-Acre scene the poem goes on to higher concerns and more abstract issues. But we must not forget that the poem owes much of its power to Langland's passionate reaction to the people and the problems of his age. *Piers Plowman*, George Kane observes, "is a living text with a content of direct concern."[37] One of those concerns was famine and its ramifications: Piers's concern—and Langland's—"if the lond faille."

This paper originally appeared in the *Yearbook of Langland Studies* 4 (1990): 86–104, copyright 1990 by Colleagues Press Inc., by whose permission it is reprinted here with slight alterations and updating (see esp. notes 9 and 25). Middle English quotations from *Piers Plowman* and other texts in the original article have been replaced here by summary or paraphrase for the convenience of the reader, and the passus ("book") and line numbers follow the passage.

Notes

1. Megan Vaughan, *The Story of an African Famine: Gender and Famine in Twentieth-Century Malawi* (Cambridge: Cambridge University Press, 1987), p. 1.

2. The texts used here will be for A: George Kane, ed., *Piers Plowman, the A Version: Will's Visions of Piers Plowman and Do-Well* (London: Athlone Press, 1960); for B: A. V. C. Schmidt, ed., *William Langland, the Vision of Piers Plowman: A Complete Edition of the B-Text* (London: J. M. Dent & Sons, 1978); and for C: Derek Pearsall, ed., *Piers Plowman by William Langland: An Edition of the C-Text*, York Medieval Texts, 2nd series (London: E. Arnold, 1978; Berkeley: University of California Press, 1979). The text concentrated on will be B (ca. 1379), with some reference to A (ca. 1370) and C (ca. 1385–86). Some scholars have argued in recent years that the incomplete text of A in MS Bodley 851 is a separate early sketch or version of the poem preceding even A. See William Langland, *Piers Plowman: The Z Version*, ed. A. G. Rigg and Charlotte Brewer (Toronto: Pontifical Institute of Mediaeval Studies, 1983), pp. 1–36, for a statement of the case followed by the text in MS Bodley 851, labeled "Z." I shall refer to Z on one occasion below.

3. Wilhelm Abel, *Agricultural Fluctuations in Europe, from the Thirteenth to the Twentieth Centuries*, trans. Olive Ordish (New York: St. Martin's Press, 1980), pp. 35–93; John Hatcher, *Plague, Population and the English Economy: 1348–1530*, Studies in Economic and Social History, Prepared for the Economic History Society (London: Macmillan, 1977), passim; Ian Kershaw, "The Great Famine and Agrarian Crisis in England 1315–1322," in *Peasants, Knights and Heretics: Studies in Medieval English Social History*, ed. R. H. Hilton (Cambridge: Cambridge University Press, 1976), pp. 85–132; Georges Duby, *Rural Economy and Country Life in the Medieval West*, trans. Cynthia Postan (Columbia: University of South Carolina Press, 1968), pp. 289–357; M. M. Postan, *Essays on Medieval Agriculture and General Problems of the Medieval Economy* (Cambridge: Cambridge University Press, 1973), pp. 41–48, 186–213; Ronald E. Seavoy, *Famine in Peasant Societies* (New York: Greenwood Press, 1986), pp. 70–75.

4. Kershaw, "The Great Famine," p. 93; M. M. Postan and J. Z. Titow, "Heriots and Prices on Winchester Manors," in Postan, *Essays on Medieval Agriculture*, pp. 150–85, esp. 169, 174a.

5. David Arnold, *Famine: Social Crisis and Historical Change* (Oxford: Basil Blackwell, 1988), p. 16.

6. Arnold, *Famine*, pp. 54–55. Seavoy advances the controversial theory that hunger and famine are endemic in peasant societies because peasants are governed by what he calls "the indolence ethic"; they seek to do only as much work as they believe will provide subsistence in a normal year. See his general statement (pp. 7–42), followed by an analysis from this position of "English Peasant Society from the Domesday Survey to the Mid-Eighteenth Century," and of Indonesian subsistence culture, the Indian famine of 1876–79, and the Irish Potato Famine, and

a concluding commentary. This is a stimulating study with much cogent analysis, but it has been sharply challenged by Arnold, *Famine*, (pp. 57–59). Seavoy's thesis is not germane to the discussion here and I shall not invoke it.

7. The time of year is a bit uncertain. Piers's reference to his "cart mare / To drawe afeld my donge while the droghte lasteth" (B.6.288–89) suggests early spring. The "droghte" is not prolonged drought but Chaucer's "droghte of March," that dry period in early spring when the ground is best for plowing and manuring in preparation for sowing: see A. Stuart Daley, "Chaucer's 'Droghte of March' in Medieval Farm Lore," *Chaucer Review* 4 (1970): 171–79. Though this seems a bit early for the hungry gap, it might not be if the preceding harvest had been a poor one. In any event, the period stretches through summer to the new harvest, Lammastide.

8. Bridget Ann Henisch, *Fast and Feast: Food in Medieval Society* (University Park: Penn State Press, 1976), pp. 79, 155–61. H. E. Hallam estimates that the English peasant ate between four and five pounds of bread a day, bread made of barley or maslin (a mixture of grain, usually wheat and rye; also called maincorn). It was the main source of calories, approximately 5,000 a day, and protein: "The Worker's Diet," in *The Agrarian History of England and Wales* 2: *1042–1350*, ed. H. E. Hallam (Cambridge: Cambridge University Press, 1988), pp. 825–45. See also H. E. Hallam, *Rural England, 1066–1348*, Fontana History of England (Glasgow: Fontana Paperbacks, 1981), pp. 65–67. (The unacceptability of peas and beans in English peasant diet that we see in *Piers Plowman* is a separate problem and needs investigation.)

9. J. Z. Titow, "Evidence of Weather in the Account Rolls of the Bishopric of Winchester, 1209–1350," *Economic History Review*, 2nd ser. 12 (1960): 360–407, esp. 360–63. The bishop of Winchester's holdings were scattered throughout several counties: Berkshire, Buckinghamshire, Hampshire, the Isle of Wight, Oxfordshire, Somerset, Surrey, and Wiltshire. For a map showing the distribution of the estates with an itemized list, see J. Z. Titow, *Winchester Yields: A Study in Medieval Agricultural Productivity* (Cambridge: Cambridge University Press, 1972), pp. 38–39.

10. J. Z. Titow, "Le climat à travers les rôles de compatabilité de l'évêché de Winchester (1350–1450)," *Annales: Économies, Sociétés, Civilisations* 25 (1970): 312–50, esp. 314–24.

11. Titow, "Climat," graph, p. 342[a–b].

12. Hatcher, *Plague, Population and the English Economy*, p. 51; Abel, *Agricultural Fluctuations*, p. 44, fig. 7; A. F. Butcher, "English Urban Society and the Revolt of 1381," in *The English Rising of 1381*, ed. R. H. Hilton and T. H. Aston (Cambridge: Cambridge University Press, 1984), pp. 84–111, esp. 100.

13. *The Brut, or The Chronicles of England*, ed. F. W. D. Brie, vol. 2, Early English Text Society, orig. ser. 136 (London: Kegan Paul, 1908), p. 304. For "dere somer" see MED, *dere* adj. (1), 3. (c) [*dere*] *sesoun* [*dere*] *time*, "a period of scarcity or high prices"; [*dere*] *somer*, "a summer of scarcity"; [*dere*] *yer*, "a scarcity, dearth." See also John Capgrave, *John Capgrave's Abbreviacion of Cronicles*, ed. Peter J. Lucas, Early English Text Society 285 (Oxford: Oxford University Press, 1983), p. 168.

Capgrave may have the wrong year. His description of the drought in 1353 (the twenty-seventh year of Edward III) parallels the *Brut*'s account for 1354 in details and phrasing. The chroniclers are not reliable on dates, especially for weather. The sometimes conflicting dates given by C. E. Britton for weather in this period (1338–70) based on reports in the chronicles, and his discussion of specific disagreements make this clear. Britton, *A Meteorological Chronology to* A.D. *1450*, Meteorological Office, Geophysical Memoirs 70 (London: HMSO, 1937): 3–5, 139–46. For cautionary articles on the use of chronicles for information on weather in the Middle Ages, see W. T. Bell and A. E. J. Ogilvie, "Weather Compilations as a Source of Data for the Reconstruction of European Climate during the Medieval Period," *Climatic Change* 1 (1978): 331–48; and P. Alexandre, "Histoire du climat et sources narratives du moyen âge," *Moyen Age* 80 (1974): 101–16. Emmanuel Le Roy Ladurie's *Times of Feast, Times of Famine: A History of Climate since the Year 1000*, trans. Barbara Bray (New York: Doubleday, 1971; reprint New York: Farrar, Straus and Giroux, 1988), is a pioneer effort but is somewhat superseded by the recent work of paleoclimatologists such as H. H. Lamb. Also, Ladurie's findings deal primarily with the Continent, not England. There is an expert discussion of general weather patterns from roughly 900 on in H. H. Lamb, "The Early Medieval Warm Epoch and Its Sequel," *Paleogeography, Paleoclimatology, Paleoecology* 1 (1965): 13–37, and in Lamb, "An Approach to the Study of the Development of Climate and Its Impact on Human Affairs," in *Climate and History: Studies in Past Climates and Their Impact on Man*, ed. T. M. L. Wigley, M. J. Ingram, and G. Farmer (Cambridge: Cambridge University Press, 1981), pp. 291–309. esp. 301–5, with a useful bibliography, pp. 307–9. Exact dates are not critical for the discussion here, only the fact of bad weather, bad harvests, and crop failures within the indicated span of time.

14. Ancel Keys et al., *The Biology of Human Starvation*, 2 vols. (Minneapolis: University of Minnesota Press, 1950) 2: 1250, list famines in England in Appendix IV, "Some Notable Famines in History": 1314, 1316, 1321, 1335, 1341, 1353, 1355, 1358, 1369, 1390. The list is not documented. Seavoy, *Famine in Peasant Societies* (pp. 72–74), mentions as famine years 1315–16 and 1321–22, and as "local famines" 1353 and 1369–70.

15. Arnold, *Famine*, pp. 6–7.

16. Kershaw, "The Great Famine," pp. 95–98.

17. Postan and Titow, "Heriots and Prices," p. 169.

18. Arnold, *Famine*, pp. 12–13.

19. See Bennett's note on B.6.324 ff.: J. A. W. Bennett, ed., *Langland's Piers Plowman: The Prologue and Passus I–VII of the B Text* (Oxford: Clarendon, 1972), pp. 214–15; and Pearsall's on C.8.343–54: *C-Text*, p. 160. But if, as Bennett suggests, the year 1362 is alluded to, it is incorrect to say "no convincing reason for an allusion [to that year] suggests itself," for that was a famine year, though a year already past.

20. Arnold, *Famine*, p. 19.

21. The descriptions and discussions of famines in Ireland, China, India, Ethiopia, and elsewhere in the last century and a half provide a useful introduction to the reality and the tragedy of famine. See Arnold, *Famine*, pp. 17–18; W. R. Aykroyd, *The Conquest of Famine* (London: Chatto & Windus, 1974), pp. 30–103; Cecil Woodham-Smith, *The Great Hunger: Ireland, 1845–1849* (New York: Harper & Row, 1962), passim; Vaughan, *Story of an African Famine*, pp. 21–49; Amartya Sen,

Poverty and Famine: An Essay on Entitlement and Deprivation (Oxford: Clarendon, 1981), pp. 52–153 (much analysis, some description); Seavoy, *Famine in Peasant Societies*, pp. 127–345 (largely analysis from his theoretical perspective). Sen lists "some absorbing accounts of the phenomenon of famine in different parts of the world and some comparative analysis" (p. 39, n.1). Though Keys's two-volume laboratory study of human starvation is very technical, the chapter on "Behavior and Complaint in Natural Starvation" contains interesting material (*Biology of Human Starvation* 2: 783–818). Useful also is Pitirim A. Sorokin, *Hunger as a Factor in Human Affairs*, trans. Elena P. Sorokin and ed. T. Lynn Smith (Gainesville: University Presses of Florida, 1975).

The literature of famine is, as one might imagine, enormous; the references here merely scratch the surface. But they offer a helpful introduction to the subject, and many of them have bibliographies invaluable for anyone who wishes to pursue the subject in greater depth.

22. *Annales Johannis de Trokelowe*, in *Johannis de Trokelowe et Henrici de Blaneforde . . . Chronica et Annales*, ed. Henry Thomas Riley, Rolls Series (London: HMSO, 1866), pp. 69–127.

23. On the destruction of livestock in these years, see Seavoy, *Famine in Peasant Societies*, p. 72.

24. Arnold, *Famine*, p. 13. In the hurried, tacked-on conclusion to the earliest version of the poem, the A-text (ca. 1370), the personification Hunger reappears as an agent of death, together with Fever (A.12.63–87). Hunger dwells with Death and is in fact on his way to kill Life (63–66). Hunger's role as a "killer" underlines the terrifying associations of the word and of the experience it points to. And the description of the half-starved dreamer's physiological reaction after eating (A.12.67–71) is evidence of the narrator's familiarity with the phenomenon of near-starvation, whoever he was. (Langland was certainly not the author of the final lines of A.12, lines 99–117. He was probably but not certainly the author of the first ninety-eight lines. See Kane, ed., *A Text*, pp. 51–52. For a recent discussion, see Anne Middleton, "Making a Good End: John But as a Reader of *Piers Plowman*," in *Medieval English Studies Presented to George Kane*, ed. Edward Donald Kennedy, Ronald Waldron, and Joseph S. Wittig (Wolfeboro, N.H.: Boydell & Brewer, 1988), pp. 243–66.

25. "Whii werre" (*The Simonie*), in *The Political Songs of England, from the Reign of John to That of Edward II*, ed. Thomas Wright, Camden Society 6 (London, 1839): 323–45.

26. R. E. Kaske, "The Character Hunger in *Piers Plowman*," in *Medieval English Studies*, ed. Kennedy, Waldron, and Wittig, pp. 187–98.

27. Here is a list of references to eating in the B-text. Not all references may have been noticed, and not all will be accepted. But it is offered as a start: (Pr.): 30, 40–43, 59, 190, 226–30; (1): 24, 36–37, 153–54, 193; (2): 96–97, 181; (3): 78–84, 194; (5): 74, 92, 119, 121, 153, 156, 171, 175, 187, 191, 286, 372–85, 390, 434–39, 493, 494, 603; (6): 19, 53, 69, 135–37, 145, 157–58, 172–301—Hunger (specifically 179–82, 186, 193–95, 198, 206, 213–17, 224, 251, 258–67, 279, 280–86, 291–301, 303–11, 320); (7): 30, 83–84, 119, 121, 122–24, 142; (9) 60–61a, 69, 80–81, 152; (10): 56–57, 58–59, 83, 94, 98–101, 359; (11): 112–14, 189–95, 216, 277; (12): 198–201, 264; (13): 23–110, 124, 216–19,

399–404, 422–24, 437, 443; (14): 10, 29–30, 38, 50, 56–69, 74–80, 134, 156–61, 177, 250; (15): 178–80, 205–7, 217, 255, 271, 276–317a, 341–42, 399–409, 413, 427a, 430–32, 439–40, 460–63, 467–77, 480–83, 573–74, 588–90; (16): 9, 10–11, 125–26, 228–29, 244; (17): 266–67; (18): 194–95, 200–201, 286–88, 332; (19): 108–9, 115, 126–27, 285, 386–93, 410–11, 417–18; (20): 3, 4–15.

Both Jill Mann, "Eating and Drinking in 'Piers Plowman,'" *Essays and Studies*, n.s. 32 (1979): 26–43, and A. V. C. Schmidt, "Langland's Structural Imagery," *Essays in Criticism* 30 (1980): 311–25, have discussed food references in *Piers Plowman*, but their emphasis is on the metaphysical, not the literal.

28. Michel Mollat, *The Poor in the Middle Ages: An Essay in Social History*, trans. Arthur Goldhammer (New Haven, Conn.: Yale University Press, 1986), pp. 158–90, 193–250.

29. Sorokin, *Hunger*, pp. 158–59; Vaughan, *Story of an African Famine*, pp. 33–34, 45; Arnold, *Famine*, pp. 91–95.

30. Quoted in "Introduction," Arnold, *Famine*, p. viii.

31. Arnold, *Famine*, p. 7. For more extended statements on famine as a primarily social phenomenon see Frances D'Souza, "Social Security and an Analysis of Vulnerability," and J. P. W. Rivers, "The Nutritional Biology of Famine," in *Famine*, ed. G. Ainsworth Harrison (Oxford: Oxford University Press, 1988), pp. 1–56, 57–59.

32. Derek Pearsall, "Poverty and Poor People in *Piers Plowman*," in *Medieval English Studies*, ed. Kennedy, Waldron, and Wittig, pp. 167–85, esp. 175.

33. Mollat, *The Poor in the Middle Ages*, pp. 158–62; Sen, *Poverty and Famine*, passim.

34. Sen, *Poverty and Famine*, p. 1.

35. Arnold, *Famine*, p. 43.

36. Konrad Burdach, *Der Dichter des Ackermann aus Böhmen und seine Zeit* (Berlin: Weidmann, 1926–32), pp. 282, 294, 297, 341, 358–59; Stephen A. Barney, "The Plowshare of the Tongue: The Progress of a Symbol from the Bible to *Piers Plowman*," *Mediaeval Studies* 35 (1973): 261–93.

37. Kane, ed., *A Text*, p. 115.

PART IV

Artistic Representations

Michael Camille

12. "When Adam Delved": Laboring on the Land in English Medieval Art

In that spurious "sequel" to the *Canterbury Tales*, *The Tale of Beryn*, which purports to relate the subsequent adventures of the pilgrims, is a comic passage describing the efforts of the Miller and the Pardoner, after finally arriving at the great shrine of Saint Thomas, to interpret the stained glass in the clerestory windows of Canterbury Cathedral. These "lewd sotes . . . like lewde gotes" make some wild guesses at the identity of a series of figures representing the ancestors of Christ. These had been designed in the late twelfth century and were now far above them, both spatially and cognitively. The Pardoner and the Miller argue whether the attribute of one is a weapon or an agricultural tool:

> "He bereth a balstaff" quod the toon, "& els a rakis ende."
> "Thou faillist quod the Miller "þow hast nat wel þy mynde."[1]
>
> [He's carrying a quarterstaff," said the one, "or else a rake handle."
> You're slipping," said the Miller, "you're losing your mind."]

While this satire on the illiterate's misperception of religious art had been commonplace for centuries, it should be a warning to all would-be iconographers, suggesting that the medievals well understood how visual interpretation is dependent upon the various expectations of beholders.

One of the best preserved panels of glass from this clerestory series, which begins the sequence of Old Testament figures at the northwest extremity of the choir and which during the Middle Ages would not, I imagine, have presented such a puzzle to peasant perceptions, even if they were unable to read the four clear letters spelling out his name, depicts Adam delving (Figure 30). Here the figure's attribute is undebatably agricultural. Adam's powerful figure, emphasizing with unusual clarity the muscular

Figure 30. Adam delving. Stained glass panel, Canterbury
Cathedral, north choir clerestory.

effort necessary to penetrate the earth, is dated ca. 1178–80. Art historians have discussed it as a "high point" in the "classical" trend in figurative art in the late twelfth century and traced its iconographical genealogy back to early Christian prototypes of Genesis illustration available in Canterbury.[2] In what follows I am more interested in looking at it like a "lewde gote" rather than as a connoisseur or Christian iconographer and in exploring what this image can tell us about the tools and techniques and perceptions of human labor in the Middle Ages.

> Cursed is the ground for thy sake; in toil shalt thou eat of it all the days of thy life; thorns also and thistles shall it bring forth to thee; and thou shalt eat of the herb of the field; in the sweat of thy face shalt thou eat bread, till thou return unto the ground. (Genesis 3: 17)

These often repeated words of the village preacher to his Sunday flock would be recalled by those pilgrims looking up at this massive figure of human strain and effort. In Genesis labor was Adam's curse, just as Eve's was subservience and childbirth, God's pronouncement being that he should "till the ground from whence he was taken" (*operaratur terram de qua sumptus est*: Genesis 3: 23). This earth, to which he must also now return in death, is as carefully visualized in the glass as the figure of the worker—its different strata of turned earth and subsoil. In this primeval post-lapsarian cycle, the earth is both the origin and the end of human life. The painful separation from the paternal Father and from the ease of Eden is indicated by the bare foot of the half-naked figure, clad only in a sheep-skin loincloth, who pushes with his body weight against the hard surface of the green earth. This tool, an extension of Adam that allows him to turn over the small strip of soil into semicircular clods, which are pictured in a band of lighter green in the window, the Miller or any other pilgrim of Chaucer's day would have recognized and named. They would have called a spade a spade. This object is constructed with some accuracy by the artist, using a lighter yellow glass for the wooden shaft and handle and a silvery blue for the metal shoe.

Adam's iron-shod tool is, strictly speaking, historically out of place at this point in the Biblical narrative since only later does one of Adam's descendants, Tubalcain, teach men the use of metals (Genesis 5: 22). Such an anachronism did not worry the designers of the window since their aim was to show Adam's archetypal labor as concretely as they could. Medieval artists and audiences represented and perceived Biblical stories and persons as happening in their own time and space. This is proved not only in the visual arts but also in the homey colloquialism of the English and French mystery plays where the story of the Fall is told in terms of domestic conflict and feudal treachery. "Why did you overstep my prohibition?" God asks Adam after discovering his transgression, in the twelfth-century French play, the *Ordo Representacionis Ade*. "You are my serf and I am your Lord!" (*Tu es mon serf e jo ton sire!*)³

That medieval artists represent Biblical subjects in terms not only of current ideology but also of visible technology can be shown by comparing the scene of Adam delving at Canterbury with its equivalent in contemporary Italian mosaic cycles at Monreale and the Palatine Chapel at Palermo (Figure 31). In these examples Adam's tool is the heavier mattock or axe.⁴ The difference in medium between the Palermo and Canterbury images— the arrangement of multiple mosaic tesserae as opposed to blown glass flattened into sheets and held together by leads—is related to divergent architectural traditions in northern and southern Europe, their contrasting needs and conditions linked to climate and geography. Similarly the difference in tools reflects the different agricultural conditions of hot, dry Sicily versus temperate, wet Kent.⁵ An axe does appear in the Canterbury window however, hanging prominently from the bough of the tree (Figure 30). This tool was crucial for grubbing and digging out roots and stumps. Early medieval examples and southern European twelfth-century depictions of the subject stress Adam's clearance of forest and undergrowth with this heavier tool or sometimes with a mattock. In two earlier prototypes for Adam delving, illustrated by English artists in the Anglo-Saxon versions of the Genesis story—Aelfric's Pentateuch (Figure 32) and the Caedmon Genesis (Figure 33)—both produced at Canterbury in the early eleventh century, the spade, not the axe or mattock, is the tool shown, despite the fact that scholars see these cycles as dependent on Italian prototypes.⁶ England was a country whose clearance for an agrarian economy stretched far back before Roman times and where the spade as a tool of cultivation rather than of clearance would be more relevant. Thus the Canterbury Adam, despite his "primitive" animal skin, is delving not into the untouched ecosystem outside Eden but into land that even in the twelfth century was among the oldest cultivated earth in Europe.

The word "spade" was by this period likewise layered with centuries of use. Derived from the Frisian/Anglo-Saxon *spada*, it describes a tool that had not changed since Roman times—a flattish end blade with a rounded or pointed extremity tipped with an iron shoe and held by a wooden handle with a grip or cross-piece at the upper end. These early depictions

Figure 31. Expulsion and labor of Adam and Eve. Mosaic, Palatine Chapel, Palermo, north wall.
Photo: University of Chicago, Epstein Archive.

are corroborated by a number of metal shoes that have been recovered and analyzed by archaeologists. The spade was distinguished from the shovel, which did not require a metal piece and was used for scooping and raising materials rather than cutting into the earth. Medieval shovels, sometimes with detachable wooden blades, are depicted in illuminated manuscripts in the context of building, particularly mixing mortar, and in specialized industries like glass-making.[7] The spade, however, is always shown in the context of its primary function of delving. The word "delve," also Germanic in origin, meant to dig with a spade (the French-derived "dig" appearing in English only later in the fourteenth century), and in some English counties to delve still means to go two spades deep into the soil.[8] In the fifteenth-century *Corpus Christi Play* a disconsolate Adam talking to Eve, uses both words to describe the couple's bleak future on the "land":

Figure 32. Expulsion and angel teaching Adam and Eve to delve. London, British Library Cotton MS Claudius B.IV, Pentateuch, fol. 7v. By permission of the British Library.

Figure 33. Adam and Cain delving. Caedmon Genesis. Oxford, Bodleian Library, MS Junius ii, p. 49. By permission of the Bodleian Library.

But let us walk forth into the land,
With right great labour our food to find,
With delving and digging with my hand,
Our bliss to bale and care to pind.[9]

An unusual feature of the Canterbury Adam is his putting his foot across the whole edge of the spade. Other depictions in manuscripts, glass, and sculpture as well as the archaeological evidence suggest that the spade usually had only a single ledge for foot pressure (Figure 33). This type also is displayed in the thirteenth-century Carrow Psalter, in the scene where Adam and Eve are given their respective implements by the angel (Figure 34). Notice the same means of depicting the soil by semicircular mounds as in the window in Figure 30; it evokes the never-ending labor that stretches before the disobedient couple and their descendants. The one-sided spade also is visible in a figure, close in all other respects to the pose of the Canterbury Adam, in an illumination of the labor of the month for March, in an eleventh-century calendar cycle also produced at Canterbury (Figure 35).[10] In these two examples the curved form of the nailed-on metal shoe is stressed. Thorold Rogers, in his still invaluable study of prices, found seven references to an object recorded by the Latin term *vanga* or *vanga ferrata* in the earlier part of the period 1259–1440, priced at about 2 ½d. each and four later references at about 5 ¾d. Sometimes they are listed separately, the wooden frame at one price and the iron frame to fit it at another.[11] All iron blades were rare and probably used only on furnace sites, where intense heat made them necessary. Like the ploughshare the *vanga ferrata* was another instance of the crucial role of the village blacksmith in the agrarian economy. Adam cannot delve without Tubalcain's invention. Likewise in medieval iconography the smith is a powerful figure of magic and strength, from Wayland the Smith on the eighth-century whalebone Franks Casket to Tubalcain at his forge as pictured in the margins of the Psalter of Queen Isabella.[12]

In the seasonal labors of the year the spade was associated with winter, as in calendar illustrations for February and March (Figure 35). At this time there was much ditching and digging to be done in preparation for sowing the spring grains of oats and barley. Drainage ditches had to be dug in the fields, even during harvest in a wet season; in addition, peat turves and marl pits had to be dug.[13] In different areas of the country this labor would have had different associations. In the fens, for example, where turving was crucial to the economy, an experienced "digger" could move 2,000 peat

Figure 34. Adam and Eve being given their tools by an angel. Carrow Psalter. Baltimore, Walters Art Gallery, MS 34, fol. 22v. By permission of the Walters Art Gallery.

Figure 35. Soil culture, digging, raking, and sowing. Calendar page for March. London, British Library, Cotton MS Tiberius B.CV, fol. 4r. By permission of the British Library.

blocks per day and was widely respected, although most men were eventually crippled by these working conditions. This least-liked activity was also the first of the real labors of the months after the winter respite according to the popular rhyme:

Januar   By thys fyre I warme my handys;
Februar   And with my spade I delfe my landys.[14]

This association with the winter makes Adam's near-naked delving depicted in the Canterbury window an even more pitiable sight. While we are wont to look on this muscular figure as a heroic type, his straining, bony body would have brought a shudder to any onlooker who knew the feel of fields in February.

In contrast to the delving manual worker, the image of the ploughman, such as that in the fourteenth-century *Luttrell Psalter*, is, as I have discussed elsewhere, an image of peasant power and status. The technological and economic advantage that the plough gave its owner can be compared with that provided by the early harvesting machinery to early twentieth-century farmers. In the *Rentier de Messire Jehan de Pamele-Audenarde*, a rare example of an illustrated manorial document, which was made for a late thirteenth-century French lord, the ploughman with his horse-drawn plough is juxtaposed with an image of *vi fourkes* on the same folio (Figure 36). These men are referred to in the text below as *manouvriers*—manual workers of lesser status and wages than the *arator* with his elaborate contraption. The plough was not only an expensive item (valued on one estate in 1341 at 12s. 10d., not counting the oxen, compared with a few pence for the spade), it had elevated symbolic associations. It indicated feudal service performed by tenants, and in everything from Aelfric's *Colloquy* to *Piers Plowman* it served as a symbol of the productivity of the peasant in an ordered "good society."[15]

The spade's associations were more mundane, literally, in that they suggested *mundus*, the earth. What it did was less spectacular but perhaps even more crucial. As John Langdon suggests, the spade was "even more important than the plough or cart" to the manorial economy because of its multipurpose use by "all levels of society."[16] While not every villein had his own plough but had to depend on borrowing or sharing this vital piece of equipment, he would certainly have owned a spade, as is shown in lists of necessary equipment prescribed by later medieval writers of agricultural treatises. Whereas the image of the plough always prompts the question of

Figure 36. Digging and ploughing. Rentier de Messire Jehan de Pamele-Audenarde. Brussels Bibliothèque Royale Albert I, MS 1175, fol. 156v.

the relationship between man and the means of production (whose plough is it, whose oxen and horses, whose ridge and furrow is being created?) the image of the delver is one of self-sufficiency. This is perhaps why the scene of Adam's primal labor became the paradigmatic image of peasant labor and not the ploughman and his team. Only in medieval Scandinavia do we find frequent depictions in wall paintings of Adam ploughing, and these are mostly fifteenth-century representations, linked to the pretensions of the higher yeomen class of peasants.[17] In the illustrations to the Anglo-Saxon Caedmon Genesis the first man to push a plough is Tubalcain, inventor of metal; later, Noah is shown ploughing with his son before his drunkenness.[18]

Ground-breaking tools were an essential aspect of the agrarian economy not only because the seeds had to be buried but also in the carving out of territories, in the construction of moats and ditches that divided, in addition to the fields, the tofts in the village. Here there may be a relationship between the popularity of the delving Adam in twelfth- and thirteenth-century English art and the contemporary expansion of arable and settlement.[19] It is a commonplace that the plough's creation of ridge and furrow sculpted the still visible contours of the English countryside. However, to continue the artistic metaphor, this subtle brushwork of lines would have been impossible without the canvas or ground of fields and strips—the spaces created by the spade.

The only medieval depiction of drainage digging I have come across occurs in an eleventh-century Oxford manuscript of the Anonymous Life of Saint Cuthbert, who not only built his own cell on the bleak island of Farne but grew his own food, aided by ravens and other creatures (Figure 37). Instead of hindering him, as they did most medieval farmers, verminous birds and beasts miraculously serve the saint. In one scene he is depicted holding a one-sided spade, illustrating how

> on a certain day on his island, he was digging and trenching the land [*sulcabat terram*] (for at first, for two or three years before he shut himself in behind closed doors, he laboured daily and gained his food by the work of his hands [*opere manuum suorum*] knowing that it is said: "He that will not work, neither shall he eat").[20]

For a monk like Cuthbert, Adam's curse of labor was incorporated into his penitential renunciation of the world.

An image related to the isolated father of humankind that we see in the Canterbury window, which became increasingly popular in the subsequent century, was that showing both Adam and Eve at their labors. This was common in wall paintings, in sculptural programmes such as the west front of Wells Cathedral (Figure 38), and in numerous manuscript illustrations. It was also enacted in the mystery plays, where a tableau representing Adam and Eve bemoaning their fall from grace emphasizes their respective tools. Adam in the Chester Play cries: "With this spade that yee may see / I have dolven. Learne yee this at mee, / howe yee shall wynne your meate." Eve's pendant refrain is, "I suffer on yearth for my misdeede; / and of this wooll I will spyn threede by threede."[21]

This was an image that divided male from female work. If Adam's was manual labor, Eve's was the labor of birth: "Childrenn must I beare with woo." There is no reference in Genesis to Eve's spindle and distaff. The

Figure 37. Digging saint aided by birds. Life of St. Cuthbert. Oxford, University College MS 165, fol. 63.

development of this attribute that came to stand for all women in medieval iconography, from the housewife chasing the fox and hen in images of the fable to the printed "Gospel of Distaffs," which records the gossip of village crones, has much to tell us about attitudes toward women in the agricultural world that cannot be discussed here. What is important is that such images provided ideological models for men and women in the audience. Precisely because such images are often ideological we must not take them as accurate depictions of reality. Historians have recently shown how women in fact shared a great deal in the heavy agricultural work—gardening, poultry, and pigs being their specialties and, of course, harvesting, as is depicted in the *Luttrell Psalter*. The image of Eve with her distaff may have some relevance considering the increasing role of women in domestic textile production in the later Middle Ages, when this activity became a

Figure 38. Adam and Eve labor. Wells Cathedral, west front.

source of extra income. In the archetypal image, however, it represents her position as clother and feeder of her children, as *inside* rather than *outside* the house.

One of the most moving depictions of this subject, which emphasizes the domestic site of Eve's work, placed inside some kind of cave, occurs in the early fourteenth-century Holkham Bible Picture Book, made for a rich layman. Adam wears even less than in the Canterbury scene, and both he and Eve look back longingly from their labors to what they have lost

(Figure 39). Above them looms the angel with the flaming sword, who bars the way to the gates of Paradise in a spatial and symbolic configuration that surely refers to the two ruling powers, the authority of ecclesiastical (angelic hierarchy) and secular (the sword) lordship over the peasant. The inscription in Norman French makes special mention of Adam's tool and notes that the angel had commanded him to use it: "Adam une beche prist/ Sicum le aungel ly avoit dit" ("Adam took a spade, just as the angel had said") and that Eve must "vesture fere" ("make clothes"). The grazing goats suggest another aspect of the agrarian economy—a dichotomy between arable farming and animal husbandry that is expanded in a subsequent picture in the manuscript, which shows Adam's descendants at their various labors.[22] Of their offspring, Abel, the meek victim, becomes the shepherd whose offering God accepts while the murderer Cain, *agricola*, becomes the "cursed husbandman" whose fruits God rejects and whose sins include "forcing the earth with his plough." Cain's curse, even heavier than that borne by his father, is that "when thou tillest the ground it shall not henceforth yield unto thee her strength" (Genesis 4: 11–12).[23]

Labor was always both curse and blessing in medieval Christian thought. It was necessary for the survival of the human race and yet it was imposed from above, by the Lord (both of the world and of the peasant's holding). This curse was carried, however, by only one group in medieval society—the villein or peasant farmer, whose bent-over body made him the descendant of the cursed Cain and the primal delver, Adam. Behind all these images, too, is the anxiety of laboring in a precarious system, one which on numerous occasions during the thirteenth and fourteenth centuries made Cain's curse quite concrete. If the image of the ploughman can be seen as one optimistic of the future rewards of labor, that of Adam delving evokes more ominously the basic struggle for survival, especially during the famines that ravaged England from 1315 to 1321, exactly the period when the Holkham Bible Picture Book was painted.

An unusual illustration of the negative association of the delving figure occurs in a treatise on the spiritual life by Anselm of Canterbury (ca. 1220). It depicts the contrast between liberty and servitude, the illustrator representing both concepts by small human exemplars in the bottom margin. Liberty is a knight on his horse, being handed a gift by his lady; servitude is a man with a spade at the edge of the page (Figure 40). As a "low" tool of the very poorest peasants, delving can be seen here as indicating not just low but unfree status in the manorial hierarchy.

One immediately thinks of that popular catchphrase of the Uprising

Figure 39. Expulsion and labor of Adam and Eve. *The Holkham Bible Picture Book*. London, British Library, MS Add. 47682, fol. 4v. By permission of the British Library.

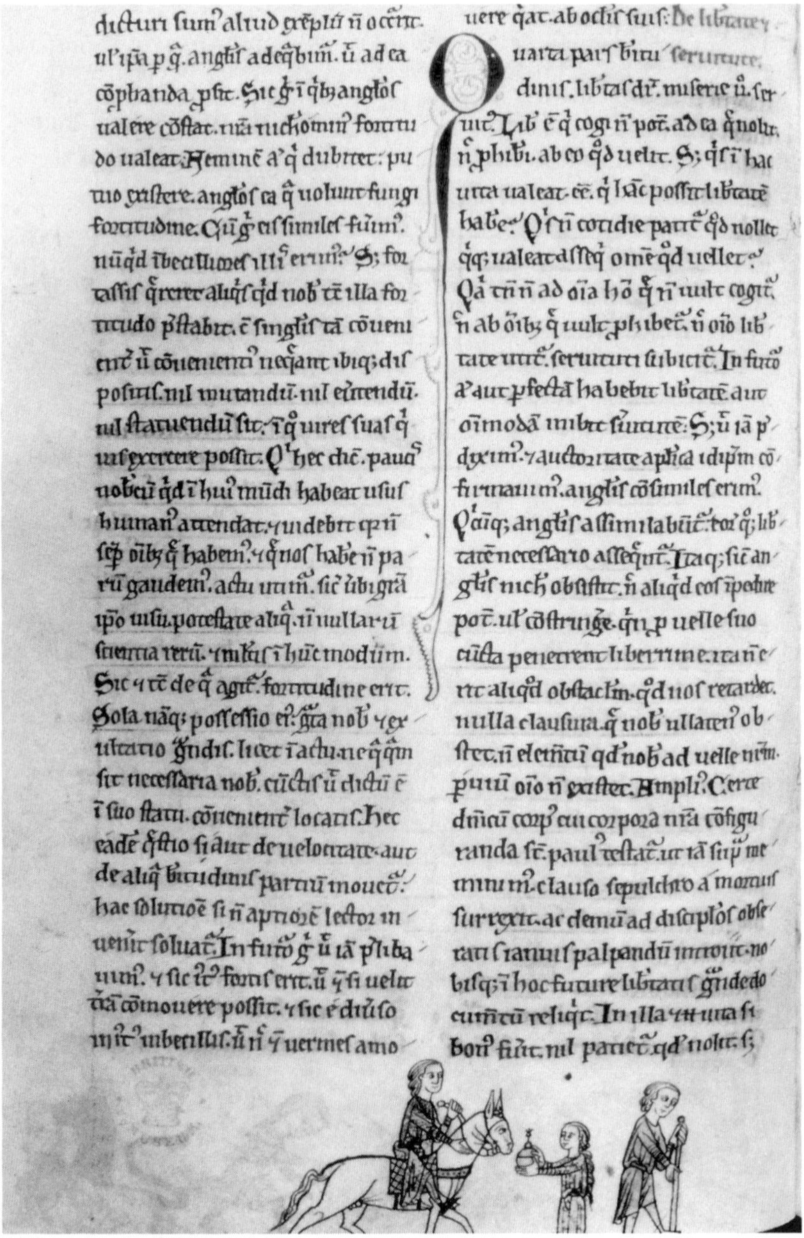

Figure 40. Liberty and servitude. Anselm, *De similitudinibus*. London, British Library, Cotton MS Cleopatra C.XI, fol. 14v. By permission of the British Library.

of 1381 (the Peasants' Revolt as it used to be called) that was, according to some accounts, shouted by John Ball in rousing the folk to revolt:

> When Adam delved and Eve span
> Who was then the gentleman?[24]

This rhyme scholars now believe to be as old as the Canterbury window and to have been circulating in oral tradition long before the late fourteenth century. Similarly, what used to be a "peasants' revolt" is now more accurately seen as one fired just as much by alienated urban craftspersons and workers. Thus it is all the more interesting that this varied disaffected group saw in the images of Adam and Eve that have been discussed here a radical Edenic equality. The church, which for the most part created and ordered such images as the Holkham manuscript, which was painted by a Dominican priest, had a different take on the first workers. Clerics argued that the social order itself had originated with the Fall. As Alexander Neckham put it, "If man had not sinned there would be no difference of degree for a degree is a lapse from the norm."[25] Another English scholastic, Alexander of Hales, described how the transgression of Adam and Eve necessitated the institution of authority from above: "Natural law ordains the equal freedom of all in the state of original nature (*naturae institutae*); but according to the state of fallen nature (*naturae corruptae*) it ordains that subjection and lordship are necessary for the constraint of evil."[26]

Adam's first labor, as depicted in the Holkham Bible or in stained glass and wall paintings that peasants could see themselves on their day of rest, did not so much represent the primal communism of their first parents but more their subjection in sin. One fifteenth-century preacher urged his listeners to follow the model of their sinful first parents: "yt ys the wylle of Gode that we schuld labour and put our body to penaunce for to fle synne. Thus dyd Adam and Eve, to example of all tho that schuld come after them."[27]

In terms of the "mental set" of the peasant, the delving Adam would have stood in a particular place in the hierarchy of labor. It is a fallacy to lump all those who are neither *oratores* nor *bellatores* into a homogeneous peasant class. As is becoming increasingly clear, there were subtle distinctions of duties, skills, and labors that depended upon each peasant's free or unfree status, manorial obligations, landholding, and region, and that distinguished different strata among the land laboring population. Well-documented for medieval France and probably also the case in England

Figure 41. The first two nightmares of Henry I. *Chronicle of Florence and John of Worcester* Oxford, Corpus Christi College MS 157, p. 382.

was a distinction between the *laboureurs* and the *manouvriers*. The *laboureurs* had sufficiently substantial lands to be able to keep plough oxen, while the *manouvriers* were those poorer peasants who worked only with their own hands.

Although the model of *oratores*, *bellatores*, and *aratores* signaled the third group as "ploughmen," the iconography of this triple order nearly always depicts the third group with the attribute of the spade; indeed I would argue that the spade distinguishes the particular status of the *manouvrier* in English medieval art. This is true in one of the earliest depictions of the three orders of society, in the Chronicle of Florence and John of

Worcester ca. 1130 (Figure 41). Here, the first of King Henry I's three nightmares, in which he is attacked by the *tres ordines* complaining of high taxation, involves a group of angry rustics. They come to the King's bedside waving their implements (*rusticanis instrumentis*). The most prominent of these, jutting out into the text space, is the spade. Just as the knight has his sword in the same position below, this instrument is the emblem of the peasant's authority. In this, the first representation of "revolting peasants" in Western art, we must not forget that the whole thing, spades and all, is a dream vision, interpreted by the king's physician seated on the left.

In a thirteenth-century French miniature of the three orders, or *les trois manières de gens*, the three conventional roles studied in their textual order by Duby and others are more tightly demarcated (Figure 42).[28] The gesticulating priest controls the Latin Word, the central knight serves his protective role with a large heraldic shield, while the third order, the only one of the three who stands behind and is pushed to the edge of the letter, holds his spade. One always has to ask for what audience such images were made. The Worcester Chronicle was made for monks and portrays peasants as a monstrous rabble. Similarly here, for a noble French-speaking reader, the central knight usurps the place of the provider and pushes the producer to the side. The great irony in dealing with both the visual and verbal testimony of medieval peasant experience is that there is no evidence of the three orders that presents the place of the laborer from the laborer's point of view. Medieval art is hegemonic and made *for* only two of the three orders. If ploughing the page was a metaphor for the scribe, the man who held the spade did not hold the pen.

William Langland's *Piers Plowman* is often seen as presenting a more sympathetic view of peasant experience, but in fact its vision of order is highly conservative. In the margins of an early fifteenth-century manuscript of the work, now in Oxford (Figure 43), is an image of a man digging into the lines:

Yblessed be al tho that here bylyue biswinketh
Throw eny lele labour as thorw lymes and handes.
Labores manuum tuarum quia manducabis.[29]

[Blessed be all those who work for their food
By any honest labor with their bodies and hands.
For thou shalt eat the labor of thy hands (Psalms 128)]

p luſ q de nule autre are

C ele en a la nobilite

d eſ · uj · maneres de genſ que

L i phlloſofe poſerent au mode

leiple
regne
ore a
parts
Anſi q
ele fu
ia diſ
a athe
neſ q
ſiet en
grece

b ne are de grm nobloe

a phlloſofe q loeſ furent

Q les autreſ enſegn durent

R e poſerno celone lor ſenſ

F orſ q · uj · maneres de genſ

Figure 42. The three orders of men. *Image du Monde*. London, British Library, MS Sloane 2453, fol. 85. By permission of the British Library.

Figure 43. Peasant digging, detail. *Piers Plowman*. Oxford, Bodleian Library, MS Douce 104, fol. 39r. By permission of the Bodleian Library.

Langland refers to the trope of manual labor once again as blessed, as did the preachers of the time. Only if everyone keeps his place in the divinely sanctioned social order, from ploughman, to knight, to priest, will the world's ills be solved. Once again we can see how the image of the laboring Adam served to ratify a system sanctioned by the Creator and set in motion at the beginning of human time.

The Russian historian Aron Gurevich, in his study of medieval attitudes toward wealth and labor, points out the contradictions inherent in medieval clerical attitudes toward work. Although labor was considered inseparable from the human condition, he points out that, theologically speaking, the better life was the contemplative, not the active, life: "Christ, the clergy pointed out, did not labour. Having gathered his disciples around him he encouraged them to give up their earthly occupations and become not labourers but 'fishers of men.'"[30] But most peasants could not depend on their loaves and fishes to multiply miraculously. Gurevich shows how later in the Middle Ages the church had to alter its position on work and present it to a predominantly agricultural audience as penance and payment towards the ultimate reward of heaven—where all labors would cease. It was crucial that the notion of labor, with its low, slavish status carried over from the ancient world, be sanctified and legitimated by both church authority and the aristocratic powers that depended upon the food and income from their human and animal possessions.

The images of the delving Adam and of Adam and Eve at work served this purpose of reminding the medieval masses of their duties in the divinely ordained social order. Rather than illustrating the cynical question posed by John Ball's supporters—"When Adam delved and Eve span / Who was then the gentleman?"—the image reminded onlookers of the fall of man into labor. Through their own sins Adam and Eve had transgressed, and only through the penance of suffering on earth could they regain the kingdom of heaven. As Master Thomas Wimbledon harangued the vast crowds who came to hear him preach at St. Paul's Cross:

> Herfore everich man see to what state God hath cleped him, and dwell he therein by travile according to his degree. Thou that art a laborer or a crafty man, do this truelly. If thou art a servant or a bondman, be suget and lowe, in drede of displeasing of thy Lord.[31]

My final example is a full-page picture from a thirteenth-century Bible Picture Book that I discovered at the Art Institute of Chicago. It was made in France in the thirteenth century, probably for a layperson since it has captions in the vernacular. This is the clearest example of the Christian sanctification of labor one could hope to encounter, since it shows Adam being instructed in delving, not by an angel but by God himself (Figure 44). The caption reads: *Adam aprend labourer terre*. In the wall paintings at Sigena created in Spain in the twelfth century, probably by an English artist, and in the eleventh-century Canterbury manuscript of Aelfric's Pentateuch (Figure 32) Adam is taught to dig by an angel. Adam being taught

Figure 44. Adam taught to delve by Christ. Art Institute of Chicago, Bible Picture Book. By permission of the Trustees of the Art Institute of Chicago.

by the archangel Michael occurs in the Latin *Vitae Adae et Evae*, one of a number of popular apocryphal texts about humanity's first parents that circulated in the Middle Ages. But the only textual source for the idea of the delving Deity comes in the Jewish apocryphal text the Hebrew Book of Jubilees, where the Lord declares, "We gave him work and instructed him to do everything that was suitable for tillage." This instruction was given while Adam was in the garden of Eden, and once expelled, the text tells us, "he tilled the land as he had been instructed in the garden."[32] I have also located the idea nearer to thirteenth-century France in the recounting of the Genesis story in the popular *Histoire Ancienne*. These vernacular manuscripts contemporary with the Chicago picture describe how "Our Lord gave them the instruments with which they would labor and showed them how to use them."[33] This context accords with my theories of how verbal and visual vernacularization in medieval culture work to materialize and humanize categories of experience compared with the earlier, more hierarchical, abstracted forms of Latin.

Returning finally to the delving Adam of the Canterbury window, I think it is no surprise that he appears so Christ-like, with the long hair and even the facial type of the one who will eventually redeem him. Typologically, of course, Christ is the "new Adam" who releases humankind from Adam's curse of death. The idealization of this figure, the softly rippling green earth and the blue water alongside that suggest irrigation and fertility, the nearby tree (the tree of Life that will become the tree of crucifixion), which provides a convenient hook for Adam's other tool, all suggest that this is a complex typological image like others in the Canterbury programme.[34] The onlooker in the cathedral only had to follow the ancestors of Christ around the clerestory to find, directly opposite Adam on the south wall, the Virgin and Child. But I have strayed too far from the perceptions of "lewde" folk and must reiterate in conclusion that this important and hitherto neglected image in the history of humanity's conquest of the earth, like most monumental images of the period, does not so much "reflect" medieval experience as construct it.

As Alfred W. Crosby puts it, "The Gothic cathedral, sublime product of the twelfth century, was more than a sign of rebirth. It marked the first birth of a society of remarkable energy, brilliance, and arrogance. Such societies are often expansionistic."[35] We often forget that cathedrals stand in the midst of enormous stretches of rich and fertile arable land. The image in the cathedral in the final analysis also refers to the enormous agricultural and economic resources necessary to erect a building of this lavish scale—the prebendery lands and their revenues, thousands of tenants, customary labors, tithes and taxes collected by the prior and monks of Christchurch. Their manors and revenues stretched all over southern England and brought in an income of £1,460 in 1207.[36] As lords of the land in a more proprietorial sense than that lonely monk Cuthbert eking out his existence centuries before, they depended upon the toil of thousands of Adams and Eves. Adam, his spade shining in its yellow and silver in the clerestory window, in this sense does not delve only for his livelihood nor even his own grave. He digs the very foundations of the great cathedral around him.

Notes

1. *The Tale of Beryn with a Prologue of the Merry Adventure of the Pardoner with a Tapster of Canterbury*, ed. F. J. Furnivall and W. G. Stone (London: Published for the Chaucer Society by N. Trubner, 1887), p. 6. (Also published by the Early English Text Society, extra ser. 105 (London: Kegan Paul, 1909).

2. For the "classical" sources of the stained glass see Madeline Harrison Caviness, *The Early Stained Glass of Canterbury Cathedral, circa 1175–1220* (Princeton, N.J.: Princeton University Press, 1977), and for a detailed description of the iconographic sources, pp. 113–14, as well as the same author's *The Windows of Christ Church Cathedral, Canterbury*, Corpus Vitrearum Medii Aevi, Great Britain 2 (London: Oxford University Press, 1981), p. 18. A vast new work on Genesis imagery, Hans Martin von Erffa, *Ikonologie der Genesis. Die christlichen Bildthemen aus dem Alten Testament und ihre Quellen* (Munich: Deutscher Kunstverlag, 1989), includes more bibliography on Adam, pp. 233–35. For a theological reading see B. D. Naidoff, "A Man to Work the Soil: A New Interpretation of Genesis 2–3," *Journal for the Study of the Old Testament* 2–3 (1978): 2–14.

3. *The Play of Adam: Ordo representacionis Ade*, ed. Carl J. Odenkirchen (Brookline, Mass.: Classical Folia Editions, 1976), pp. 90–91. For the "contemporizing" aspect of the drama, see V. A. Kolve, *The Play Called Corpus Christi* (Stanford, Calif.: Stanford University Press, 1966). I have not yet looked through prop accounts for references to tools.

4. For the Monreale mosaic see Ernst Kitzinger, *The Mosaics of Monreale* (Palermo: S. F. Flaccovio, 1960), pl. 17. Caviness (*Windows of Christ Church Cathedral*, p. 18) sees the fleece loincloth and spade, as opposed to the mattock, forming a distinct Canterbury type.

5. An exception to this is a thirteenth-century window in the Elisabethkirche, Marburg, showing the mattock. For the iconography of tools in medieval art see Siegfried Epperlein, *Der Bauer im Bild des Mittelalters* (Leipzig: Urania-Verlag, 1975), and Perrine Mane, *Calendriers et techniques agricoles: France-Italie, XIIe–XIIIe siècles* (Paris: Sycomore, 1983), which reproduces the February digger sculpted on the Parma baptistery (pl. 164).

6. See Israel Gollancz, *The Caedmon Manuscript of Anglo-Saxon Biblical Poetry: Junius XI in the Bodleian Library* (London: Roxburghe Club, 1927) and Elz-

bieta Temple, *Anglo-Saxon Manuscripts, 900–1066* (London: Harvey Miller, 1976), for Junius ii. It has two images of Adam with a spade on pp. 45 and 46, in which he also holds a bag of seeds. On p. 49 is the unusual image of Adam digging together with his son Cain, later depicted in stone at Lincoln; see George Zarnecki, *Romanesque Lincoln: The Sculpture of the Cathedral* (Lincoln: Honywood, 1988), fig. 56. The Pentateuch scenes are also fascinatingly original as described in C. R. Dodwell and Peter Clemoes, eds., *The Old English Illustrated Hexateuch*, Early English Manuscripts in Facsimile 18 (Copenhagen: Rosenkilde & Bagger, 1974). The Anglo-Saxon gloss added inside the frame of fol. 7v (fig. 3) between the figures suggests a later misreading of the image, since it describes the burying of Abel.

　　7. On medieval wooden shovels and spades see C. A. Morris, "Early Medieval Separate-Bladed Shovels from Ireland," *Journal of the Royal Society of Antiquaries of Ireland* iii (1981): 50–69, and "A Group of Early Medieval Spades," *Medieval Archaeology* 24 (1980): 205–10.

　　8. See "delve" and "dig" in the *Oxford English Dictionary*.

　　9. *The Corpus Christi Play of the English Middle Ages*, ed. R. T. Davies (London: Faber; Totowa, N.J.: Rowman and Littlefield, 1972), p. 85.

　　10. For the important theme of calendar illustration see James Carson Webster, *The Labors of the Months in Antique and Mediaeval Art to the End of the Twelfth Century* (Princeton: Princeton University Press, 1938), and Mane, *Calendriers et techniques agricoles*. A broad view is presented by Henrik Specht, *Poetry and the Iconography of the Peasant: The Attitude to the Peasant in Late Medieval English Literature and in Contemporary Calendar Illustration* (Copenhagen: Department of English, University of Copenhagen, 1983), pp. 31–44.

　　11. James E. Thorold Rogers, *A History of Agriculture and Prices in England* 1: *1259–1400* (Oxford: Clarendon, 1866; reprint Vaduz: Kraus, 1963), pp. 540–41. See also John Langdon's excellent section on hand tools in "Agricultural Equipment," in *The Countryside of Medieval England*, ed. Grenville Astill and Annie Grant (Oxford: Basil Blackwell, 1988), pp. 86–107, esp. 95–99, which has further bibliography on visual representations as well as archaeological material.

　　12. For the association of smith and farmer see Ian H. Goodall, "The Medieval Blacksmith and His Products," in *Medieval Industry*, ed. D. W. Crossley (London: Council for British Archaeology, 1981), pp. 51–62. An iconographic and social study of the smith as metalworker/artist/magician in medieval society is much needed. For the early Anglo-Saxon casket see P. W. Souers, "The Wayland Scene on the Franks Casket," *Speculum* 18 (1943): 104. The fourteenth-century Psalter of Isabella of England (Munich, Staatsbibl. Cod. gall. 16, fol. 19v.) has a marginal image of Tubalcain at his forge.

　　13. See M. Patricia Hogan, "The Labor of Their Days: Work in the Medieval Village," *Studies in Medieval and Renaissance History* n.s. 8 (1986): 94.

　　14. Rossell Hope Robbins, ed., *Secular Lyrics of the Fourteenth and Fifteenth Centuries*, 2nd ed. (Oxford: Clarendon, 1956), p. 62.

　　15. Michael Camille, "Labouring for the Lord: The Ploughman and the Social Order in the Luttrell Psalter," *Art History* 10 (4) (1987): 423–54; John Langdon, *Horses, Oxen and Technological Innovation: The Use of Draught Animals in English Farming from 1066 to 1500* (Cambridge: Cambridge University Press, 1986).

The ploughman, however, could be called upon to "to dig, make enclosures . . . remove earth or dig trenches to dry the land and drain off the water" according to *The Seneshaucy*; see Dorothea Oschinsky, ed., *Walter of Henley and Other Treatises on Estate Management and Accounting* (Oxford: Clarendon, 1971), p. 283.

16. Langdon, "Agricultural Equipment," p. 95.

17. C. Ostling, "The Ploughing Adam in Medieval Church Paintings," in *Man and Picture: Papers from the First International Symposium for Ethnological Picture Research in Lund 1984*, ed. Nils-Avrid Bringéus (Stockholm: Almqvist & Wiksell, 1986), pp. 13–19.

18. Oxford, Bodleian Library, MS Junius 11, pp. 54 and 77.

19. For the expansion of the thirteenth century see M. M. Postan, "Medieval Agrarian Society in its Prime, Pt. 7: England," in *The Cambridge Economic History of Europe 1: The Agrarian Life of the Middle Ages*, 2nd ed., ed. Postan (Cambridge: Cambridge University Press, 1966), pp. 548–632, and the recent discussions in H. E. Hallam, ed., *The Agrarian History of England and Wales, 2: 1042–1350* (Cambridge: Cambridge University Press, 1988). A useful theoretical overview of current controversies can be found in a review of the latter work by Kathleen A. Biddick, "Malthus in a Straightjacket? Analyzing Agrarian Change in Medieval England," *Journal of Interdisciplinary History* 20 (1990): 623–35.

20. Bertram Colgrave, *Two Lives of Saint Cuthbert* (1939; reprint New York: Cambridge University Press, 1985), pp. 100–101.

21. *The Chester Mystery Cycle*, ed. R. M. Lumiansky and David Mills, Early English Text Society, suppl. ser. 3 (Oxford, 1974): 33.

22. W. O. Hassall, "Introduction and Commentary," The Holkham Bible Picture Book (London: Roxburghe Club, 1954), pp. 64–65 for the toil (fol. 4v.) and 68–69 (fol. 6) for the image of Cain and his descendants. The theme of Eve spinning is just as important and deserves to be studied as an image inculcating the domestic place and space of women. For now, see the few comments by Philippe Verdier, "Woman in the Marginalia of Gothic Manuscripts and Related Works," in *The Role of Women in the Middle Ages*, ed. Rosmarie Thee Morewedge, Papers of the Sixth Annual Conference of the Center for Medieval and Early Renaissance Studies, State University of New York at Binghamton, May 6–7, 1972 (Albany: State University of New York Press, 1975), p. 129.

23. The theme of the "cursed husbandman Cain" is discussed in Camille, "Labouring for the Lord," pp. 433–34.

24. For this poem see G. R. Owst, *Literature and Pulpit in Medieval England*, 2nd rev. ed. (Oxford: Basil Blackwell, 1961), pp. 290–91, and J. A. Raftis, "Social Change versus Revolution: New Interpretations of the Peasants' Revolt of 1381," *Social Unrest in the Late Middle Ages*, ed. Francis X. Newman, Papers of the Fifteenth Annual Conference of the Center for Medieval and Early Renaissance Studies (Binghamton, N.Y.: State University of New York Press, 1986), p. 17.

25. "Si igitur non peccasset homo, nihil esset gradus; est namque gradus elongatio a temperantia." Alexander Neckham, *De naturis rerum*, ii, 156, ed. Thomas Wright, Rolls Series 34 (London: Longman, 1863): 250. See George Boas, *Essays on Primitivism and Related Ideas in the Middle Ages* (Baltimore: Johns Hopkins University Press, 1948), pp. 83–85.

26. Quoted by Bede Jarrett, *Social Theories of the Middle Ages, 1200–1500* (Boston: Little, Brown, 1926), p. 9.

27. This sermon is quoted in Owst, *Literature and Pulpit*, p. 555.

28. This miniature illustrates the *Image du Monde* by Gossouin of Metz, a popular cosmological compendium; see Oliver H. Prior, *L'Image du Monde de Maître Gossouin* (Lausanne: Payot, 1913), p. 77. Elsewhere in the manuscript illustrations of this treatise, the figure with the spade reappears in a cosmological image of the centrifugal universe with *terre* at its center, signified often by a little figure digging; see Prior, *Image du Monde*, p. 158. This does not put the laborer at the center of the universe, however, but at its basest rung. For the history of the theme of the *trois manières de gens*, see Georges Duby, *The Three Orders: Feudal Society Imagined*, trans. Arthur Goldhammer (Chicago: University of Chicago Press, 1980).

29. Derek Pearsall, *Piers Plowman by William Langland: An Edition of the C-Text*, York Medieval Texts, 2nd ser. (London: E. Arnold, 1978), passus VIII, ll. 260–61. For the digger in the Bodleian manuscript with more examples of later English manuscript images of the spade, see Kathleen L. Scott, "The Illustrations of *Piers Plowman* in Bodleian Library MS. Douce 104," *Yearbook of Langland Studies* 4 (1990): 42–43. Scott also draws my attention to a half-page miniature depicting Lady France chastising the three orders, including a bearded peasant with a prominent spade in an English fifteenth-century miniature, University College, Oxford MS 85, fol. 1, reproduced in Margaret S. Blayney, ed., *Fifteenth-Century English Translations of Alain Chartier's "Le Traité de L'Esperance" and "Le Quadrilogue Invectif,"* Early English Text Society 270 (Oxford: Oxford University Press, 1974), frontispiece.

30. A. J. Gurevich, *Categories of Medieval Culture*, trans. G. L. Campbell (London: Routledge & Kegan Paul, 1985), p. 260.

31. Quoted in Owst, *Literature and Pulpit*, p. 551.

32. R. H. Charles, *The Apocrypha and Pseudepigrapha of the Old Testament in English* 2 (Oxford: Oxford University Press, 1913): 16. This is cited by Otto Pächt as the source for the fresco at Sigena, Spain (perhaps by an English artist), where the angel digs in the same strong pose as Adam in the Canterbury window: "A Cycle of English Frescoes in Spain," *Burlington Magazine* 103 (1961): 169 and fig. 9. For the influence of the *Vita Adae* on Genesis cycles in thirteenth-century stained glass at Tours and sculpture at Rouen, see Linda Morey Papanicolaou, "The Iconography of the Genesis Window of the Cathedral of Tours," *Gesta* 20 (1981): 179–89.

33. Mary Coker Joslin, ed., *The Heard Word, a Moralized History: The Genesis Section of the Histoire Ancienne in a Text from Saint-Jean D'Acre*, rev. ed. (University, Miss.: Romance Monographs, 1986), p. 87.

34. Caviness, *Early Stained Glass*, pp. 116–17.

35. Alfred W. Crosby, *Ecological Imperialism: The Biological Expansion of Europe, 900–1900* (Cambridge: Cambridge University Press, 1986), p. 44.

36. For the agricultural wealth of the most important ecclesiastical center in England see William Page, ed., *Victoria History of the County of Kent* 2 (London: Oxford University Press, 1926): 114–15, and also R. A. L. Smith, *Canterbury Cathedral Priory* (Cambridge: Cambridge University Press, 1943).

Jane Welch Williams

13. The New Image of Peasants in Thirteenth-Century French Stained Glass

Art-historical studies of medieval peasant imagery are scarce and tend to be more concerned with artistic tradition and agricultural practices than with the historical circumstances surrounding the creation of particular images.[1] This chapter emphasizes the importance of such historical research to the analysis of medieval peasant imagery by showing how the few depictions of peasants in thirteenth-century French cathedral windows as if they were donors had very different meanings at different times and in different places and contexts.

The traditional venues for medieval peasant imagery prior to the thirteenth century were biblical or astrological.[2] In cycles of the birth of Christ, anonymous shepherds looking like medieval peasants hear the news of Christ's birth and come to see him.[3] In cycles of the labors of the months, anonymous peasants perform their yearly round of agricultural work accompanied by astrological signs, typically in church calendars and around the margins of church portals. When artisans painted peasants in stained glass for church windows, they copied these peasant images.

In the thirteenth century, for the first time, peasants appeared as if they were donors in the bottom margin of a few French stained glass windows, traditionally the location of upper-class donor imagery. This is the first public appearance of medieval peasants in a new social position in art. Throughout medieval art, upper class individuals identified by inscriptions or blazons and usually shown kneeling in prayer appeared as donors at the base of the art works they donated to churches. However, the peasants are differentiated in the manner of their representation; unlike the upper classes, peasants are shown laboring. Their customary representation accords with their designation as "those who work" in the ecclesiastical conception of the three orders of medieval society.[4]

At Chartres Cathedral, in two adjacent ambulatory lancets, vineyard workers appear in the bottom register. One window illustrates the Life of the Virgin,[5] and the other illustrates the Signs of the Zodiac and the Labors of the Months.[6] A clerestory oculus at Chartres depicts peasants ploughing the earth who are identified as donors by inscription.[7] In a choir clerestory Genesis window at Tours Cathedral, four scenes of peasants ploughing appear in the bottom register.[8] At Le Mans Cathedral, a window in the choir contains images of vineyard workers beneath scenes from the life of the patron Saint Julian and Christ.[9] These are the only thirteenth-century windows of which I am aware that represent peasants in this new context.[10]

It is questionable whether peasants really donated the windows. Our meager information suggests that a big lancet in the early thirteenth century cost around 30 livres equal to 7,200 deniers.[11] The average income of a laborer was 1 to 2 deniers per day, and so a window might have cost the equivalent of at least 3,600 days of peasant labor.[12] The few windows depicting peasants in the donor location in northern French Gothic cathedrals appeared along with many more windows depicting tradespeople in their lowest register; the latter also may not have been intended to indicate donation.[13]

Vineyard Workers and Ploughmen at Chartres

In fact, peasants did not donate the two lancets in the south ambulatory at Chartres. These are the earliest of the peasant windows, possibly installed soon after 1208, but surely by 1220.[14] In both windows, a large medallion on the left side of the lowest register depicts vineyard workers. In the Window of the Life of the Virgin, two workers dressed in calf-length winter clothes and hoods prune the vine between them with *serpes* or billhooks (Figures 45 and 46).[15] At the base of the adjacent Window of the Signs of the Zodiac and the Labors of the Months, five vineyard workers tend vines (Figure 47).[16] This scene is confused by sloppy repairs, but one can discern on the left a worker hoeing, and, close by, four other workers pruning a vine between them. In small square scenes at the bottom corners of the window border, single vineyard workers are depicted in profile carrying hoes.[17]

In both windows, the large vineyard scenes appear opposite a nobleman on horseback carrying a shield painted with the arms of Champagne.[18] The nobleman addresses a group of supplicating monks, accompanied by secular figures. An inscription beneath the nobleman in the Zodiac Window now reads: COMES : TEOBALD(US) DAT . . . HO(C) VESPO VINERUS

Figure 45. Bottom register, Life of the Virgin window. Chartres Cathedral, bay 16, c. 1208–20. Author's photo.

Figure 46. Detail, vineyard workers. Chartres Cathedral, bay 16. Author's photo.

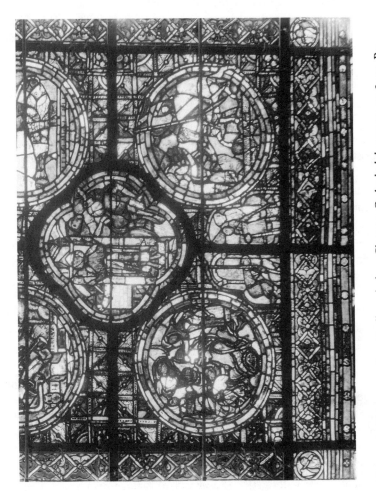

Figure 47. Bottom section, Zodiac window. Chartres Cathedral, bay 17 ca. 1208–20. By permission of ARS/SPADEM.

AD PRECES COMITIS PERTICEENSIS (Figure 47).[19] The inscription has lost letters and coherence as a result of repairs, necessitating some imaginative reconstruction to read: "Count Thibaut gave the vineyard by Vespres to the prayers of the count of Perche." Vespres, or Vepres, was a road to Bouglainval, near Maintenon.[20] Bouglainval was one of the cathedral chapter's prebend headquarters, which already had ten parishes by the end of the twelfth century, and was heavily planted in vines.[21] The implication is that the vineyard given to the church by Count Thibaut at the death of his friend, the count of Perche, was intended to pay for prayers in his memory.

An unusual picture appears in the middle of the bottom register in the Zodiac Window, between the scene of vineyard workers and the image of the Count Thibaut (Figure 47). On the left of the picture, in an arcaded space suggesting a church, a monk in a long cloak and hood bends far over toward the ground to pull at a cord, apparently ringing a church bell. Opposite him, two men stand with staffs. One wears a red hooded cape, and the other a white *cotte*, or short work dress. They seem to be the same people who appear in the adjacent scene hailing the count. They are probably intoning a bell in memory of the dead count of Perche. The occasion of the gift seems clear, but one is not sure to which church the gift was given—the cathedral, one of its parish churches, or a monastery. Nor is it clear from the inscription who might have paid for the windows that attest to the gift of a vineyard.[22]

However the inscription and imagery might be completely explained, they definitely indicate that the peasants working on vines signify a vineyard donated to the church, as so many vineyards were in the twelfth and thirteenth centuries. In the Zodiac window, just above the donation imagery, March appears illustrated by the usual peasant pruning a vine. Toward the top, September is represented by the harvesting and trampling of grapes, and October by the storing of new wine. The cycle is unusual in that three of the Labors illustrate viticulture,[23] apparently to honor the donation of the vineyard memorialized at the bottom.

An oculus in the nave clerestory at Chartres depicts a peasant guiding a wheeled plough with two handles while another peasant whips the horses and guides them with a stick (Figure 48).[24] A third man, much larger in scale, appears in the oculus above the plough's axle, standing beyond the plough. He looks backward toward the ploughman, gesturing toward him with his right hand; his left hand holds the cord fastening his cape. This gesture and the man's attire were features reserved for the upper classes in contemporary imagery. This third man seems to direct the work; hence his

Figure 48. Oculus of ploughmen. Chartres Cathedral, north nave clerestory, bay 160, c. 1205–24. Author's photo.

Figure 49. Detail, Medieval Model Book. Vienna Nationalbibliothek Cod. 507, fol. 1.

relatively large scale in the scene is symbolic of his superior status. His pose and location closely resemble a tradition preserved in an early thirteenth-century model book, now in Vienna (Figure 49).[25] In the window, how-ever, the plough is more complex. Two stanchions that rise on either side of the plough's wheel base closely resemble the superstructure of a turn-ear type of plough, which is more carefully depicted in a thirteenth-century manuscript now in Brussels (Figure 36, p. 258).[26]

In the Chartres oculus, the stained-glass artist seems to have conflated two traditional representations, one showing an advanced type of plough[27] and one showing the peasants supervised by a superior. This is a proud ploughing image that portrays the local use of the latest ploughing technology. The details of the plough are carefully drawn, and the horses denote advanced agricultural power. In the diocese of Chartres in the twelfth century, the change from the ox to the horse for ploughing had taken place.[28] As was customary in medieval art, the animals are reduced in number to two, although between four and eight animals were commonly used to pull a plough.[29] An enigmatic red jug stands at the bottom of the picture.

The little oculus contains an inscription in the middle of the scene that is not readable from the cathedral floor: "NOVIGE MEI DA(N)T HA(N)C VITREAM,"[30] or "My Nogentians gave this window." There were several Nogents in the diocese. Nogent-le-Rotrou in the land of the Count of Perche[31] and Nogent-le-Roi in the land of the king of France[32] were not located in prime wheat land, where ploughing was characteristic of the agricultural life.[33] Although we cannot be sure, the reference was most likely to Nogent-le-Phaye, located just southeast of Chartres at the edge of the Beauce wheat fields.[34] It was the administrative center of one of the cathedral chapter's former provostships.[35] The chapter's provosts had greedily mismanaged four large sections of the chapter's farms until 1193, when the chapter took back direct control of them, doubling its income.[36] After 1193, Nogent-le-Phaye constituted four and one-half prebends, directed by a mayor who oversaw the taxation of five parishes within his administration.[37] It is probably peasants of Nogent-le-Phaye and their mayor who appear in the window.

The rural mayors themselves were originally serfs whom the chapter appointed to direct the exploitation of its properties. The mayors' positions became hereditary over time. They had a valuable fief and powerful authority, and therefore came to be nicknamed "village roosters."[38] Of the twenty-seven mayors mentioned in local charters, fifteen were knights or squires.[39] The chapter's peasants in Beauce, on the other hand, remained serfs; they did not obtain their "liberty" through purchase until the end of the Middle Ages.[40] It was most likely the mayor who actually collected the money for the little oculus from the peasants in Nogent-le-Phaye. This might explain the red jar at the base of the window; it may have been symbolic of the collections of small amounts of money from a large number of chapter serfs.[41]

Ploughmen at Tours

In the choir of the St. Gatien Cathedral at Tours, at the base of a clerestory lancet known as the Genesis window, installed after 1255, four teams of peasants are shown ploughing (Figure 50).[42] The stained-glass artists here used two different ploughing scenes in alternation. In one model, two oxen pull the plough; in the other, two horses. The artists drew the ploughs in very abbreviated manner, without distinguishing their components in any detail, emphasizing their rustic nature by showing the handles contrived from forked branches of a tree. In the first model, a peasant holds the forked handle with both hands behind two oxen that are yoked to the axle of a wheeled plough (Figure 51). A second peasant stands behind the oxen, prodding them forward with a long pole while glancing backward and gesturing as if to communicate with the ploughman. Although their faces look as if they have been repainted, it is not insignificant that they both appear distressed. In the second model, horses with horse collars pull the plough, and the second peasant behind the horses carries a switch over his shoulder, suggesting that horses were more willing, easily directed work animals (Figure 52). These peasants' faces look much less stressed.[43]

Scholars have claimed that the window was donated by these ploughmen[44] or by a confraternity dedicated to Adam as the patron saint of laborers.[45] There is no inscription in the window, and no documentation to substantiate either claim.[46] On the other hand, the imagery of ploughing associated with the story of Adam has a long, significant medieval tradition. Above the plough teams, five registers of medallions depict scenes from Genesis. Adam is frequently shown in Genesis cycles ploughing after his expulsion from the Garden of Eden, to indicate his laboring in punishment for his sin. A striking example appears on the bronze doors of San Zeno in Verona, where Adam guides the plough, his son Cain pulls it, and Eve spins.[47] The tradition was known at Tours, since the monks of the Abbey of St. Julien owned the Ashburnham Pentateuch and used its illustrations as models for the murals in their Romanesque church.[48] The Genesis page from the Ashburnham Pentateuch depicts both Adam and Cain ploughing.[49] But this does not appear in the cathedral glass at Tours, where, instead, Adam tills and Eve spins and, in another panel, Adam sows and Cain reaps.[50] Hence the ploughmen, in a sense, replace the usual depictions of Adam and Cain ploughing, to represent the eternal punishment of men after the Fall. Women's punishment is epitomized by Eve spinning, and later in the window by her birthing. But the window, like the Biblical text, is a male-centered narrative.

Figure 50. Bottom section, Genesis window. Tours Cathedral, after 1255. Author's photo.

Figure 51. Detail, ploughmen with oxen. Genesis window, Tours Cathedral. By permission of ARS/SPADEM.

Figure 52. Detail, ploughmen with horses. Genesis window, Tours Cathedral. By permission of ARS/SPADEM.

The upper registers of the Genesis window depict the history of Cain and the Chastisement of Lamech. At the top in three cusped roses, three unidentified, sainted bishops appear enthroned, holding croziers and performing a blessing. Linda Papanicolaou, in a study of the iconography of the Genesis window, has concluded that the window was derived from a Parisian model partly based on apocryphal elaborations of Genesis in *The Book of Adam and Eve*[51] and on the *Historia Scholastica* of Peter Comestor.[52] She suggests that the ploughing scenes refer to the generations after Lamech, who suffered the seventy and seven-fold vengeances he predicted.[53] Since both the idea of ploughing as punishment and the repetition of the generations after Lamech had long been incorporated into the pictorial cycles of Genesis,[54] it is possible that the archbishop and canons ordered repeated images of ploughmen with these notions of punishment in mind. The images themselves do not look like something a successful peasant might commission. The simplicity of the ploughs and the decorative tree additions suggest the city dweller's romanticized concept of the countryside. The scenes are not the kind of proud, detailed image that appears in the oculus at Chartres, where the commission likely came from a country dweller.

But the inference of donation remains, since the ploughmen appear in the bottom register, separate from the Biblical scenes. The bishop and canons at Tours needed donations to help repair war damage as well as to rebuild their cathedral. Battles between the English and French had left Tours destitute in the early thirteenth century. Rebuilding the cathedral of Tours, which had begun in 1168, was repeatedly delayed by the interruptions of war and lack of funds, and was not completed until the fourteenth century.[55] The fenestration of the new choir got under way around 1255,[56] after Saint Louis stopped demanding war taxes from the churches.[57] Local abbeys, especially St. Martin of Tours, repressed communal organization and constrained their peasants in serfdom.[58] Despite evidence of peasant aggrandizement in income and status elsewhere, the local devastation, debts, and repression suggest that local peasants would have been unable to accumulate sufficient surplus to pay for a window.[59] The situations of local peasants did not change until later in the thirteenth century.[60]

While we cannot say who the peasants at the base of the Genesis window might be, we can elaborate on these images as signifiers of property. Tours, like Chartres, was surrounded by wheat fields.[61] The measurement of wheat fields was calculated by the plough. Theoretically, *carrucata* or *aratrum* designated the extent of land that one plough team could work in one year. South of Paris, in the twelfth century, the *carrucata* in documents was 42 hectares 80 ares, or some 106 acres, less than its original Carolingian size.[62] As land values increased, the standard land measurement shrank. Lynn White has calculated that, considering the improvements in medieval ploughing methods, a team of oxen would have been able to work 400 acres of crops in a three-field system where 200 acres were fallow. Horses could work twice as much land as oxen, but were expensive to buy and to maintain.[63] The horses here add to the impression of a very large and rich agricultural domain—perhaps several domains. Adding to the importance of wheat lands toward the mid-thirteenth century was the increasing price of wheat, as a result of the diminishing availability of marginal lands on which to expand cultivation.[64] It is possible that the three sainted bishops at the top of the window symbolize dioceses whose wheat lands provided the funds for the window,[65] or areas where the chapter of St. Gatien owned large wheat fields.[66] Considering the economic distress and social repression at Tours, the images of peasants, like those at Chartres, are most likely symbols of property.

Vineyard Workers at Le Mans

A change occurs in the meaning of this unusual peasant "donor" imagery at Le Mans, where the donation of a large window by local vineyard workers is documented. Construction began on the new choir of the cathedral of St. Julian at Le Mans under Bishop Hamelin in 1217.[67] After delays in construction, Bishop Geoffroy finally fixed the date of the consecration of the new choir on the day after Easter, April 20, 1254.[68] According to the description of the translation of Saint Julian to the new cathedral of Le Mans at that time, vineyard workers who arrived late at the celebration pledged to donate a window. Their pledge is recorded in the *Actus Pontificum Cenomannis* written in the late thirteenth century:

> It is pleasing to add, concerning the head vineyard keepers and the cultivators of vines, who, upon seeing the candles of the others, on whose model they had done nothing, said to each other, talking amongst themselves: Others have made light for the moment; let us make windows to illuminate the church in the future. They garnished it with an entire window composed of five lancets in which they are represented attending to the work of their trade. It is important to say in praise of the other inhabitants, that besides sumptuous lights, they also gave windows where they appear with the signs of their professions.[69]

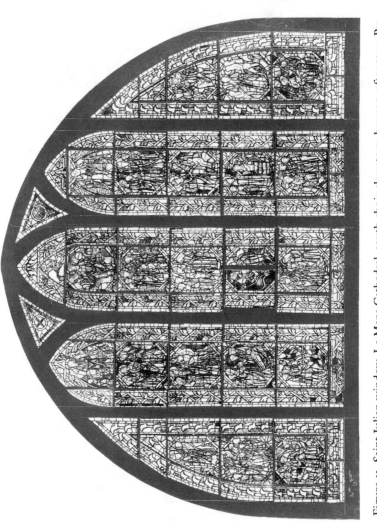

Figure 53. Saint Julian window. Le Mans Cathedral, south choir clerestory, bay 107, after 1254. By permission of ARS/SPADEM.

Figure 54. Detail, vineyard workers. Saint Julian window, Le Mans Cathedral. By permission of ARS/SPADEM.

This is, to my knowledge, the first written evidence of the donation of a church window by agricultural workers in France. The vineyard workers appear in four medallions, which seem to show three moments of viticulture: Two scenes depict peasants pruning vines in winter and in spring, and the other two show peasants, after the harvest, tasting the new wine.[70] The rest of the window depicts the history of Saint Julian in eleven medallions interspersed with four scenes of Christ and the Virgin Mary (Figure 53).

The first scene of vineyard workers depicts two men wearing *cottes*, who tend a vine (panel 2, Figures 53 and 54). An older man, differentiated by his beard and hood, trims back a leafed branch with a *serpe*; behind him,

Figure 55. Detail, vineyard workers. Saint Julian window, Le Mans Cathedral. After André Bouton, *Le Maine: Histoire économique et sociale* 1 (Le Mans: Monnoyer, 1962).

a younger man wearing a cap tills the soil with a hoe. The second scene shows a similar pair of men pruning a vine that is bare of leaves (panel 8, Figures 53 and 55). The older man in front bends far over to prune a branch, while the younger man leans over him to move another branch out of the way. Behind them in the blue background of the medallion two huge white *serpes* appear suspended in the air.[71] Similar upright *serpe* heads form a heraldic pattern in the upright borders of the two outside lancets of the five-lancet window. This is a unique design, likening the vineyard workers' tool to the upper class coat-of-arms motifs that appear frequently in other clerestory windows at Le Mans.[72]

The two wine-tasting scenes appear as pendants at the ends of the lowest register. The scenes take place in vineyards after the wine harvest. The bare branches suggest winter, as do the long robes of the men. In the

Figure 56. Detail, wine tasting. Saint Julian window, Le Mans Cathedral. By permission of ARS/SPADEM.

left medallion, the largest man in the center holds a pitcher on his knee with his left hand and gestures in conversation to the man to his left (Figure 56). This man looks like the older man in the pruning scenes, bearded and hooded, but he wears a full-length robe. He holds a goblet with his left hand and touches the larger man next to him on the arm. These two are engaged in a conversation, presumably about the wine, which the young man on the left seems to be tasting from a small vial. The other wine-tasting scene on the right of the lowest register depicts four men seated in a vineyard in winter (Figure 57). Here the two men in front look very much like the two vine trimmers. The older man wears gloves and holds a *serpe* with his left hand while reaching out for a drink, while the younger man prepares to pour out wine into a goblet.[73]

    Such scenes of wine-tasting are not found in depictions of the Labors of the Months.[74] At Le Mans the wine-tasting scenes imply the vineyard workers' change from simple laborers to men who enjoy the wine they have

Figure 57. Detail, wine tasting. Saint Julian window, Le Mans Cathedral. By permission of ARS/SPADEM.

produced. This change is underlined by the repeated images of pruning tools like blazons in the borders.

We have no clue about who these men were beyond the cryptic "clausorii et cultorii vinearum" in the *Actus Pontificum Cenomannis*. A *clausorius* was the manager of an enclosed vineyard, or *clos*. Enclosed vineyards were usually old, top-quality implantations held by lay and ecclesiastical lords. Wine production promised much greater income in good years relative to grain culture.[75] In the thirteenth century, the most valuable old vineyards usually had managers appointed by their noble owners, and tenant farmers who were assisted by day laborers.[76] At Le Mans, vines were planted all around the town, but larger vineyards were located to the southeast. Hence, it seems likely that these vineyard managers arrived late at the cathedral consecration because they had come from vineyards beyond the suburbs of Le Mans.

These vineyard workers may have come from their own vineyards. In this area many contracts of *complant*—that is, contracts for new vine plantings on noble lands—show peasants receiving a large part of the new vineyards in reward for their efforts.[77] It is believable that such enterprising vineyard workers at Le Mans could have paid for the big window.[78] In the course of the thirteenth century, the previous norm, whereby the peasant with a contract of *complant* who planted new vines received one-half of the vineyard, gradually changed in favor of the peasant. Increasingly, the peasant received the whole vineyard and the property owner received only one-half of the first harvest and then a fixed amount of wine yearly after that.[79] This window, if it does portray enterprising vineyard laborers who made contracts of *complant*, as I believe, presented an image charged with the suggestion of diocesan social and economic advancement.[80]

The political implications of the window are also significant. Not by chance, it seems, the window combines images of Saint Julian, the patron saint of the cathedral, with images of Christ at a time when the cathedral bishop, Geoffroy of Loudun, was trying to establish an apostolic legend of the saint. Saint Julian had long been worshiped as the founder of the cathedral, but in the thirteenth century the idea that he was an apostle of Christ was widely disseminated and was confirmed or suggested in the imagery of three thirteenth-century stained glass windows dedicated to the saint in cathedrals at Le Mans, Tours, and Angers.[81] The claim that Saint Julian was an apostle of Christ apparently met a need for the elevation of episcopal prestige and personal legitimacy in response to Bishop Geoffroy's repeated humiliation in local conflicts. The beleaguered bishop traveled three times to Italy to seek the pope's intervention on his behalf.[82] Hence the progressive social position of the vineyard workers contributed to the political amplification of the bishop.

There is another, only partly understood, political meaning to the donation imagery in the choir. It has to do with the tensions between the bishops of Le Mans and Charles of Anjou, who received the royal apanage of Touraine, Maine, and Anjou on his majority in 1246.[83] Meredith Lillich discovered that several of the coats of arms in the windows, presumed to indicate donations of windows, belonged to local nobles who joined the forces of Charles d'Anjou in his short-lived Flemish campaign of 1254. She reasoned that the windows must have been given on the nobles' return from the campaign, and conjectured that the bishop hastily planned the translation of Saint Julian's body to occur on April 24, 1254, to create a patriotic event that would send the local armed forces off in glory.[84]

Attractive as this idea might be, it is partly contradicted by the description of the translation in the *Actus Pontificum Cenomannis*: "Moreover, when the prince [Charles of Anjou] wanted to drag the citizens into the army, threatening their physical well-being and loss of their possessions, they chose rather to be subject to both dangers than to miss seeing the longed-for translation of the most holy patron."[85] The bishop had obtained from the pope 100 days' pardon and indulgences for all attending the translation, but this did not soothe the count of Anjou's fury. Indeed, animosity escalated between Count Charles of Anjou and Bishop Geoffroy. The year after the translation, Bishop Geoffroy died in Rome while seeking Pope Alexander IV's support, after he refused to comply with the count of Anjou's demand for homage.[86] In the years immediately following, antagonism between the count and the new bishop became so exacerbated that the count's officers forced the bishop to flee from his diocese.[87] It is not yet clear how the knights who fought for the count came to donate windows to the bishops, who fought against the count. But the donation imagery of nobles, knights, monks, tradesmen, and vineyard workers in the choir makes an ideological statement of local support for the cathedral at a time when the cathedral bishops engaged in protracted altercations—not just with the count but also with local monastic houses and the archbishop of Tours.[88] Hence it was politically astute that Bishop Geoffroy de Loudun had his own image placed in the center lancet of the new choir together with eight small images of his coat of arms, surrounded by all these images of local allegiance to him and his cathedral.[89]

Conclusion

The new images of peasants as donors had very different meanings. In the first example, at Chartres, the peasants working on vines evidently represented a vineyard given over to the church. In the second example at Chartres, the ploughmen represented five parishes in Beauce from which their mayor must have extracted funds for the oculus. At Tours, the images imply punishment for sin inherent in the depictions of ploughing associated with Genesis. In view of the socially repressive policies of ecclesiastical lords in the region, the donors were probably bishops. Le Mans seems to be the one instance where documentation and agricultural entrepreneurship combine to argue in favor of peasant donation. At Le Mans, the images of vineyard workers pruning vines and tasting wine appear in the programmatic statement of public support in a cathedral whose bishops were frequently the losers in local conflicts. Political confrontation and economic crisis were common to Chartres and Tours as well. The peasant images at Chartres and Tours also take their place in a broad ideological panorama that represents all classes of society as obeisant benefactors of the church.[90]

The inclusion of peasants in new artistic contexts in thirteenth-century cathedral windows reveals that ecclesiastical lords were beginning to value the labor of peasants, who could no longer be thought of as immutable elements of nature. It is not by chance that these rare instances when peasants break through a social barrier in art occur in cathedrals whose resources were agriculturally based. The churchmen were aware of changing social formations in the countryside. But it was precisely in areas where ecclesiastical lords sought to hold back the trend of peasant enfranchisement, at Chartres and Tours, that peasants were first given artistic stature in place of legal status. The bishops and canons felt the need to represent publicly the proper role of peasants to labor in support of the church. They ordered artists to single out agricultural workers for scrutiny in public places and present them as cleaned up, cheerful figures doing their work as if it were easy—which it was not—and as if ploughing and pruning were beautiful, even enviable occupations. It does not last, this idealized image of the prosperous peasant. In the fourteenth century, artistic representation returns peasants to their monthly labors and to marginalized drolleries in luxurious manuscripts for noble patrons.

In the contemporary debate over the transition from feudalism to capitalism, traditional concepts of secular Malthusianism and economic transformation have marginalized the role of peasants in medieval social change. In opposition to this, Robert Brenner has argued that the relationship between peasants and lords is precisely the crucial factor in the relatively retarded development toward capitalism in France compared with England.[91] I think that this study offers support to Brenner's materialist insistence on the vital economic factor of agricultural labor in social change. The obscurity of French peasants in medieval historical records and medieval art is evidence of their feudal suppression, not their unimportance.

In a separate discourse, Marxist art history, empowered by a critical stance toward the relationship between medieval art production and its agricultural base, has been increasingly marginalized by literary and semiological theoreticians for its alleged failure to see the multiplicity of social interactions and irretrievably complex historical variables mediating the relationship between levels of the social organization, known in Marxist analysis as base and superstructure. The peasant imagery discussed here offers examples of striking diversity in the meaning of apparently similar representations emerging from historical research. Ploughmen and pruners in the early thirteenth century were not donors, but rather signifiers of property. Representation of agricultural entrepreneurship, precocious at Le Mans, gave way soon to a deeper entrenchment of lordly privilege in art. The unusual peasant imagery has much to tell about the stress in the social hierarchy, but its modern interpretation without historical basis remains far from accurate, as this demonstration, I hope, has shown. I would like to think that these examples argue for the return of history to art history and for the revitalization of a materialist stance in scholarly work.

This article was presented as a lecture at two stages of its development, first at the *Rusticus* Conference at the University of British Columbia in 1986, and secondly at a session on "The Margins of Medieval Art," College Art Association Annual Meeting, New York, 1990. I am grateful to the University of Arizona for providing a subvention for the illustrations for this essay. Thanks also are due to the Medieval Studies Committee for including me in the Workshop in 1986 and supporting my travel there, and to Jerrilynn D. Dodds, for including me in her Session at CAA. I am also grateful to Barbara Abou-El-Haj for her inspiration, Jeryldene Wood for her encouragement, and Cynthia Hibbard and John Day for their help.

Notes

1. See James C. Webster, *The Labors of the Months in Antique and Mediaeval Art to the End of the Twelfth Century* (Princeton, N.J.: Princeton University Press, 1938); Perrine Mane, *Calendriers et techniques agricoles (France-Italie, XIIe–XIIIe siècles)* (Paris: Sycomore, 1983); Perrine Mane, "Comparaison des thèmes iconographiques des calendriers monumentaux et enluminés en France, aux XIIe et XIIIe s.," *Cahiers de Civilisation Médiévale* 29 (1986): 257–64; Christine Östling, "The Ploughing Adam in Medieval Church Paintings," *Man and Picture: Papers from the First International Symposium for Ethnological Picture Research, Lund, 1984*, ed. Nils-Avrid Bringéus (Stockholm: Almqvist & Wiksell, 1986), pp. 13–19. Jonathan Alexander, while writing primarily about broad ideological issues in the imagery of the *Très Riches Heures*, an early fifteenth-century illustrated calendar, touches briefly on peasant conditions at the time and calls for more research on this issue: "*Labeur* and *Paresse*: Ideological Representations of Medieval Peasant Labor," *Art Bulletin*

72 (1990): 436–52, esp. 449–50. The chapters in this book by Michael Camille and Bridget Ann Henisch are excellent exceptions to the rule, as is the article by Michael Camille, "Labouring for the Lord: The Ploughman and the Social Order in the Luttrell Psalter," *Art History* 10(4) (1987): 423–54.

2. For example, in early medieval Genesis cycles, Adam and Eve labor like medieval peasants in punishment for disobeying God. Engelbert Kirschbaum et al., eds., *Lexikon der christlichen Ikonographie* 1 (Rome: Herder, 1968): 42–70; Louis Réau, *Iconographie de l'art chrétien* 2, pt. 1 (Paris: Presses Universitaires de France, 1956): 77–93. See also the discussion in Alexander, "*Labeur* et *Paresse*," pp. 437–38.

3. See Gertrud Schiller, *Iconography of Christian Art*, trans. Janet Seligman (Greenwich, Conn.: New York Graphic Society, 1971) 1: 84–88.

4. A "principle of necessary inequality" emerged from the first medieval written declarations of the proper Christian social order. Georges Duby, *The Three Orders: Feudal Society Imagined*, trans. Arthur Goldhammer (Chicago: University of Chicago Press, 1980), p. 59. Jonathan Alexander, "*Labeur* et *Paresse*," also noticed the differentiation of the imagery of the three orders in his study of the *Très Riches Heures*.

5. Yves Delaporte, *Les Vitraux de la cathédrale de Chartres* (Chartres: E. Houvet, 1926), I, 223–26, Bay 16, plates XLIV–XLVII, color plates VII and VIII (hereafter Delaporte and Houvet). For a complete bibliography on the windows, see Louis Grodecki et al., *Les Vitraux du centre et des pays de la Loire*, Corpus Vitrearum, Recensement des Vitraux Anciens de la France 2 (Paris: Éditions du Centre National de la Recherche Scientifique, 1981): 31–32, Bay 28 (hereafter *CVR* 2). For a bibliography on medieval stained glass, see Madeline Harrison Caviness, *Stained Glass Before 1540: An Annotated Bibliography* (Boston: G. K. Hall, 1983).

6. Delaporte and Houvet, pp. 227–33, plates XLVIII–LI; *CVR* 2: 31–21, Bay 28.

7. Delaporte and Houvet, Bay 160, p. 508, plate CCLXIX; *CVR* 2: 41, Bay 131.

8. Linda Morey Papanicolaou, "The Iconography of the Genesis Window of the Cathedral of Tours," *Gesta* 20 (1981): 179–89, and Papanicolaou, "Stained Glass Windows of the Choir of the Cathedral of Tours," unpublished Ph.D. dissertation, New York University, 1979. See also Jean Jacques Bourassé and Canon Manceau, *Verrières du choeur de l'église metropolitaine de Tours* (Paris: V. Didron, 1849), pp. 65–66; Henri Boissonnot, *Les Verrières de la cathédrale de Tours* (Paris: Frazier-Soye, 1932), pp. 25–28; Emile Mâle, *The Gothic Image: Religious Art in France of the Thirteenth Century*, trans. Dora Nussey (New York: Harper, 1972), p. 205; Francis Salet, *La Cathédrale de Tours* (Paris: H. Laurens, 1949), p. 35.

9. André Mussat, *La Cathédrale du Mans* (Paris: Berger-Levrault, 1981), Bay 107, p. 113; Ambroise Ledru, "La cathédrale du Mans," *Inventaire general des richesses d'art de la France: Province, Monuments religieux* 4 (1907): 256–57; Louis Grodecki, "Les vitraux de la cathédrale du Mans," *Cahiers Archéologiques de France*, CXIXe session Maine (1961), pp. 59–99, esp. 84–85; *CVR* 2: 251, Bay 107.

10. The only possible addition is the Lubinus window at Chartres Cathedral. See Jane Welch Williams, *Bread, Wine and Money: The Windows of the Trades at Chartres Cathedral* (Chicago: University of Chicago Press, 1993), pp. 73–95.

11. A record from Soissons states that the king gave a lancet for 30 livres, and

a record from Limoges lists a window price of 23 livres, which might mean that it was smaller or that prices varied. Louis Grodecki, *Les Vitraux de Saint-Denis: Étude sur le vitrail au XIIe siècle*, Corpus Vitrearum Medii Aevi, France, Études 1 (Paris: Centre National de la Recherche Scientifique, 1976): 27 and n. 19.

12. In reality, peasants normally did not receive daily wages, making their ability to accumulate the cost of a window even more unlikely.

13. On the trade windows of Chartres, see Williams, *Bread, Wine and Money*.

14. The windows are dated 1217–20 in *CVR* 2: 31; John James suggested 1208–9 or after. *The Contractors of Chartres*, 2nd ed., 2 vols. (London: Croom Helm, 1979–81) 1: 282–85. See also Williams, *Bread, Wine and Money*, pp. 15–17.

15. Delaporte and Houvet, p. 223 and plate XLIV.

16. Ibid., p. 227 and plate XLVIII.

17. Ibid., p. 230.

18. Ibid., pp. 223, 227.

19. Ibid., pp. 228–29. COMES TEOBALD(us) DAT must mean "Count Thibaut gave." We might deduce from this that Count Thibaut was the donor of the window, not the vineyard workers. But HO—presumably for hoc—and VESPO VINERUS (the latter misconstrued Latin) can be translated as "this vineyard by Vespres," evidently referring to vines given by the count. Hence the count gave vines, not necessarily the window. The inscription ends with AD PRECES COMITIS PERTICEENSIS, or "to the prayers of the count of Perche," to indicate that the vineyard was given in memory of the count of Perche. Delaporte believed that the count, not the vineyard workers, donated the window, but two pages later offered another hypothesis: Delaporte and Houvet, p. 230.

20. Lucien Merlet, *Dictionnaire topographique du Département d'Eure-et-Loir* (Paris: Imprimerie Impériale, 1861), p. 187.

21. E. de Lépinois and Lucien Merlet, eds., *Cartulaire de Notre-Dame de Chartres*, 3 vols. (Chartres: Garnier, 1861–65), 2: 287 (hereafter *CND*). Also found in the medieval documents as Bogleinval, Bugleinval, Buglainvallis, Bouglainvallis, Buglainvilla. *CND* 3: 242. The region had many vineyards, since the thirteenth-century *Polypticon* of Chartres Cathedral lists among the Bouglainval prebend dues a payment for matins of 15 livres from the wine tithes; ibid., p. 291. The prebend income included 20 sous from a monastery pressorium and income from five arpents of vines; ibid., p. 343. Also in the prebend of Bouglainval was the town of Medium-Vicini where the chapter had a press, a cellar, and three arpents of vines; ibid., p. 344. Other wine income in the prebend list further indicates a large number of vineyards; ibid., pp. 345–46.

22. Delaporte thought that the inscription along with the adjacent images might be explained as perpetuating the memory of a gift of vineyard property by the count to a religious establishment of monks who were the donors of the window; ibid., p. 230.

23. Perrine Mane found that the more usual labor cycles included only the trimming of vines in March and the trampling of grapes in September. According to his calculation, only 15 percent of labor cycles of the twelfth and thirteenth centuries show viticulture three months of the year, and only 23 percent show the filling

of a barrel. *Calendriers*, pp. 186–98. The twelfth-century archivolt reliefs on the west facade at Chartres have only two wine-producing scenes, March pruning and September treading of grapes in a vat. Webster, *Labors of the Months*, p. 157. In fact, of all five medieval representations of agricultural calendars at Chartres, only the Zodiac window contains three scenes of wine production; see Marcel J. Bulteau, "Étude iconographique sur les calendriers figurés de la cathédrale de Chartres," *Société Archéologique d'Eure-et-Loir: Mémoires* 7 (1882): 197–224, esp. 222–23.

24. The date of the oculus is controversial: The Corpus Vitrearum dates it to 1205–15, while others suggest a date after 1223. *CVR* 2: 41; Williams, *Bread, Money and Wine*, pp. 15–16.

25. Vienna, Nationalbibliothek Cod. 507, fol. 1. The image is reproduced in Robert W. Scheller, *A Survey of Medieval Model Books* (Haarlem: Erven F. Bohn, 1963), catalogue no. 9, fig. 34. See also Alexander, "*Labeur* et *Paresse*," fig. 17; *Reiner Musterbuch*, facsimile and commentary by Franz Unterkircher, Codices Selecti 64, 64* (Graz: Akademische Druck- und Verlaganstalt, 1979).

26. Brussels, Bibliothèque Royale Albert I, Ms. 1175, known as "Rentier d'Audenarde." The manuscript is a polyptique (revenue list) of a Flemish landlord named Messire Jehan de Pamele-Audenarde, which is enlivened by pen drawings. The manuscript is dated to approximately 1275. See Léo Verriest, *Le Polyptyque illustré dit "Veil Rentier" de Messire Jehan de Pamele-Audenarde (vers 1275)* (privately published, Brussels, 1950), pp. xiii–xciii.

27. Gérard Sivéry, *Terroirs et communautés rurales dans l'Europe occidentale au moyen âge* (Lille: Presses Universitaires de Lille, 1990), pp. 17–27.

28. André Chédeville, *Chartres et ses campagnes (XIe–XIIe siècle)* (Paris: Klincksieck, 1973), p. 79; Marc Bloch, "Les problèmes du peuplement Beauceron," *Mélanges Historiques* 2 (Paris: S.E.V.P.E.N., 1963): 646. On the advantages of horse power, see Lynn White, Jr., "The Discovery of Horse Power," in White, *Medieval Technology and Social Change* (New York: Oxford University Press, 1970), pp. 57–69; Guy Fourquin, *Le Paysan d'Occident au Moyen Age* (Paris: F. Nathan, 1972), p. 89; John Langdon, *Horses, Oxen and Technological Innovation* (Cambridge: Cambridge University Press, 1986), p. 160.

29. H. G. Richardson, "Historical Revision No. c: The Medieval Plough-Team," *History* n.s. 26 (1941–42): 287–296, esp. 288–89.

30. Delaporte and Houvet, plate CLX.

31. Chédeville, *Chartres et ses campagnes*, pp. 39–40. The area was known for its wool textile production; ibid., pp. 71, 448, 450, 455.

32. Ibid., pp. 42–43. This area has a very thin layer of silt on a rocky base, unfit for wheat production, and is adjacent to the forest of Yveline; ibid., pp. 58, 67. Philip Augustus confiscated the land from the Montfort family in 1200. See Roger Durand, "Le domaine de Nogent-le-Roi au XVIIe siècle," *Société Archéologique d'Eure-et-Loir: Mémoires* 15 (1915–22): 117–38, esp. 117.

33. Chédeville, *Chartres et ses campagnes*, pp. 56–59. There also were a castle and a priory with Nogent placenames, and three other towns. On the castle of Nogent-le-Rotrou, see ibid., pp. 270, 274–76, 410; on the priory of Saint-Denis at Nogent-le-Rotrou, see ibid., pp. 132, 196 (n. 176), 204, 225, 270, 328, 348, 365

(n. 190). Delaporte mentions also the possibilities of reference to Nogent-sur-Eure and Nogent-sous-Coucy. Delaporte and Houvet, p. 508. Topographic dictionaries of the region were published in the cathedral *cartulaire* by Lépinois and Merlet and in a separate volume by Merlet; *CND* 3: 229–317, and Merlet, *Dictionnaire topographique*. Nogent-sur-Eure was a village on the banks of the Eure River, undoubtedly planted with vines along the river and mostly with grain in the adjacent cleared lands. It was the center of one of the Chapter's provosts who became corrupt and was dismissed. *CND* 1, no. 56: 154–55, and Louis Amiet, *Essai sur l'organisation du Chapitre Cathédrale de Chartres du XIe au XVIIIe siècle* (Chartres: F. Laine, 1922), 112–18. The provosts were replaced with mayors (*CND* 1: 237). It was in the land held primarily by the Count of Perche (*CND* 1, no. 110: 218–19); but the area paid grain yearly to the chapter's granary (*CND* 2: 292). It could have been this Nogent referred to in the window, but this is less likely since it was not as grand a center of the chapter's wheat production as was Nogent-le-Phaye. The other possibility, Nogent-sous-Coucy, is much more obscure, since it is unmentioned in the topographic dictionaries of the region.

34. The prebend of Nogent-le-Phaye grew rapidly in the thirteenth century due to donations. *CND* 2, no. 227: 88 and n. 2.

35. *CND* 3: 285.

36. *CND* 1, intro.: c, and charters no. 86 (1171), 188–90; no. 119 (1193), 225–27; and no. 128 (1195), 248–49.

37. *CND* 2: Polypticon, 289–90.

38. Chédeville, *Chartres et ses campagnes*, pp. 386–89. On this phenomenon, see also Monique Bourin and Robert Durand, *Vivre au village au moyen âge: Les solidarités paysannes du 11e au 13e siècles* (Paris: Messidor/Temps Actuels, 1984), pp. 113–14.

39. Chédeville, *Chartres et ses campagnes*, p. 289.

40. Ibid., p. 369.

41. This could have been a *taille*, or special tax charged to peasants by their lord for his special needs. See Bourin and Durand, *Vivre au village au moyen âge*, pp. 199–200.

42. See note 8.

43. The tree motif in the ploughing scenes repeats as the Tree of Life in the scene of Adam and Eve tasting its fruit in the next register. One wonders if exegetical metaphor is suggested here.

44. Papanicolaou, "Stained Glass Windows," pp. 87–88.

45. This is especially suggested since the imagery follows the apocryphal text of the *Book of Adam and Eve*, or *Vita Adae*; Papanicolaou, "Genesis Window," p. 179. Boissonnot thought the window was donated by a confraternity of laborers. *Verrières de Tours*, p. 25.

46. The search by Chauvigne in local archives for occupational corporations did not reveal any such confraternity. See Auguste Chauvigne, "Histoire des corporations d'arts et métiers de Touraine," *Société d'Agriculture, Sciences, Arts et Belles-Lettres du Département d'Indre-et-Loire: Annales* 62 (Tours: La Société, 1884): 254–79.

47. See Waltraud Neumann, *Studien zu den Bildfeldern der Bronzetür von San*

Zeno in Verona (Frankfurt: Haag und Herchen, 1979), plate 30; and Ursula Mende, *Die Bronzetüren des Mittelalters, 800–1200* (Munich: Hirmer, 1983), plate 73.

48. André Grabar, "Fresques romanes copiées sur les miniatures du Penta-teuque de Tours," *Cahiers Archéologiques* 9 (1957): 329–41; Annabelle Simon Cahn, "A Note: the Missing Model of the Saint-Julien de Tours Frescoes and the Ashburnham Pentateuch Miniatures," *Cahiers Archéologiques* 16 (1966): 203–7.

49. Bibliothèque Nationale, MS. nouv. acq. lat. 2334, fol. 6r. On the Ashburnham Pentateuch, see Oskar L. Gebhardt, *The Miniatures of the Ashburnham Pentateuch* (London: Asher, 1883).

50. Reading from bottom to top and from left to right, as the window is intended to be read, these scenes are panels 12 and 15. For a reproduction of the entire window in painted color, see Boissonnot, *Verrières de Tours*, plate IV.

51. Papanicolaou, "Genesis Window," pp. 179–89.

52. Ibid., pp. 179–89, especially 183.

53. Ibid., p. 186, n. 39, and Papanicolaou, "Stained Glass Windows," p. 89.

54. See, for example, the Aelfric Paraphrase cycle (British Museum, MS. Cotton Claudius B IV) and M. R. James, *Illustrations of the Book of Genesis, Egerton 1894* (Oxford: Roxburghe Club, 1921).

55. Étienne Giraudet, *Histoire de la ville de Tours* (Tours, 1873; reprint Marseille: Laffitte, 1977) 1: 103–32.

56. Papanicolaou, "Genesis Window," p. 179.

57. Papanicolaou, "Stained Glass Windows," p. 24. On the war taxes, see also Giraudet, *Histoire de Tours* 1: 109–10.

58. Ibid. 1: 105–9, 124. The monks of the great abbey of Marmoutier were endeavoring to build an enormous Gothic church, but their project was interrupted by the violent attack of the count of Blois, who seized the abbey and its revenues in 1237; the monks did not recover their rights until after 1253. The atmosphere of crisis and despair seems to have been widespread around Tours during the mid-thirteenth century. Charles Lelong, *L'Abbaye de Marmoutier* (Chambray les Tours: C. L. D., 1989), 33–34.

59. There is no local evidence to support Papanicolaou's idea that the ploughers in the window might have been "village capitalists" who owned their own draft animals. "Stained Glass Windows," p. 87.

60. Only in the last quarter of the thirteenth century did the situation change as fiefs were allowed to be sold. This opened the way for laborers to acquire noble inheritances, but the loosening of social rank was caused by economic hardship and seems primarily to have benefited merchants. Giraudet, *Histoire de Tours* 1: 110–11.

61. See Roger Dion, *Le Val de Loire: Étude de géographie régionale* (Tours: Arrault, 1934; reprint Marseille: Laffitte, 1978), p. 306 and fig. 38. Paul Fenelon, *Atlas et geographie des pays de la Loire* (Paris: Flammarion, 1978), plan 44, shows 25 to 50 percent of the area around Tours still planted in cereals. The local churches received large tithes in wheat. Louis de Grandmaison, ed., *Cartulaire de l'archévêché de Tours*, Mémoires de la Société Archéologique de Touraine, 2 vols. (Tours: L. Péricat, 1892–94), 2: nos. 309–11.

62. M. Guérard, *Cartulaire de l'abbaye de Saint-Père de Chartres* (Paris: Crapelet, 1840) 1: clxviii–clxix.

306 Jane Welch Williams

63. Fourquin, *Paysan d'Occident*, pp. 89–91; White, "Discovery of Horse Power," p. 62.

64. John Day, "Crise de féodalisme et conjoncture des prix à le fin du moyen âge," *Annales: Économies, Sociétés, Civilisations* 34 (1979): 305–18.

65. For example, St. Julian of Le Mans.

66. The canons were acquiring wheat lands through purchase and confirming rights to other agricultural lands at the time. See Grandmaison, ed., *Cartulaire de l'Archévêché de Tours* 1: no. 76 (1265), no. 77 (1265), no. 84 (1257–70).

67. Gabriel Fleury, *La Cathédrale du Mans* (Paris: H. Laurens, 1925), p. 14. For a history and description of the construction, see Frances Salet, "La cathédrale du Mans," *Congrès Archéologique de France*, CXIXe session Maine (Paris, 1961), pp. 18–58.

68. Ambroise Ledru, *La Cathédrale Saint-Julien du Mans* (Mamers: G. Fleury et A. Dangin, 1900), p. 235.

69. The whole passage reads: "Addere libet de clausoriis et cultoribus vinearum, qui videntes cereos aliorum, ad quorum exemplum nichil fecerant, inter se mutuo loquentes aiebant: Alii fecerunt momentaneum luminare; faciamus vitreas, que illuminent ecclesiam in futurum. Fecerunt autem formam integram quinque vitreas continentem, in quibus ipsi per officia depinguntur. Nec a laudibus civium putavimus esse tacendum, quia, etsi luminaria fecerint sumptuosa, nichillominus fecerunt vitreas, in quibus per officia sunt depicti." "Gesta Domini Gaufridi de Loduno Episcopi," in *Actus Pontificum Cenomannis in urbe degentium*, ed. G. Busson and A. Ledru, Archives Historiques du Maine 2 (Le Mans, 1902), p. 491.

70. The panels seem to have been placed out of normal order. Reading from left to right and from bottom to top, the two vine-pruning scenes are panels 2 and 8, and the unusual scenes of wine-tasting are in panels 1 and 5 at either end of the lowest register. It may be that there were originally five scenes of viticulture across the bottom register, and that one of the other scenes, possibly the single scene of the Harrowing of Hell, was added during a repair campaign. Or perhaps the two wine-pruning scenes originally were in the outer two lancets with the wine-tasting scenes, which would seem logical since it is these lancets that are surrounded with repeated *serpe* heads.

71. This curious suspension of tools above workers' heads has a long history in European art, going back to the funeral stone relief carvings of late Roman tradesmen. Stained glass artisans repeated the iconographic custom in some of the windows of the trades at Chartres—for example, the stone carving scenes in the St. Cheron window depict tools suspended over the carvers' heads; Delaporte and Houvet, pp. 337–44 and plate CXXIII.

72. For example, the axial clerestory window with the image of Bishop Geoffroy de Loudun as donor repeats the image of his blazon eight times.

73. The faces of the two men seated behind them are only partly visible. I am indebted to the careful description of these scenes by André Bouton, *Le Maine: Histoire économique et sociale* 1: *Des Origines au XIVe siècle* (Le Mans: Monnoyer, 1962), p. 356.

74. Similar scenes sometimes appear in cycles of Noah, which show Noah's sons giving him wine of their first harvest after the flood. See, for example, the

window of Noah at Chartres. Delaporte and Houvet, Bay 64, pp. 409–11 and plate CLXXXI.

75. Because of the relative profitability of wine production, the independent vintners were heavily taxed. See, for example, Edouard Perroy, *La Terre et les paysans en France aux XIIe et XIIIe siècles* (Paris: Société d'Édition d'Enseignement Superieur, 1973), pp. 144–45. For the history of the production of wine in France, see Roger Dion, *Histoire de la vigne et du vin en France des origines au XIXe siècle* (Paris: Flammarion, 1959).

76. Georges Duby, *Rural Economy and Country Life in the Medieval West*, trans. Cynthia Postan (Columbia: University of South Carolina Press, 1968), p. 140.

77. Bouton, *Maine*, p. 344; Robert Latouche, "Un aspect de la vie rurale dans le Maine au XIe et au XIIe siècle: l'établissement des bourgs," *Moyen Age* 47 (1937): 44–64, esp. 58.

78. I have not found prices for wine and vineyard work for the second half of the thirteenth century in Maine, but in 1335 a muid of wine was worth 12 livres, daily workers pruning vines earned 2 sous 6 deniers per day to 3 sous, and a grape harvester earned 8 deniers. See André Joubert, "La vie agricole dans le Haut-Maine," *Revue Historique et Archéologique du Maine* 19 (1886): 288–89.

79. Roger Grand, *Le Contract de complant depuis les origines jusqu'à nos jours* (Paris: L. Tenin, 1917), esp. pp. 40–41; see also Constant Gauducheau, "Une coutume du moyen âge venue jusqu'à nous: les vignes à complant," *Aquitaine* 14(100) (1980): 329–33. These contracts were also called *méplant*. Duby, *Rural Economy and Country Life*, p. 139.

80. There might have been a deliberate reference to the improvement in the wines enjoyed by the landlords who had established contracts of *complant*, since the original impulse for the contracts seems to have come from the poor quality of wine sometimes presented to landlords by dishonest serfs; Gauducheau, "Coutume du moyen âge," p. 330.

81. Saint Julian's apostolic legend was based on the model of Saint Martial. Linda Morey Papanicolaou, "A Window of St. Julian of Le Mans in the Cathedral of Tours: Episcopal Propaganda in Thirteenth-Century Art," *Studies in Iconography* 7–8 (1981–82): 35–64.

82. For the history of Bishop Geoffroy de Loudun, see Ledru, *Cathédrale Saint-Julien*, pp. 233–42; also Busson and Ledru, *Actus Pontificum Cenomannis*, pp. 486–505. The thirteenth-century bishops at Le Mans also were humiliated by a group of cathedral canons who were trying to supersede episcopal authority on the model of the canons at Chartres Cathedral. See L. Hubert, "Enquête de 1245 relative aux droits du chapitre Saint-Julien du Mans," *La Province du Maine* 3 (1923): 172–81; Julien Chappée, Ambroise Ledru, and Louis-J. Denis, "L'exemption et les privilèges du chapitre du Mans," *Enquête de 1245 relative aux droits du chapitre Saint-Julien du Mans* (Paris: Champion, 1922), pp. cviii–cxvi.

83. This ambitious prince launched wars in Naples, the Balkans, and elsewhere in an attempt to establish his own empire. See François Lebrun, *Histoire des pays de la Loire* (Toulouse, 1972), p. 156.

84. Meredith Lillich, "The Consecration of 1254: Heraldry and History in the Windows of Le Mans Cathedral," *Traditio* 38 (1982): 344–52.

85. "Preterea, cum cives ipsos, sub interminatione pene corporum et amissionis rerum, in exercitum trahere vellet princeps, elegerunt magis utrique periculo subjacere, quam diu desideratam translationem patroni sanctissimi non videre." Busson and Ledru, *Actus Pontificum Cenomannis*, p. 491.

86. Ibid., pp. 241, 503.

87. Paul Piolin, *Histoire de l'église du Mans* 4 (Paris: Julien, Lanier, 1858): 405–411.

88. Ibid., pp. 422–23.

89. Ledru, "Cathédrale du Mans," p. 253.

90. On Chartres, see Williams, *Bread, Wine and Money*, pp. 19–36. On Tours, see Giraudet, *Histoire de Tours* 1: 103–11.

91. In 1976, Robert Brenner published an article in the journal *Past and Present* that attacked demographic determinism in the interpretation of agrarian social transformation. The article stimulated a debate in subsequent literature about the nature and conditions of the transition from feudalism to capitalism in western Europe. The heated and ideological combustion of this debate led to the publication in 1985 of the key arguments by the main protagonists in *The Brenner Debate: Agrarian Class Structure and Economic Development in Pre-Industrial Europe*, eds. T. H. Aston and C. H. E. Philpin (Cambridge: Cambridge University Press, 1985). Brenner argued, I believe correctly, that class relations between peasants and lords were the key to different transitions toward capitalism. The thirteenth century saw increasingly rigid controls over the peasantry in northern France generally. The old feudal relationship between peasants and local lords broke down in France slowly, giving way to individual proprietorship. The process was already under way in the thirteenth century, especially in the Paris basin. The windows are the first indications of this change toward local "liberation" in the more conservative, poorer provinces of central France. But smallhold peasantry freed from local lords' surplus extraction did not lead to the accumulation and specialization prerequisite to capitalist production—just the opposite. The center of profits exploitation shifted to the monarchy. Petty-holders aimed at reproducing themselves within a local market economy, without motivation for or capability of expansion and innovation. Hence these images are ironic, marginal evidence of the continuance of peasant marginalization in the late stages of feudalism. For the continued debate on the importance of the peasantry to the transition to capitalism, see Harvey J. Kaye, "From Feudalism to Capitalism: The Debate Goes On," *Peasant Studies* 13(3) (1986): 171–80, and Robert A. Denemark and Kenneth P. Thomas, "The Brenner-Wallerstein Debate," *International Studies Quarterly* 32 (1988): 47–65.

Bridget Ann Henisch

14. In Due Season: Farm Work in the Medieval Calendar Tradition

When a medieval artist was told to illustrate a calendar, he knew exactly what he was expected to provide. It made no difference whether he was working in wood or in stone, tracing the design for a stained-glass window, or brushing gold onto a sheet of vellum. He reached into his store of patterns and pulled out, not twelve scenes, or emblems, one for each month of the year, but twenty-four. One illustration showed a characteristic occupation for the month, and the other displayed the month's dominant zodiac sign. The artist then proceeded to group his pictures in any number of configurations, of which the simplest and most straightforward was the pair of compartments, as can be seen on a page for June, in a psalter made in northern France toward the end of the twelfth century, which shows a man mowing in one frame, and a crab, the zodiac sign, in the other.[1]

The presence of the occupation scene is readily understood. The sequence of twelve activities, almost always drawn from the countryside and the farm, represents the annual, endlessly repeated, cycle of necessary, basic tasks that put food on the table: pruning and ploughing, sowing and reaping; the fattening up of livestock, and the slaughter.

The presence of the zodiac sign needs a little more explanation. The zodiac is the narrow pathway across the sky in which the sun, the moon, and the principal planets seem to move throughout the year.[2] It is divided into twelve equal sections, or signs, each named after a constellation whose position once, long ago, lay within it. The sun passes through one of these sections each month, as it makes its progress from one year's end to the next. Because the sun was all-important to society, its movements were studied with the greatest attention, and it was only natural and fitting that the twelve divisions of the calendar should be marked with the zodiac signs, as reminders of the sun's journey through the sky, as well as with the scenes of man's essential duties, as he bustled about his work down below.

One more pair of scenes, from a later, fifteenth-century French manuscript, offers a crude and cheerful representation of July, with a man cutting grain in one compartment, and Leo the lion flourishing his tail among the stars, next door (Figure 58).[3] Over the centuries, the tradition of calendar illustration became comfortably established. And just for that reason the artist could play with it, presenting the same familiar scenes in a variety of conventions, from the use of isolated figures set against a plain or patterned background[4] to groups of people moving in a fully developed landscape.[5]

The calendar tradition had very long roots, tapping into the classical past. In the medieval period, in western Europe, we begin to find traces of it from the ninth century onward; by the twelfth century it had become firmly established, and was to grow especially strong and popular in France, England, and Flanders. As the Middle Ages drew to a close in the early sixteenth century, the convention still showed great vitality, with many splendidly rich examples in the Books of Hours made in Flanders for an international market.[6]

The name often given to the tradition is "The Labors of the Months," but in fact by the end of the medieval period it had become a cycle of occupations rather than labors, because so many pleasures of the seasons had been tucked into the scheme, from snowball fights in January[7] to dancing and dicing in December (Figure 59).[8] But, however frivolous the details of any particular calendar might become, always at the core there was the round of activity on the land intended to pile provisions high in the larders of society.

The cycle might be shown anywhere, up on a roof boss, or down on a misericord, half-hidden in the shadows beneath a choir-stall. It could decorate the pages of a Book of Hours, for the private pleasure of a private owner, or be carved as a public statement for all to see, around the great west doorway of a church.[9] No matter where it appeared, whether in solemn majesty or as a light-hearted frivolity, it was the embodiment of a deeply felt, long-held belief that life on earth was an unending round of work, shaped and driven by the year's unending round of seasons.

It was an accepted truth that man's fall from grace had led to the punishment of incessant toil and struggle in the world beyond the gates of Paradise. The terrible words of God to Adam in the third chapter of Genesis summed up the situation: "In the sweat of thy face shalt thou eat bread, till thou return unto the ground; for out of it was thou taken: for dust thou art, and unto dust shalt thou return" (Genesis 3: 19). Nature herself had been corrupted by human sin, and the corruption showed in her contrariness and lack of cooperation with human efforts: unpredictable weather, difficult soils, rampaging weeds, ravening wildlife. Adam had been God's

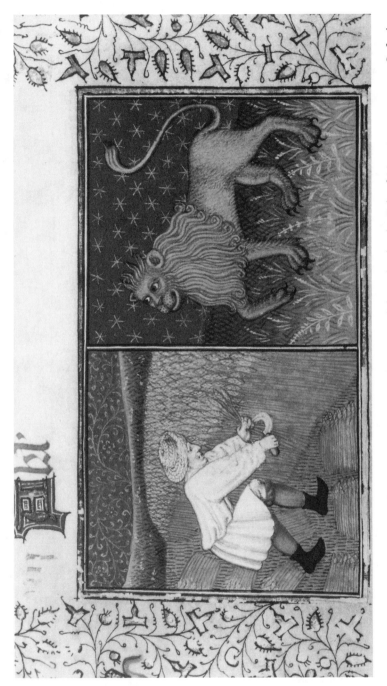

Figure 58. Reaping, July. Book of Hours, made in France for the English market by the Falstof Master, ca. 1440–50. Oxford, Bodleian Library, MS Auct. D. inf. 2.11, fol. 7r. By permission of the Bodleian Library.

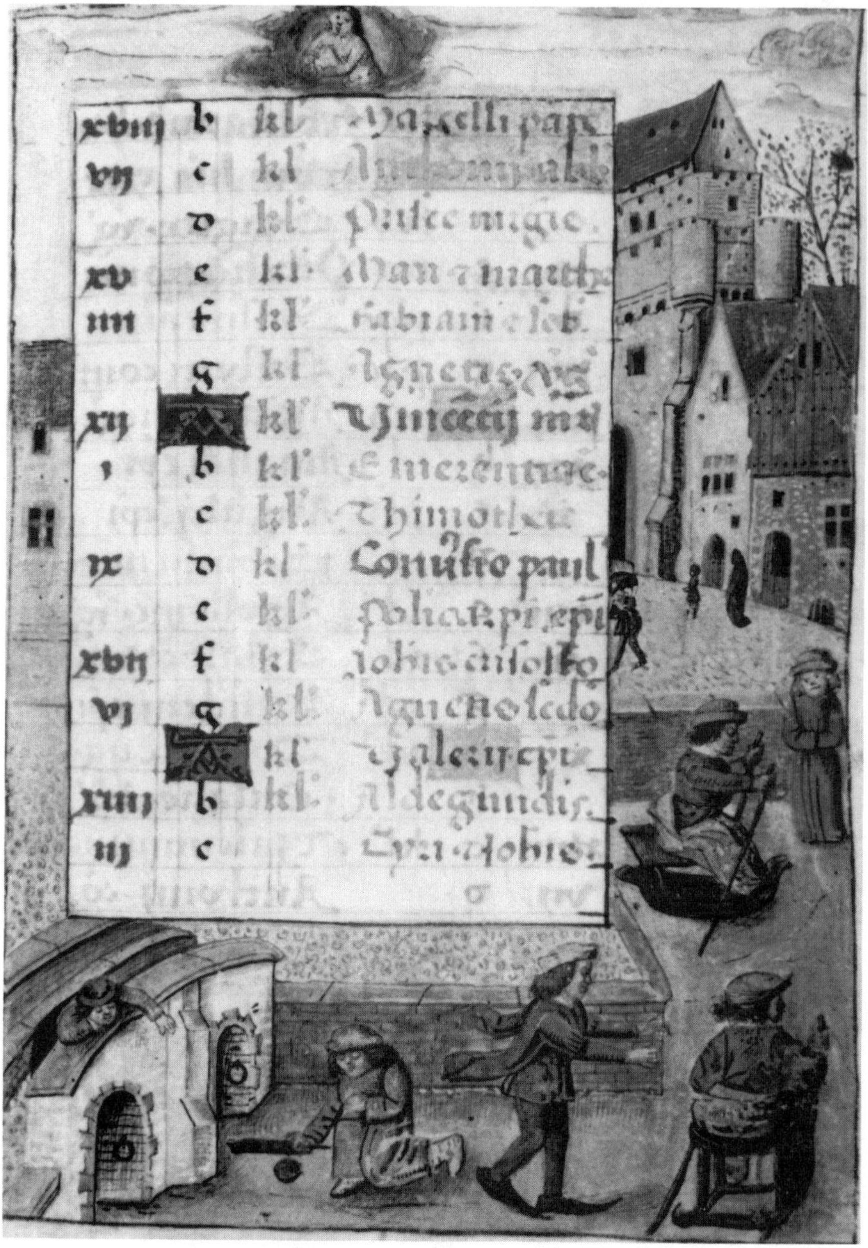

Figure 59. Winter sports, January. Book of Hours, Flemish, ca. 1490. London, British Library MS Add. 18852, fol. 2. By permission of the British Library.

first gardener in Paradise, but his undemanding round of duties in that blessed enclosure was but a poor preparation for the realities he had to face when, thrust out into the hostile world, he was handed a spade as a parting present from a reproachful angel.[10] It was a very rude awakening.

Although the curse of unending toil had been laid on the whole of society, not everyone was expected to toil in the same way. Although all people had to face a verdict after death on the life they had led on earth, all were not rewarded for their efforts in quite the same way while still living in the world. According to a very simplified, shorthand scheme, which remained popular as a teaching tool for centuries despite its obvious limitations, society was served by three groups: those who looked after its spiritual needs, those who defended it against injustice, and those whose job it was to feed it: the church, the governing class of kings, lords, and knights, and the laborers. It was a coarse but convenient grid, laid over the teeming complexities of real life, to create a bold, easily memorized platitude.[11]

Of these three groups, all were necessary, but some were more equal than others. Just as history is always written by the victors, so rules are drawn up by those already in position to derive most benefit from them. The relations between two, the church and the secular government, showed an endless jockeying for real power in the world throughout the period. The third group, of laborers, was regarded, by and large, as a necessary evil. It was worked very hard, punished harshly for ordinary misdemeanors and ruthlessly for any stirrings of revolt. It was also held in some contempt. Then, as now, there was no strong desire felt by those with some comfort and authority in their own way of life to change places with anyone in obviously less agreeable conditions. Voluntary poverty, accepted as a spiritual education, was one thing. Ordinary, grinding poverty imposed by circumstance offered no such rewards and often had a dishearteningly bad effect on the character of its victim. Preachers pointed out, frequently, that as much pride, and greed, and anger, lurked in a peasant's heart as in that of the most arrogant baron.[12]

The poor, in short, were not very attractive. Not in clothes, in appearance, in habits, in situation. That remarkable man, Henry Grosmont, duke of Lancaster, who was not only one of Edward III's great magnates and military commanders but also a devout layman who was able to write his own manual of devotion, confessed there, quite frankly, that he did not like the smell of the poor. He was sorry for it, he prayed for forgiveness, but there it was: He found it most disagreeable.[13] The same kind of attitude is to be found in another aristocratic author, Joinville, the biographer of Saint Louis, king of France. He loved and honored his master as a saint, but was appalled when Louis insisted on following Christ's example to the letter and actually knelt to wash the feet of some poor men on Maundy Thursday. Joinville's rigid disapproval is recorded in his own book[14] and remembered in an early fourteenth-century illustration of the scene.[15]

The lot of the poor was sometimes described with compassion, and the figure of the honest workman was sometimes held up for imitation, but even in such cases the emphasis was on the harshness of the peasant's life, the courage and obedience with which he shouldered his heavy burden. Remarkably little was ever said in praise of the good sides of that life. It is very rare to come upon this kind of remark, set down in a schoolboy's exercise book in the late fifteenth century: "It is a great pleasure to be in the contrey this hervest season . . . to se the Repers howe they stryffe who shal go before othere."[16]

To turn from the written record to the calendar pictures is to step into a very different world. In comments about the peasant and life on the land, the three notes most often struck, whether in sermons, in manuals for priests, or in the secular literature of the age, are contempt, criticism, and compassion. Not one of these is sounded in the labors of the months tradition. In the calendar cycle, man seems to have exchanged one paradise for another. Admittedly, he is always busily at work throughout the year, but in circumstances never to be matched this side of heaven.

The emotional tone of the cycle is noticeably calm. This is one of the few places in medieval art where serenity, not suffering, is the order of the day. The pictorial presentation of the passing seasons is pierced with no sense of sin, no sense of paradise lost. The harmonies are disturbed by no fear of death, no forebodings of disaster. There is no hint of effervescent high spirits, but everywhere we look there is an air of quiet purposefulness and confidence (see Figure 60).[17] The figures know what they are doing; no one is getting in anyone else's way, no quarrels flare up. No one is in despair about ever finishing the job in time. The work may be back-breaking, but it is never heart-breaking.

One reason for this happy state of affairs is that the weather is always accommodating, always appropriate. No untimely drought shrivels the new growth of springtime; no sudden hailstorm flattens the harvest. Nature provides the right weather, at the right time; man takes the right action to reap best advantage from ideal conditions.[18] Peace of mind is further guaranteed by the fact that not only the weather but also the equipment is in

Figure 60. Ploughing and sowing, September. Da Costa Hours, Simon Bening and others, Bruges, ca. 1515. New York, Pierpont Morgan Library, MS M.399, fol. 10v.

perfect shape. The necessary tools for any job are always in working order. We never see a broken ploughshare or a rusty billhook, and there is never any sign of an accident, even when an enormous scythe slices through the air, alarmingly close to a set of vulnerable toes.[19]

In the world set forth in medieval literature, it is not hard to find distinctly unflattering descriptions of the peasant's physical appearance:

His hosen overhongen his hokschynes. on everiche a side,
Al beslombred in fen. as he the plow folwede . . .
This whit waselede in the fen. almost to the ancle.[20]

[His stockings hung down round his legs,
All splattered with mud, as he followed the plough.
He was mired in mud, almost up to his ankles.]

Alternatively, it is not hard to find the peasants presented as objects of pity, as in an early fourteenth-century English poem on the daily miseries they had to face, miseries summed up in a somber last line: "Might as well die straightaway, as struggle on like this" ("ase god in swynden anon as so forte swynke.")[21] In the calendar world, the impression of peasants and their life is quite different. The figures going about their work may not be strikingly handsome, but they are sturdy, trim, capable. They have had enough to eat. They are dressed not in rags and tatters but in appropriate clothes, warm in winter,[22] loose and easy in summer.[23] They are shown at just the right age: young enough to have energy and strength, old enough to have experience (see Figures 61 and 62).

Ideas of death and mortality are kept firmly at bay. It is true that there is one branch of the calendar tradition in which the seasons are linked with the ages of man, and death comes with winter, but as it is not relevant to the theme of agriculture there is no need to consider it here. Apart from this special case, death plays remarkably little part in the cycle. The human figures show no sign whatever of advancing age or dwindling energy, and the only death that occurs in the entire year is in November or December, when animals are slaughtered for the meat supply. Almost always, in the calendar tradition, the animal chosen is the pig. Even here the idea of death is controlled and colored by the theme of the tradition as a whole: the promise of life's ever-returning, ever-renewing cycle. Death is accepted with composure. Pigs are killed to fill the larder in December, but as we

Figure 61. Pruning, February. Da Costa Hours, Simon Bening and others, Bruges, ca. 1515. New York, Pierpont Morgan Library, MS M.399, fol. 3v.

Figure 62. Mowing, June. Book of Hours, made in France for the English market by the Falstof Master, ca. 1440–50. Oxford, Bodleian Library, MS Auct. D. inf. 2.11, fol. 6r. By permission of the Bodleian Library.

look at the scene we hear no squeals of agony, see no blood stains, smell no sweat.[24] We know, and the pigs know, that, in the calendar cycle at least, they are absolutely safe and indestructible. Come next November, they will be resurrected, to root for acorns once again.[25] On this point there is a yawning gap between the treatment of time's passage in art and in literature. In medieval poetry, the haunting question is always: "Who wot nowe that ys here / Where he schall be anoder yere?"[26] In calendar art that question is never raised, because it is quite unnecessary. Everyone knows what will happen next year: exactly the same round of seasons and the same round of activities as in the present one.

Calendar scenes are small. They are ornaments, whether decorating a page, a lead font, or a stone porch. This limitation in size takes the figures one step further from reality. There is a doll's house air to many examples and, even in the severe medium of stone, the little figures often look more like pixies than like people.[27] One of art's mysterious powers is the ability to draw pleasure from pain. Just as Samuel Beckett's prose gives a mesmerizing beauty to disintegration and decay, so the artists in the Labors tradition transformed the mud and misery of demanding work into satisfying harmonies. The medieval mastery of line and pattern creates from everyday movements, in everyday jobs, the choreographed rhythms of a dance, as in a carving of a man sowing seed.[28]

Wood and stone offer the spectator the satisfactions of contour and texture, of actions and gestures caught by the artist and modeled by light and shadow. In manuscript examples, bewitching harmonies of color soften the rigors of work and add a bloom of beauty to the most humdrum activities (see Figure 63).[29] Refinement of line and the precious pigments chosen for the scene give an early fifteenth century illustration of manure being poured around a vine-stock, in early March, an elegance strangely at variance with its subject matter (Figure 64).[30]

In the same way, a typical calendar scene of harvesting is all bright gold against a bright blue sky.[31] Its smooth perfection of surface, and its serenity of tone offer not a hint of the discomfort of an actual day spent cutting the grain. Gertrude Jekyll, the great English garden designer, touched on the truth of the matter when she once spoke about her memories of holidays on a farm when she was a little girl, and remarked: "Anyone who has never done a day's work in the harvest-field would scarcely believe what dirty work it is. Honest sweat and dry dust combine into a mixture not unlike mud."[32]

The idea of life as a round of unremitting toil is softened in some

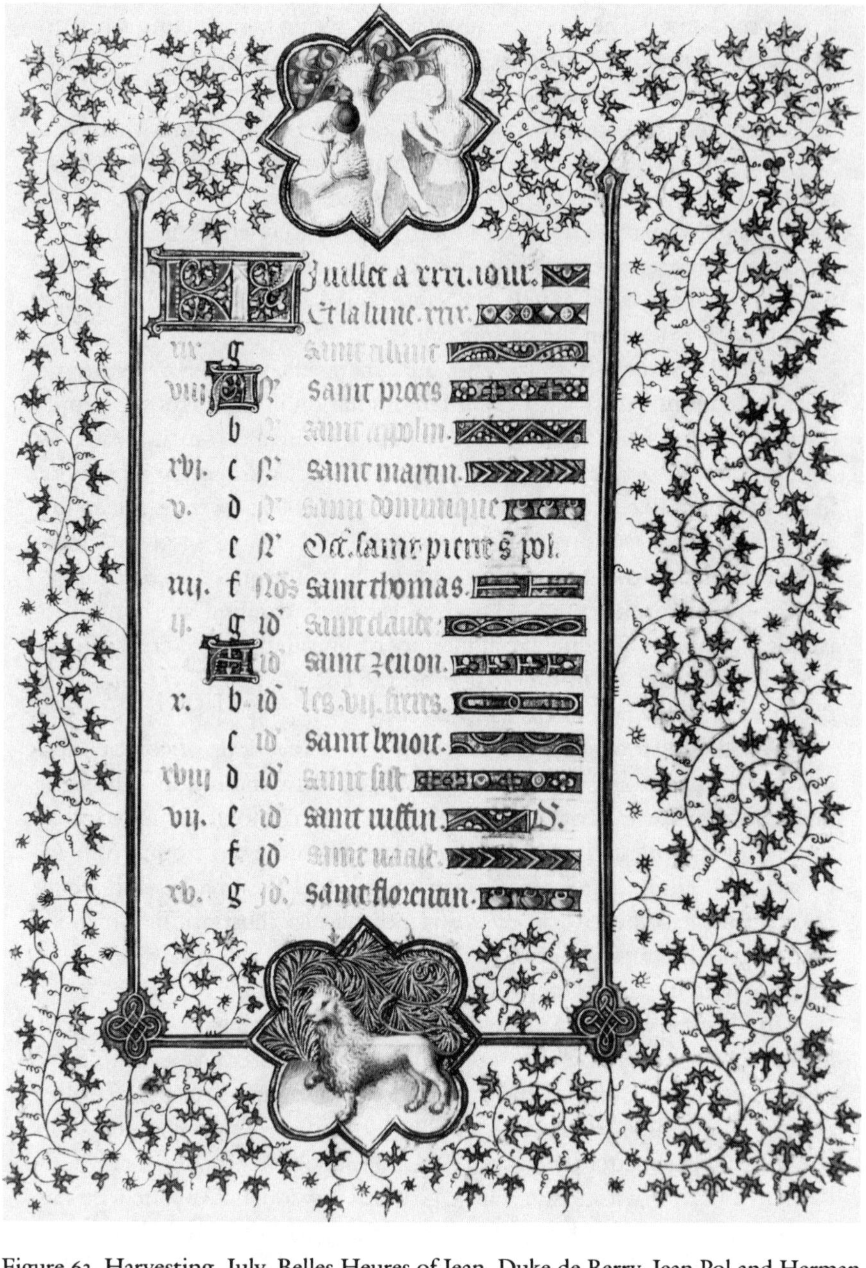

Figure 63. Harvesting, July. Belles Heures of Jean, Duke de Berry, Jean Pol and Herman de Limbourg, French, ca. 1408. New York, Metropolitan Museum of Art Cloisters Collection, MS 54.1.1, fol. 8.

Figure 64. Pouring manure around vines, March. Belles Heures of Jean, Duke de Berry, Jean Pol and Herman de Limbourg, French, ca. 1408. New York, Metropolitan Museum of Art Cloisters Collection, MS 54.1.1, fol. 4.

calendar scenes by yet another element: the element of enjoyment. Tiny details, caught by the artist, add a sweetness or a zest to the yearly round. A cart-horse is offered a tidbit after hauling a heavy load.[33] A peasant in an enormous vat presses grapes, while he helps himself to a cluster.[34] Workers look forward to a picnic lunch in the harvest field (Figure 65).[35] It is significant that the touch of relaxation or pleasure here and there never interferes with the work at hand, never breaks the rhythm of purposeful activity. It is not an interruption and never antisocial. It is never a protest against the rules of the game, a sullen gesture made against the system. The relaxation comes at appropriate times. One of the traditional images of Sloth, in manuals on the sins drawn up for preachers, is the laborer sitting idle by his plough.[36] In the calendar tradition this sin is avoided, because workers relax only when a particular job has been finished, or in ways which do not affect the task in hand, like munching grapes while still treading the vat. Just as no one in a calendar scene is ever shown stealing from the crop, sneaking home with a few ears of grain, so no one steals time. Work moves to the rhythm of the seasons. Pleasure moves in counterpoint and fills the natural pauses in the measure; it never disrupts the dance.

Beneath the smooth, deceptively simple surface of the cycle lurk many surprises. The biggest surprise of all, in a medieval work of art, is that there are no obvious religious overtones. Occasionally, a religious scene is chosen as the occupation of a month as, for example, the distribution of ashes on Ash Wednesday, the first day of Lent, in a few calendar pages for March.[37] At the core of every cycle, however, lies the agricultural story of the year, and there no religious touches of any kind are to be found. There is never a hint of divine intervention, or a turning for help or consolation to the Virgin Mary or to some local saint. There is no scene that shows the offering of harvest tithes to the church, no blessing of the fields at Rogation-tide by the parish priest. Work goes on quite outside the framework of belief, doctrine, or discipline.

There is another missing ingredient. Nowhere to be found is any sense of social context. Figures are hard at work, but they are not shown in any recognizable community. They are busy and, apparently, independent. Very rarely is there to be seen any figure of authority directing operations, or any hint of coercion. The occasional exception, as in an early fourteenth-century English scene of an overseer in the harvest field, only goes to prove the rule.[38] Every activity seems to be free, planned and carried out by the peasants themselves. Usually, masters and peasants, when shown together in the same picture, seem to inhabit entirely separate worlds, as in

Figure 65. Harvest picnic, August. Book of Hours, Flemish, early sixteenth century. London, British Library MS Add. 24098, fol. 25b. By permission of the British Library.

an August scene where lords and ladies ride out hawking in the foreground while, in the far distance, the harvest is gathered in.[39] It is only in some very late examples, produced at the end of the fifteenth or beginning of the sixteenth century, that it is possible to find orders being given and received. In a few gardening scenes for early spring it is made quite clear that the garden belongs to an owner, and the gardeners work under watchful, proprietorial eyes (see Figure 66).[40]

While there is scarcely a trace of an order given or obeyed in the calendar tradition, there is no suggestion at all of resentment, let alone actual rebellion, against the system itself. Peasants had ample grounds for grievance throughout the medieval period, and every now and again violent protest flared up, to be met in due course with even more violent retribution. When Froissart described the Peasants' Revolt of 1381 in England, he put into the mouth of John Ball, a leader of the rebellion, a speech that is a mosaic of traditional complaints against the high and mighty: "They have the wines, and spices, and the good bread; we have the rye, the husks and the straw, and we drink water. They have shelter and ease in their fine manors, and we have hardship and toil, the wind and the rain in the fields. And from us must come, from our labor, the things which keep them in luxury."[41]

Rulers were uneasily aware that their thrones rested on tinder-boxes, and the thought of peasant turmoil often troubled their dreams. In the mid-twelfth century, Henry I of England had a nightmare so upsetting that it was not only recorded but illustrated in a contemporary chronicle. Henry dreamt that maddened peasants pressed around his bed, menacing him with their pitchforks and their scythes.[42] Of such explosive anger and pent-up fury, or indeed of any breakdown in the social system, not a hint scratches the smooth surface of the calendar tradition.

The principles of selection that shaped that tradition remain shrouded in mystery. Of all the myriad, varied jobs to be done on the land, only a handful ever found their way into the calendar cycle. Considering that the tradition flourished most vigorously in northern Europe, in France, Flanders, and England, it is a little surprising that one of the most frequently represented tasks for early spring is the pruning of vines,[43] and the almost invariable one for early autumn is some aspect of the grape harvest.[44] Does this preoccupation with the vine stem from the tradition's root in the classical, Mediterranean world? Or is it due to a quite different fact of life? Wine played a central role in the service of the Mass observed in every Christian country, no matter how far it lay to the north. As a result, there was a need

Figure 66. Gardening, March. Da Costa Hours, Simon Bening and others, Bruges, ca. 1515. New York, Pierpont Morgan Library, MS M.399, fol. 4v.

to produce some kind of wine for the Church, however thin or however acid, in every region of the Christian world. Whatever the reason may be, the calendar's emphasis is always on the grape and the vine. Beer-making and cider production are never shown.

Why is the core of the calendar year the growing of grain: breaking ground, sowing seed, harvesting the ears, winnowing the grain from the chaff?[45] Is this because bread is a staple of life in the West? If so, it is surprising that no place is found for that other symbolic staple, salt, nor for the process of salt-making. To descend from the level of symbol to that of mundane reality, why is there never any hint of interest in the backbone of the medieval diet: dried peas, dried beans, and cabbage?

Considering how important sheep farming was in the economy of Europe throughout the medieval period, it is puzzling that it appears only now and then as an occupation in the calendar cycle, and never becomes a regular feature (Figure 67). Is the reason that the sheep was prized as a source of wool, the raw material of the cloth trade,[46] and so sheep farming does not fit with perfect propriety into a cycle concerned above all else with the production of food? Or were there reservations about the pastoral life itself? The shepherd with his sheep was an isolated figure, and his hours and conditions of work set him a little apart from ordinary village life. In certain important ways the shepherd's life was more primitive than life on the farm. Certainly it did not depend to the same degree on organization and cooperation, on people moving together as a team, and so it was perhaps less satisfactory than farm life as an image of society productively at work. Gerald of Wales, in the late twelfth century, thought the Irish were inferior to the English and French for several reasons, of which one was that they showed no interest in arable farming. He remarked: "They have not progressed at all from the primitive habits of pastoral living. Man usually progresses from the woods to the fields, and from the fields to settlements and communities of citizens, but this people despises work on the land."[47] Several of Gerald's contemporaries made the same point about other people living on the fringes of the known, civilized world.[48]

There is one other problem, with no easy answer. Why do women appear, more and more frequently, in the calendars of the later medieval period, the fifteenth and early sixteenth centuries (see Figures 68 and 69)?[49] It would be rash to assume any sudden surge of support for women's rights to be behind the new fashion! Women had always, by long-established custom, labored side by side with men in village farm work. Is their absence from most calendar cycles due to the fact that so many scenes are

Figure 67. Sheep shearing, April. Book of Hours, Flemish, ca. 1490. London, British Library MS Add. 18852, fol. 4v. By permission of the British Library.

Figure 68. Hunting for acorns, November. Playfair Book of Hours, made in France (Rouen) for the English market, late fifteenth century. By courtesy of the Victoria and Albert Museum.

Figure 69. Shepherd and shepherdess, with courtly couple, April. Book of Hours, Flemish, ca. 1490. London, British Library MS Add. 18852, fol. 5. By permission of the British Library.

presented within small medallions, or confining frames, inside which there was simply no room for more than one figure? Or does their entrance on the scene have more to do with changed economic conditions caused by the many plagues that swept through Europe at frequent intervals after the mid-fourteenth century? Was women's contribution needed—and felt— more keenly in this period of labor shortages created by the high rate of illness and death? Or was that contribution acknowledged at last, quite simply, on aesthetic grounds, with the discovery that to include the figure of a woman was to add a new interest, a new charm of line and detail, to a very old scene? [50]

The characteristic arrangement and appearance of the calendar cycle hide more than they reveal, and foster a somewhat distorted view of reality. They help to sustain the illusion that everything in medieval society was on a very small scale: one field, one plough; two men, two scythes. There is scarcely a hint in the tradition, from one century to the next, of commerce or of trade, either of the great international markets that punctuated the year and drew merchants from all over Europe, or of the local ones held any week, in any town.

When an elderly husband in late fourteenth century Paris gave his young wife advice on how to plan a special dinner party, he directed her, as a matter of course, to buy her supplies at the many specialized shopping districts in the city. Beef was to be found at one group of butchers, pork at another. Delicate wafers were to be ordered from one expert, garlands and table decorations from another. And when the husband suggested a recipe for a sweet confection made from carrots, he took care to add this helpful note: "Carrots are red roots, which are sold in handfuls in the market, for a silver penny a handful." [51]

The controlled confusion of such commerce finds no foothold in the tradition, and it is rare indeed to find a scene in which any farm produce is actually being exchanged for money (Figure 70). [52] Trade requires sur-plus, but in the calendar the emphasis is all on sturdy self-sufficiency. Only in the very last stages of the medieval tradition, for example in some of the Flemish calendars made in the early sixteenth century, do we begin to catch a glimpse of large-scale operations. After the diminutive, doll's house scale of most calendar activities, it comes as a shock to see a gigantic crane, for hoisting barrels, looming in the background of an autumn scene that shows the tasting of the season's new wine [53] and realize that its presence points to an economic system considerably more sophisticated than would have been suspected from the clues provided by the tradition as a whole.

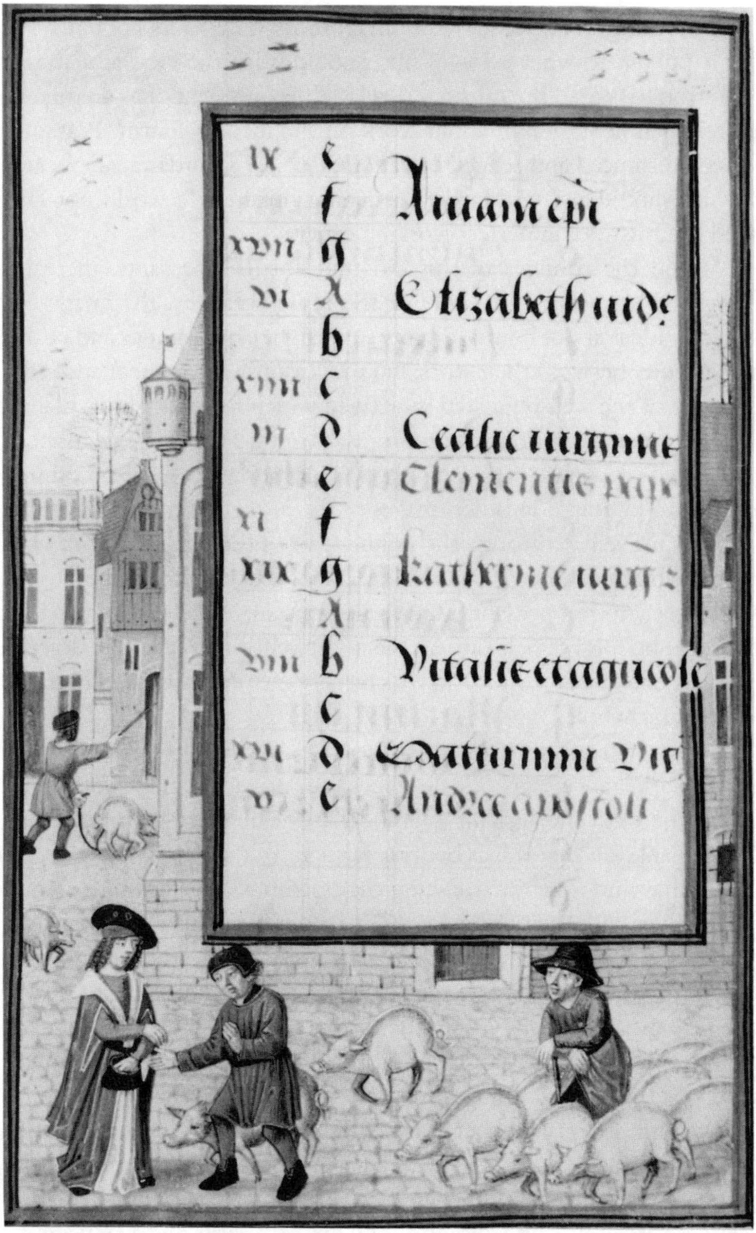

Figure 70. Pig market, November. Book of Hours, Flemish, probably Bruges, ca. 1500. Cambridge, Fitzwilliam Museum, MS 1058–1975, fol. 11v. By permission of the Fitzwilliam Museum.

Everything about that tradition, from its tone to its contents, from what it puts in to what it leaves out, should warn the viewer against the temptation to regard the calendar cycle as the equivalent of a careful, even-handed documentary film about work on the medieval farm. Peasant life has been distanced and refined by art, and society's burden made bearable by being shouldered within the sustaining dream of a world not fatally flawed but, instead, in perfect working order.

Within the confines of this old tradition the peasants, in real life so despised or disregarded, became the representatives, the image of all humanity. Ideas about human dignity shaped their appearance and bearing. The harmony between their work and the seasons was a potent and satisfying image of the well-regulated society, in which forethought in planning and skill in execution drew the appropriate reward from a responsive, and equally well-regulated, nature. The cycle's harmonies express something of the spirit to be found in other images of the properly functioning society. Its figures move not through the polluted air of the real world but within the pellucid atmosphere of an ideal model of that world. They are related not so closely to real peasants as to those honest laborers who, in an image elaborated by John of Salisbury in the mid-twelfth century, were described as the feet of a "Body Politic," in which every member was an essential part of an organic whole.[54]

In the 1370s, a translation of Aristotle's *Politics* was made for King Charles V of France. Aristotle describes four possible kinds of democracy, of which he picks the agricultural model as the best, with the disarmingly frank explanation that this is because farmers are "always busily occupied, and thus have no time for attending the assembly" and making a nuisance of themselves with their opinions.[55] The manuscript illustration of this section, labeled "Good Democracy," follows the same careful rules about the relation of work and pleasure that can be found in the calendars.[56] While some peasants are harrowing in the foreground, the men—and horses—that have done the first ploughing of the field are enjoying a well-earned picnic, one that does nothing to disrupt the rhythm of the job in hand.

The high seriousness that underlay the old calendar tradition was also, in time, sweetened and softened by a far more frivolous, but most engaging dream, the dream of a very different kind of good life. In this, the wearisome vexations and disappointments of wealth and privilege were contrasted with the pleasures of poverty. Viewed from the vantage point of high position and, perhaps, in the digestive pause after a satisfactory dinner, the simple, strenuous life in the open air, the fiber-packed diet of black bread and pure water, the untroubled dreams of the contented peasant, could seem positively enviable, and such attractions found praise in a number of elegant poems composed in the fourteenth and fifteenth centuries.[57] None of the authors, nor any of their readers, had the slightest intention of actually exchanging a comfortable life for the rigorous realities of the farm, but it became fashionable to play with the idea. It is to be suspected that the charm and grace of many calendar pictures owe something to this fancy, and were intended to please the eyes of just such patrons.

One of the most beautiful of all the calendar cycles, and certainly the one best known today, is that in the manuscript known as the "Très Riches Heures," made for the Duke de Berry, a man not noted for his love of farm life or, indeed, of peasants. In another manuscript that he commissioned, he is shown being welcomed by Saint Peter into Paradise.[58] If this happy event ever did take place, outside the Duke's fond dreams and the pages of his own manuscripts, there must be a strong presumption that he was received into heaven on the strength of his generosity as a patron of the arts, not for his generosity as a lord and master. In that role, he showed a harsh indifference toward his peasants, and a rapacious interest in the profits he could wring from their exertions. His record as a master called forth not paeans of praise from grateful subjects but resentment and rebellion throughout his vast domains.[59] For him, at least, the calendar pictures he enjoyed as he turned the pages of his Book of Hours must have woven a beautiful veil of illusion, to mask the ugly reality of the world outside his castle walls.

Notes

1. Psalter, northern France, end of the twelfth century, Paris, Bibliothèque Nationale, MS. Lat. 238, f. 3v, June, Mowing.

2. Woodcut from Textus de sphaera, by Johannes de Sacrobosco (Paris, 1538), New York, The Metropolitan Museum of Art (Dick Fund, 1934), Armillary Sphere.

3. Book of Hours, illuminated in France for the English market by the Fastolf Master, ca. 1440–50, Oxford, Bodleian Library, MS. Auct. D. inf. 2.11, f. 7r, July, Reaping.

4. Martyrologe d'Usuard, Saint-Germain-des-Près, northern France, ca. 1270, Paris, Bibliothèque Nationale, Ms. Lat. 12834, f. 64v, July, Threshing.

5. Book of Hours, Flanders, illustrated by Simon Bening, ca. 1530, Munich, Staatsbibliothek, Cod. Lat. 23638, September, Ploughing, harrowing, sowing.

6. For calendars of the early period see James C. Webster, *The Labors of the Months in Antique and Mediaeval Art to the End of the Twelfth Century* (Princeton,

N.J.: Princeton University Press, 1938). For later examples see Wilhelm Hansen, *Kalenderminiaturen der Stundenbücher: Mittelalterliches Leben im Jahreslauf* (Munich, Georg D. W. Callway, 1984).

7. Book of Hours, Flanders, illustrated by Simon Bening, ca. 1530, Munich, Staatsbibliothek, Cod. Lat. 23638, January, Snowballing.

8. Book of Hours, Flanders, illustrated by Simon Bening, ca. 1530, Munich, Staatsbibliothek, Cod. Lat. 23638, December, Dancing and dicing.

9. West doorway, Cathedral of St. Lazarus, Autun, Burgundy, ca. 1135, carved by Gislebertus.

10. Winchester Psalter, English, ca. 1150, London, British Library, Cotton MS. Nero C IV, f. 2, Expulsion of Adam and Eve from Paradise.

11. For a full discussion of the scheme see Georges Duby, *The Three Orders: Feudal Society Imagined*, trans. Arthur Goldhammer (Chicago: University of Chicago Press, 1980).

12. For examples of sermon criticism of peasant behavior see G. R. Owst, *Literature and Pulpit in Medieval England*, 2nd rev. ed. (Oxford: Basil Blackwell, 1961), pp. 365–69.

13. Henry of Grosmont, Duke of Lancaster (1310–61), *Le Livre de Seyntz Medicines* (1354), ed. E. J. Arnould, Anglo-Norman Text Society 2 (Oxford: Basil Blackwell, 1940): 13–14, 25.

14. Jean de Joinville (ca. 1224–after 1309), *The Life of St. Louis* (1309), in *Chronicles of the Crusades*, trans. M. R. B. Shaw (Harmondsworth, Middlesex: Penguin, 1963), p. 169.

15. The Hours of Jeanne d'Évreux, Paris, Jean Pucelle, ca. 1325–28, New York, The Cloisters, The Metropolitan Museum of Art, f. 148v, Saint Louis washing the feet of the poor.

16. William Nelson, ed., *A Fifteenth Century School Book* (Oxford: Clarendon, 1948), p. 5.

17. The Da Costa Hours, Bruges, illustrated by Simon Bening and others, ca. 1515, New York, The Pierpont Morgan Library, MS. M. 399, September, Ploughing and Sowing.

18. Book of Hours, Flanders, illustrated by Simon Bening, ca. 1530, Munich, Staatsbibliothek, Cod. Lat. 23638, August, Harvesting.

19. The Rohan Hours, France, 1420s, Paris, Bibliothèque Nationale, MS. Lat. 9471, f. 8v, June, Mowing.

20. W. W. Skeat, ed., *Pierce the Ploughmans Crede* (ca. 1394), Early English Text Society, orig. ser. 30 (London, 1873), ll. 426–27, 430.

21. "Song of the Husbandman" (1300), in *Historical Poems of the XIVth and XVth Centuries*, ed. R. H. Robbins (New York: Columbia University Press, 1959), no. 2, pp. 7–9, l. 72.

22. The Da Costa Hours, Bruges, illustrated by Simon Bening and others, ca. 1515, New York, The Pierpont Morgan Library, MS. M. 399, February, Pruning.

23. Book of Hours, illuminated in France for the English market by the Fastolf Master, ca. 1440–50, Oxford, Bodleian Library, MS. Auct. D. inf. 2.11, June, Mowing.

24. Stained-glass roundel, English, fourteenth century, Bilton Church, near Rugby, November, Pig-killing.

25. The Playfair Hours, made in France (Rouen) for the English market, late fifteenth century, London, Victoria and Albert Museum, MS. L. 475–1918, November, Hunting for acorns.

26. Fifteenth century carol, in *A Selection of English Carols*, ed. Richard Greene (Oxford: Clarendon, 1962), no. 27, p. 85, ll. 1–2.

27. West doorway, Cathedral of St. Lazarus, Autun, Burgundy, ca. 1135, carved by Gislebertus, Detail, March, Pruning vines.

28. Misericord, Church of St-Martin, Champeaux, France, sixteenth century, the Sower.

29. The Belles Heures of Jean, Duke de Berry, French, ca. 1408, New York, The Cloisters, The Metropolitan Museum of Art, f. 8, July, Reaping.

30. The Belles Heures of Jean, Duke de Berry, French, ca. 1408, New York, The Cloisters, The Metropolitan Museum of Art, f. 4, March, Pouring manure around the vines.

31. The Playfair Hours, made in France (Rouen) for the English market, late fifteenth century, London, Victoria and Albert Museum, MS. L. 475–1918, July, Reaping.

32. Betty Massingham, *Miss Jekyll* (London: Country Life, 1966), p. 24.

33. Book of Hours, Flanders, illustrated by Simon Bening, ca. 1530, Munich, Staatsbibliothek, codex latinus 23638, July (b), Feeding horse.

34. Martyrologe d'Usuard, Saint-Germain-des-Près, northern France, ca. 1270, Paris, Bibliothèque Nationale, MS. Lat. 12834, f. 69v, September, Grape-treading.

35. Book of Hours, Flanders, early sixteenth century, London, British Library, MS. Add. 24098, f. 25b, August, Harvest picnic.

36. Somme le Roi, Flanders, 1415, Brussels, Bibliothèque Royale, MS. 11041, f. 88v, Sloth.

37. Book of Hours, Paris, ca. 1450–60, Chantilly, Musée Condé, MS. lat. 1362, f. 3r, March, Distribution of ashes on Ash Wednesday.

38. Queen Mary's Psalter, England, early fourteenth century, London, British Library, MS. Royal 2 B VII, f. 78v, August, Overseer and harvesters.

39. The Très Riches Heures of Jean, Duke de Berry, France, the Limbourg Brothers, ca. 1413–15, Chantilly, Musée Condé, MS. 65 (1284), f. 8v, August, Hawking.

40. Book of Hours, Flanders, illustrated by Simon Bening, ca. 1530, Munich, Staatsbibliothek, Cod. Lat. 23638, March, Gardening.

41. Jean Froissart (ca. 1337–ca. 1410), *Chronicles* (ca. 1369–ca. 1400), trans. and ed. Geoffrey Brereton (Harmondsworth, Middlesex: Penguin, 1968), Book II, p. 212.

42. Chronicle of John of Worcester, England, ca. 1130–40, Oxford, Bodleian Library, MS. Corpus Christi Coll. 157, f. 382, Henry I's nightmare.

43. Bedford Hours, France, ca. 1423, London, British Library, MS. Add. 18850, f. 3, March, Pruning.

44. Book of Hours, made in France (Rouen) for the English market, late fifteenth century, Oxford, Bodleian Library, MS. Auct. D. inf. 2.11, f. 9r, September, Treading of grapes.

45. The Playfair Hours, made in France (Rouen) for the English market, late

fifteenth century, London, Victoria and Albert Museum, MS. L. 475–1918, August, Threshing and winnowing.

46. Book of Hours, Flanders, illustrated by Simon Bening, ca. 1530, Munich, Staatsbibliothek, Cod. Lat. 23638, June (a), Sheep-shearing.

47. Gerald of Wales (ca. 1145–1223), *The History and Topography of Ireland* (ca. 1188), trans. John J. O'Meara (Harmondsworth, Middlesex: Penguin, 1982), Part III, ch. 93, pp. 101–2.

48. Robert Bartlett, *Gerald of Wales* (Oxford: Clarendon, 1982), pp. 160–62.

49. The Playfair Hours, made in France (Rouen) for the English market, late fifteenth century, London, Victoria and Albert Museum, MS. L. 475–1918, October, Sowing.

50. The Très Riches Heures of Jean, Duke de Berry, the Limbourg Brothers, ca. 1413–15, Chantilly, Musée Condé, MS. 65 (1284), f. 6v, June, Haymaking.

51. *The Goodman of Paris* (ca. 1393), trans. and ed. Eileen Power (London: Routledge, 1928), sec. 2, art. IV: "How to Order Dinners and Suppers," pp. 221–47, 296.

52. Book of Hours, Flanders (probably Bruges), ca. 1500, Cambridge, Fitzwilliam Museum, MS. 1058–1975, f. 11v, November, Pig market.

53. Book of Hours, Flanders, illustrated by Simon Bening, ca. 1530, Munich, Staatsbibliothek, Codex Latinus 23638, October (a), Tasting wine and loading barrels.

54. John of Salisbury (ca. 1115–80), *The Statesman's Book (Policraticus)* (1159), trans. John Dickinson (New York: Knopf, 1928), VI, xx, p. 243.

55. Aristotle, *The Politics*, trans. and ed. Ernest Barker (Oxford: Oxford University Press, 1958), VI.4.2, p. 263.

56. Aristotle, *The Politics*, translated into French by Nicole Oresme, Paris, ca. 1372, Brussels, Bibliothèque Royale, MS. 11201–2, f. 241, Farming in a good democracy.

57. For example, Philippe de Vitry, *Franc Gontier*, Pierre d'Ailly, *Le Tyran*, and François Villon, *Les Contredis Franc Gontier*, all in *The Penguin Book of French Verse: To the Fifteenth Century*, ed. Brian Woledge (Harmondsworth, Middlesex: Penguin, 1961), pp. 216, 218, 327.

58. The Grandes Heures of Jean, Duke de Berry, ca. 1409, Paris, Bibliothèque Nationale, MS. Lat. 919, f. 96, Saint Peter welcomes the Duke de Berry at the gate of Paradise.

59. Millard Meiss, *French Painting in the Time of Jean de Berry: The Late Fourteenth Century and the Patronage of the Duke*, 2nd ed. (London: Phaidon, 1969), text vol., p. 32.

Alfred W. Crosby

15. Afterword

The disadvantage of commenting on papers on a subject about which one knows a great deal is that one is unlikely to be astonished: one arrives acquainted with the protagonists and with an outline of the plot in an inside pocket. When asked to write an afterword, one may feel obliged to say yes, but how to express any attitude warmer than the satisfaction that things are coming along as expected? Fortunately, I am not an expert on medieval European agriculture. I was invited to comment, I presume, as one who is interested in agriculture and history on a large scale, a world scale. Probably I am expected to place medieval agriculture in "the big picture," that last resort of the underinformed. But before I launch into that, let me make a few remarks about the specifics that have most impressed me in the papers included in this volume.

I am impressed with how devilishly complicated and subtle the subjects of medieval agriculture and agriculturists are, subjects that most people, even historians, I am ashamed to say, would assume to be simple. After all, it includes nothing about the vagaries of the world market, nothing about pork belly futures, nothing as mind-boggling as the shifting relationship between the prices of wheat, farm machinery, chemical fertilizers, soil degradation, and borrowed money. I was sure, to give one example of my naiveté, that the medieval farmer's tools were inferior to those that his successors used and use. I *knew* that scythes were better than sickles, just as McCormick knew that reapers are better than scythes. Now, however, I have learned that with a sickle a farmer can cut no more than the ears of grain, leaving a tall stubble for livestock to feed upon as they refertilize the land—an example of "sustainable farming" that would delight an old Yankee and a new organic farmer equally.

Now I know, thanks to Bökönyi and Brunner, that the "fall of Rome" was, agriculturally, a complicated descent that even included a few ascensions. Roman farm tools and practices were not forgotten, but were handed on, generation after generation, and the design of scythes was even improved during the darkest of the Dark Ages. On the other hand, for some reason or other, Roman animal breeding practices lapsed almost completely.

Campbell has corrected my plausible assumption that the fourteenth-century slump before the Black Death was simply ecological in nature, that it was simply a matter of too many people trying to wrest a living out of the land with primitive methods. Now I know that it was political, too, and in some regions even more political than anything else. Edward II actually *raised* taxes during the Great Famine, an act of moral and economic insanity. Campbell has shaken my faith in plausible explanations, be they those of Ricardo or Marx or the ecologists.

Frank's rereading of *Piers Plowman* has fortified my belief, derived from post-medieval studies, in the importance of starvation in history. Hunger and the fear of it, the fear expressed in this ancient document as "if the land fail," are constant factors in the history of the West well into the nineteenth century, and very complicated in their influence. For example, there is no doubt that hunger leads to disease in general, but, paradoxically, it discourages some infections because some bacteria need well-fed bodies to feed on. Hunger can lead to poor health and behaviors that do not easily occur to the well-upholstered historians of our time. I found Mary Kilbourne Matossian's book about the effect of mold on the staples of European diet impressive,[1] but had my doubts about the magnitude of that effect. Now I know about the "hungry gap," that is, that during the Middle Ages and for long after the poor were forced, in the last weeks before harvest, to subsist on the dregs of the last harvest's grains and bran, exactly the foods most likely to be moldy and to poison them with ergot and send them reeling into madness.

I am impressed with how narrow was the margin of choice of medieval peasants in matters of subsistence, but, on the other hand, I am also impressed with how humans can defy what seems to me to be good sense and reject their best opportunities to fill their bellies. Why were Europeans so slow to take up the new crops the Muslims brought into Iberia and the Mediterranean islands, especially hard wheat, sugar cane, and rice? Because of Christian bigotry? Probably, but there is more to it than that. Farmers living on an annual hand-to-mouth regimen—you eat what you raise or go hungry—dare not experiment. Anything less than a decent crop means not fiscal but nutritional disaster. But Watson tells us that an elite truly in control of the land, as in Islamic areas, can take chances. Elites are confident of their next meal and of next year's meals. And so, aristocrats, in contradiction to the egalitarian pablum fed me since birth, can be the razor edge of progress.

In their totality, these papers oblige me to acknowledge that history is not logical, but evolutionary. It is as tricky as the sequence that leads from the eohippus to the sturdy animals that carried the medieval knights. Therefore, there is no substitute for digging up the sticky, stubborn facts of what actually happened, happened in defiance of Ricardo and Marx.

And now to fulfill my obligation to provide "the big picture." Europe in the Middle Ages was "the wild west" of the band of civilized lands that stretched from the Atlantic to the China Sea. Circa the year 700 or 1000 or even 1400, European society did *not* seem destined to become the most successfully imperialistic society of all time. But it did: Queen Victoria, not Genghis Khan, was monarch of the world's most extensive empire. Yet Islamic society was more expansive than Europe's during the Middle Ages. At the beginning of that era it conquered the southern half of Christendom, and at the end, in 1453, besieged and seized Byzantium, the second Rome. The China of the Song dynasty was the most impressive industrial and commercial power on earth, and when Marco Polo saw Beijing in the thirteenth century, it took his breath away; and he was from northern Italy, the most progressive region of all Europe. So why and how did Europe triumph?

Every advanced society rests on the backs of its farmers. What can I, a self-styled world historian, see in Europe's medieval countryside that was unique? Machines? In the high Middle Ages Europe did have large numbers of water and wind mills, but they were quite recent, that is, probably more an effect than a cause of the subterranean factors we are looking for. What made Europe distinctive from the decline of Rome onward? Bökönyi's animals, big domesticated animals! The Amerindian civilizations had llamas, dogs, guinea pigs—practically nothing. China had some animals, but by far most of its large animals were people. The Muslim and Hindu peoples may have had more livestock per capita than the Chinese, but we quibble if we hesitate to say that no area of large, complex societies was so full of big animals as Europe. Europe may have had more meat per capita than any other advanced people. The Chinese peasant diet, to cite an extreme example, was almost completely vegetarian. In a world in which crises were common, especially famine, Europeans could resort to a technique available to no other advanced people. They could kill a large percentage of their biggest mammals and eat them. Where the Far Easterners were obliged to use their own excrement for fertilizer, maximizing the exchange of internal parasites, the Europeans could use the manure of their animals, which limited said exchange. It is possible that Europeans were subject to less chronic illness than most people. In an age when power was a matter of muscle, not engines, Europe had more muscle, animal muscle. (And, if I may be permitted a plausible hypothesis, utilizing animal power leads to machine-making. Animals are, unlike people, strong enough to pull heavy loads, and wagons operate on wheels, just like dynamos and jet engines.)

The most distinctive aspect of the Europeans' future would be their ability to manipulate physical reality more successfully than any others. Therefore, when I look at medieval Europe, I look for indications of actual interest in (or at least active acknowledgment of) palpable nature, rather than in the impalpable reality of saints and demons flickering behind it. I look for indications of a respect for workers and work, for the manipulation of the here-and-nowness of what can freeze your toes, fill your barns with calves and lambs, or rip the roof off your home, that is, respect for the secondary causes rather than the Prime Mover Unmoved. Most societies, including most of the most advanced, have not focused on such mundanities; or, to be more precise, the literate and artistic elites of those societies, the classes that would eventually produce and/or encourage the scientists, bankers, engineers, and entrepreneurs (the "movers and shakers"), have been too ethereal and cerebral to waste their time on materiality. What about medieval Europeans?

Dutton's Carolingians believed in storm-makers, not in updrafts of moist air, but did not accord them the deference that Africans do their rain-makers. Dozer-Rabedeau's *rusticus* was not a bipedal domesticated animal; he enjoyed a high degree of freedom of action, albeit at a low level, but was his status rising or falling? The peasants that Williams has found in stained glass seem of central importance in some examples of thirteenth-century glass, but in the dismal fourteenth century retreated to the margins. We see *in* glass "as we see through a glass," that is, "darkly."

What was medieval Europe's view of *work* in and of itself? Was it something to be done exclusively by the barely human "rustic"? Artistic and literary evidence is elusive, no better than what we can ascertain in what artists thought they saw so many centuries ago. Such evidence as Braet has examined shows the peasant as swinish, but Piers Plowman (literally a ploughman) is clearly a virtuous figure. And there is the Canterbury image of Adam and his space introduced to us by Camille. This Adam looks much put-upon, but tragic rather than vulgar. He even looks Christ-like. Yet, Braet tells us, if this magnificent figure had emerged from the fields to marry above his class, he would have become an obnoxious clown.

Jaritz preaches to us that our sources on the medieval peasantry are ambiguous, even contradictory. This is evidence in itself. "To admire the peasants," he writes, "and at the same time to condemn them . . . seems to have been rather normal." The untoward was stirring, which always stimulates incoherence.

The market! Wherever I see signs of a decline of subsistence agriculture and the rise of a market economy, bells ring in my twentieth-century head. I consider the market to be the Philosopher's Stone, the Universal Solvent, the single mechanism, which, more than any other, made us what we are. Economic determinism is a gross oversimplification of the way that societies function, but I cannot imagine a large, complex, hierarchical society (like our own) in which economic factors are not among the most powerful. Circa 1300, according to Campbell, a significant proportion of English peasants were already sliding into what would become the mainstream of modern life, that is, the market economy (and Chaucer not even born yet). The chief items being bought and sold were not spices and silks and other luxury goods, but bulk goods, farm goods, whose production, transport, and sale involved the peasant masses.

The market, like Jehovah, is a jealous god and tolerates no traditions or institutions not its own, or alters them to serve its own purposes. But Marco Polo saw market economies in the Far East, and Watson has much to tell us about the market as it functions in medieval Islamic societies. Why were the great civilizations of Asia and North Africa not transformed as Europe was? The stock answer is that outside Europe there were elites, priestly and political, that reined in the market economy and kept the merchant class in line. The Chinese emperor and his court and bureaucracy would be a good example of an elite operating to stem the tide of capitalism and its cheeky entrepreneurs. In Europe, in contrast, power was split, politically and religiously, between kings and popes; and monarchs and nobles and doges squabbled endlessly over who or which controlled what. And so the merchants were able to slip between their legs and run off to invent the modern world.

Well, fine, that sounds plausible. But have I not admitted that the lesson I have learned from this collection is that *plausible* does not necessarily mean *right*. We historians must hie ourselves back to the archives, to the facts; and if we are medievalists we must grant our most respectful attention to the peasants because in the Middle Ages the cultural center of gravity resided in the peasantry. If we are to begin to comprehend how they resisted and how they adapted to market forces, we must understand their family and land ownership systems, an inquiry that Kuchenbuch already has begun. The elites thought of the peasantry at best as capable of nothing more demanding than obedience: the illustrations in their calendars and Books of Hours, which Henisch has so carefully examined, show the villeins as barely more intelligent than the other beasts of burden. But when this horny-handed majority changed or accepted change, then the world changed. All else was the fluttering and chirping of people of ephemeral authority.

Note

1. Mary Kilbourne Matossian, *Poisons of the Past: Molds, Epidemics, and History* (New Haven, Conn.: Yale University Press, 1989).

Selective Bibliography

The following citations are drawn from the essays in this collection.

Abel, Wilhelm. *Agricultural Fluctuations in Europe, from the Thirteenth to the Twentieth Centuries*, trans. Olive Ordish. New York: St. Martin's Press, 1980.

———. *Geschichte der deutschen Landwirtschaft vom frühen Mittelalter bis zum 19. Jahrhundert*. Deutsche Agrargeschichte 2, ed. Gunther Franz. 3rd ed. Stuttgart: Ulmer, 1978.

———. *Die Wüstungen des ausgehenden Mittelalters*. 2nd ed. Stuttgart: G. Fischer, 1955.

Achilles, Walter. "Bemerkungen zum sozialen Ansehen des Bauernstandes in vorindustrieller Zeit." *Zeitschrift für Agrargeschichte und Agrarsoziologie* 34(1) (1986): 1–30.

Alexander, Jonathan. "*Labeur* and *Paresse*: Ideological Representations of Medieval Peasant Labor." *Art Bulletin* 72 (1990): 436–52.

Arié, Rachel. *Études sur la civilisation de l'Espagne musulmane*. Leiden: E. J. Brill, 1990.

Arnold, David. *Famine: Social Crisis and Historical Change*. Oxford: Basil Blackwell, 1988.

Arnold, Klaus. "Mentalität und Erziehung: Geschlechtsspezifische Arbeitsteilung und Geschlechtersphären als Gegenstand der Sozialisation im Mittelalter." In *Mentalitäten im Mittelalter. Methodische und Inhaltliche Probleme*, ed. František Graus. Vorträge und Forschungen 35. Sigmaringen: Thorbecke, 1987.

Bailey, Mark. *A Marginal Economy? East Anglian Breckland in the Later Middle Ages*. Cambridge: Cambridge University Press, 1989.

———. "*Per impetum maris*: Natural Disaster and Economic Decline in Eastern England, 1275–1350." In *Before the Black Death: Studies in the "Crisis" of the Early Fourteenth Century*, ed. Bruce M. S. Campbell. Manchester: Manchester University Press, 1991.

———. "The Rabbit and the Medieval East Anglian Economy." *Agricultural History Review* 36 (1988): 1–20.

Baker, Alan R. H. "Evidence in the 'Nonarum Inquisitiones' of Contracting Arable Lands in England during the Early Fourteenth Century." *Economic History Review* 2nd ser. 19 (1966): 518–32.

Barney, Stephen A. "The Plowshare of the Tongue: The Progress of a Symbol from the Bible to *Piers Plowman*." *Mediaeval Studies* 35 (1973): 261–93.

Bäuerliche Sachkultur des Spätmittelalters. Veröffentlichungen des Instituts für mittelalterliche Realienkunde Österreichs 7, Sitzungsberichte der österreichischen Akademie der Wissenschaften, phil.-hist. Kl. 439. Vienna: Verlag des österreichischen Akademie der Wissenschaften, 1984.

Bennett, Judith M. "The Village Ale-Wife, Women and Brewing in Fourteenth-Century England." In *Women and Work in Preindustrial Europe*, ed. Barbara A. Hanawalt. Bloomington: Indiana University Press, 1986.

———. *Women in the Medieval English Countryside: Gender and Household in Brigstock Before the Plague*. New York: Oxford University Press, 1987.

Bentzien, Ulrich. *Bauernarbeit im Feudalismus: Landwirtschaftliche Arbeitsgeräte und -verfahren in Deutschland von der Mitte des ersten Jahrtausends u.Z. bis um 1800*. Veröffentlichungen zur Volkskunde und Kulturgeschichte 67. Berlin: Akademie-Verlag, 1980.

Bergquist, Harry, and Johannes Lepiksaar. "Medieval Skeletal Remains from Medieval Lund." *Archaeology of Lund: Studies in the Lund Excavation Material* 1. Lund: Museum of Cultural History, 1957.

Berkner, Lutz K. "Inheritance, Land Tenure and Peasant Family Structure: A German Regional Comparison." In *Family and Inheritance: Rural Society in Western Europe, 1200–1800*, ed. Jack Goody, Joan Thirsk, and E. P. Thompson. Cambridge: Cambridge University Press, 1976.

Bernheimer, Richard. *Wild Men in the Middle Ages: A Study in Art, Sentiment, and Demonology*. Cambridge, Mass.: Harvard University Press, 1952.

Biddick, Kathleen. "Malthus in a Straightjacket? Analyzing Agrarian Change in Medieval England." *Journal of Interdisciplinary History* 20 (1990): 623–35.

———. "Medieval English Peasants and Market Involvement." *Journal of Economic History* 45 (1985): 823–31.

———. "Missing Links: Taxable Wealth, Markets, and Stratification among Medieval English Peasants." *Journal of Interdisciplinary History* 18 (1987): 277–98.

———. *The Other Economy: Pastoral Husbandry on a Medieval Estate*. Berkeley: University of California Press, 1989.

Biddick, Kathleen, ed. *Archaeological Approaches to Medieval Europe*. Kalamazoo: Medieval Institute Publications, Western Michigan University, 1984.

Blickle, Peter. "Les communautés villageoises en Allemagne." In *Les Communautés villageoises en Europe occidentale du moyen âge aux temps modernes*. Flaran 4. Auch: Comité Départemental du Tourisme du Gers, 1984.

———, ed. *Deutsche ländliche Rechtsquellen*. Stuttgart: Klett-Cotta, 1977.

Bloch, Marc. "Les transformations des techniques comme problème de psychologie collective." *Journal de Psychologie Normale et Pathologique* 41 (Jan.–March 1948): 104–15.

Bois, Guy. *The Crisis of Feudalism: Economy and Society in Eastern Normandy, c. 1300–1550*. Cambridge: Cambridge University Press, 1984.

———. *The Transformation of the Year One Thousand: The Village of Lournand from Antiquity to Feudalism*, trans. Jean Birrell. Manchester: Manchester University Press, 1992.

Bökönyi, Sándor. "Animals, Draft" and "Animals, Food." In *Dictionary of the Middle Ages* 1, ed. Joseph R. Strayer. New York: Charles Scribner's Sons, 1982.

———. "The Earliest Waves of Domestic Horses in East Europe." *Journal of Indo-European Studies* 6 (1–2) (1978): 17–76.

———. *History of Domestic Mammals in Central and Eastern Europe*, trans. Lili Halápy. Budapest: Akadémiai Kiadó, 1974.

Bibliography 345

45g

———. "Horse." In *Evolution of Domesticated Animals*, ed. Ian L. Mason. London: Longman, 1984.

Bolens, Lucie. *Les Méthodes culturales au moyen âge d'après les traités d'agronomie andalous: Traditions et techniques.* Geneva: Éditions Médecine et Hygiène, 1974.

Bonnassie, Pierre. *From Slavery to Feudalism in South-Western Europe*, trans. Jean Birrell. Cambridge: Cambridge University Press, 1991.

Boshof, Egon. *Erzbischof Agobard von Lyon.* Cologne: Böhlau, 1969.

Bosl, Karl. "Gesellschaftsentwicklung, 900–1350." In *Handbuch der deutschen Wirtschafts- und Sozialgeschichte* 1, ed. Hermann Aubin and Wolfgang Zorn. Stuttgart: Union Verlag, 1971.

Bourin, Monique, and Robert Durand. *Vivre au village au moyen âge: Les solidarités paysannes du 11e au 13e siècles.* Paris: Messidor/Temps Actuels, 1984.

Braet, Herman. "'Cucullus non facit monachum': Of Beasts and Monks in the Old French 'Renart' Romance." In *Monks, Nuns and Friars in Medieval Society*, ed. Edward B. King, Jacqueline T. Schaefer, and William B. Wadley. Sewanee, Tenn.: Press of the University of the South, 1989.

Brandon, P. F. "Demesne Arable Farming in Coastal Sussex during the Later Middle Ages." *Agricultural History Review* 19 (1971): 113–42.

Brenner, Robert. "Agrarian Class Structure and Economic Development in Pre-industrial Europe." *Past and Present* 70 (1976): 30–75. Reprinted in *The Brenner Debate: Agrarian Class Structure and Economic Development in Pre-industrial Europe*, ed. T. H. Aston and C. H. E. Philpin. Cambridge: Cambridge University Press, 1985.

Bridbury, A. R. "Before the Black Death." *Economic History Review* 2nd ser. 30 (1977): 393–410.

Britnell, R. H. *The Commercialisation of English Society 1000–1500.* Cambridge: Cambridge University Press, 1993.

———. "England and Northern Italy in the Early Fourteenth Century: The Economic Contrasts." *Transactions of the Royal Historical Society* 5th ser. 39 (1989): 167–83.

———. "The Proliferation of Markets in England, 1200–1349." *Economic History Review* 2nd ser. 34 (1981): 209–21.

Brown, Peter. *Society and the Holy in Late Antiquity.* Berkeley: University of California Press, 1982.

Brunner, Karl. "Nachgrabungen: Sachkultur und Kontinuitätsfragen am Beispiel der bayerischen Quellen des Frühmittelalters." In *Typen der Ethnogenese unter besonderer Berücksichtigung der Bayern* 1, ed. Herwig Wolfram and Walter Pohl. Denkschriften der österreichischen Akademie der Wissenschaften, phil.-hist. Kl. 201, Veröffentlichungen der Kommission für Frühmittelalterforschung 12. Vienna: Verlag der österreichischen Akademie der Wissenschaften, 1990.

Brunner, Karl, and Gerhard Jaritz. *Landherr, Bauer, Ackerknecht: Der Bauer im Mittelalter—Klischee und Wirklichkeit.* Vienna: Böhlau, 1985.

Brunner, Otto. *Land and Lordship: Structures of Governance in Medieval Austria*, trans. Howard Kaminsky and James Van Horn Melton. Philadelphia: University of Pennsylvania Press, 1992.

Buchda, G. "Die Dorfgemeinde im Sachsenspiegel." In *Die Anfänge der Land-*

gemeinde und ihr Wesen 2. Vorträge und Forschungen 8. Konstanz: Thorbecke, 1964.

Burgess, Glyn S. *Contribution à l'étude du vocabulaire précourtois*. Geneva: Droz, 1970.

Butcher, A. F. "English Urban Society and the Revolt of 1381." In *The English Rising of 1381*, ed. R. H. Hilton and T. H. Aston. Cambridge: Cambridge University Press, 1984.

Camille, Michael. "Labouring for the Lord: The Ploughman and the Social Order in the Luttrell Psalter." *Art History* 10(4) (1987): 423–54.

Campbell, Bruce M. S. "Agricultural Progress in Medieval England: Some Evidence from Eastern Norfolk." *Economic History Review* 2nd ser. 36 (1983): 26–46.

————. "Commonfield Origins: The Regional Dimension." In *The Origins of Open Field Agriculture*, ed. Trevor Rowley. London: Croom Helm, 1981.

————. "The Diffusion of Vetches in Medieval England." *Economic History Review* 2nd ser. 41 (1988): 193–208.

————. "Land, Labour, Livestock, and Productivity Trends in English Seigniorial Agriculture, 1208–1450." In *Land, Labour and Livestock: Historical Studies in European Agricultural Productivity*, ed. Campbell and Mark Overton. Manchester: Manchester University Press, 1991.

————. "People and Land in the Middle Ages, 1066–1500." In *An Historical Geography of England and Wales*, ed. R. A. Dodgshon and R. A. Butlin. 2nd ed. London: Academic Press, 1990.

————. "Population Pressure, Inheritance and the Land Market in a Fourteenth-Century Peasant Community." In *Land, Kinship, and Life-Cycle*, ed. Richard M. Smith. Cambridge: Cambridge University Press, 1984.

————. "Towards an Agricultural Geography of Medieval England." *Agricultural History Review* 36 (1988): 87–98.

Campbell, Bruce M. S. et al. *A Medieval Capital and Its Grain Supply: Agrarian Production and Distribution in the London Region c. 1300*. Historical Geography Research Series 30 (Historical Geography Research Group, 1993).

Campbell, Bruce M. S. and John P. Power. "Mapping the Agricultural Geography of Medieval England." *Journal of Historical Geography* 15 (1989): 24–39.

Caviness, Madeline Harrison. *The Early Stained Glass of Canterbury Cathedral, circa 1175–1220*. Princeton, N.J.: Princeton University Press, 1977.

————. *The Windows of Christ Church Cathedral, Canterbury*. Corpus Vitrearum Medii Aevi, Great Britain 2. London: Published for the British Academy by Oxford University Press, 1981.

Chambers, J. D. *Population, Economy, and Society in Pre-Industrial England*. London: Oxford University Press, 1972.

Chapelot, Jean, and Robert Fossier. *The Village and House in the Middle Ages*, trans. Henry Cleere. London: Batsford, 1985.

Chevallier, Claude-Alain, ed. *Théâtre comique du moyen âge*. Ser. 10/18: 752. Paris: Union Générale d'Éditions, 1982.

Cipolla, Carlo M. *Before the Industrial Revolution: European Society and Economy, 1000–1700*. 2nd ed. New York: W. W. Norton, 1980.

————. *Between Two Cultures: An Introduction to Economic History*, trans. Christopher Woodall. New York: W. W. Norton, 1991.

Clark, Elaine. "Debt Litigation in a Late Medieval English Vill." In *Pathways to Medieval Peasants*, ed. J. A. Raftis. Toronto: Pontifical Institute of Mediaeval Studies, 1981.

Les Communautés rurales/Rural Communities, pt. 5. Récueils de la Société Jean Bodin 44. Paris: Dessain et Tolra, 1987.

Constable, Giles. "*Nona et Decima*: An Aspect of Carolingian Economy." *Speculum* 35 (1960): 224–50.

Conze, Werner. "Arbeit." In *Geschichtliche Grundbegriffe* 1, ed. Otto Brunner, Werner Conze, and Reinhart Koselleck. Stuttgart: E. Klett, 1972.

Cooter, William S. "Ecological Dimensions of Medieval Agrarian Systems." *Agricultural History* 52 (1978): 458–77.

Crosby, Alfred W. *Ecological Imperialism: The Biological Expansion of Europe, 900–1900*. Cambridge: Cambridge University Press, 1986.

Cuisenier, Jean, and Rémy Guadagnin, eds. *Un Village au temps de Charlemagne: Moines et paysans de l'abbaye de Saint-Denis du VIIe siècle à l'An Mil*. Paris: Éditions de la Réunion des Musées Nationaux, 1988.

Daley, A. Stuart. "Chaucer's 'Droghte of March' in Medieval Farm Lore." *Chaucer Review* 4 (1970): 171–79.

Davies, Wendy. *Small Worlds: The Village Community in Early Medieval Brittany*. London: Duckworth, 1988.

Denemark, Robert A., and Kenneth P. Thomas. "The Brenner-Wallerstein Debate." *International Studies Quarterly* 32 (1988): 47–65.

Denis, B. "Le peuplement ovin de la France septentrionale avant l'introduction des Mérinos." In *L'Homme, l'animal domestique et l'environnement du moyen âge au XVIIIe siècle*, ed. Robert Durand. Enquêtes et Documents 19. Nantes: Centre de Recherches sur l'Histoire de la France Atlantique, 1993.

DeWindt, Edwin Brezette. *Land and People in Holywell-cum-Needingworth*. Toronto: Pontifical Institute of Mediaeval Studies, 1972.

Dodgshon, Robert A. *The Origin of British Field Systems*. London: Academic Press, 1980.

Doehaerd, Renée. *The Early Middle Ages in the West: Economy and Society*, trans. W. G. Deakin. Amsterdam: North-Holland, 1978.

Donkin, R. A. *The Cistercians: Studies in the Geography of Medieval England and Wales*. Toronto: Pontifical Institute of Mediaeval Studies, 1978.

Duby, Georges. *The Early Growth of the European Economy*. London: Weidenfeld & Nicholson, 1973.

————. *Rural Economy and Country Life in the Medieval West*, trans. Cynthia Postan. Columbia: University of South Carolina Press, 1968.

————. *The Three Orders: Feudal Society Imagined*, trans. Arthur Goldhammer. Chicago: University of Chicago Press, 1980.

Dufournet, J. "Du 'Jeu de Robin et Marion' au 'Jeu de la Feuillée.'" In *Études de langue et de litteratures du moyen âge offertes à Félix Lecoy*. Paris: Champion, 1973.

Durliat, Jean. "Du caput antique au manse médiéval." *Pallas* 29 (1982): 66–77.

Dutton, Paul Edward, ed. *Carolingian Civilization: A Reader*. Peterborough, Ont.: Broadview, 1993.

Dyer, Christopher. *Lords and Peasants in a Changing Society: The Estates of the Bishopric of Worcester, 680–1540*. Cambridge: Cambridge University Press, 1980.

Edwards, James Frederick and Brian Paul Hindle. "The Transportation System of Medieval England and Wales." *Journal of Historical Geography* 17 (1991): 123–34.

Ennen, Edith, and Walter Janssen. *Deutsche Agrargeschichte: Vom Neolithikum bis zur Schwelle des Industriezeitalters*. Wiesbaden: Steiner, 1979.

Epperlein, Siegfried. *Der Bauer im Bild des Mittelalters*. Leipzig: Urania-Verlag, 1975.

Erffa, Hans Martin von. *Ikonologie der Genesis: Die christlichen Bildthemen aus dem Alten Testament und ihre Quellen*. Munich: Deutscher Kunstverlag, 1989.

Farmer, David L. "Grain Yields on Westminster Abbey Manors, 1271–1410." *Canadian Journal of History* 18 (1983): 331–47.

———. "Two Wiltshire Manors and Their Markets." *Agricultural History Review* 37 (1989): 1–11.

Faucher, D. "L'assolement triennal en France." *Etudes Rurales* 1 (1961): 7–17.

Flint, Valerie I. J. *The Rise of Magic in Early Medieval Europe*. Princeton, N.J.: Princeton University Press, 1991.

Fossier, Robert. *Peasant Life in the Medieval West*, trans. Juliet Vale. Oxford: Basil Blackwell, 1988.

Fourquin, Guy. *Le Paysan d'Occident au moyen âge*. Paris: F. Nathan, 1972.

Fox, H. S. A. "The Alleged Transformation from Two-Field to Three-Field Systems in Medieval England." *Economic History Review* 2nd ser. 39 (1986): 526–48.

Frank, Grace. *The Medieval French Drama*. Oxford: Clarendon, 1954.

Freedman, Paul. *The Origins of Peasant Servitude in Medieval Catalonia*. Cambridge: Cambridge University Press, 1991.

Fumagalli, Vito and Gabriella Rossetti, eds. *Medioevo rurale: Sulle tracce della civiltà contadina*. Bologna: Il Mulino, 1980.

Galloway, J. H. *The Sugar Cane Industry: An Historical Geography from Its Origins to 1914*. Cambridge: Cambridge University Press, 1989.

Geary, Patrick J. *Aristocracy in Provence: The Rhône Basin at the Dawn of the Carolingian Age*. Philadelphia: University of Pennsylvania Press, 1985.

———. "La coercition des saints dans la pratique religieuse médiévale." In *La Culture populaire au moyen âge*, ed. Pierre Boglioni. Montréal: L'Aurore, 1979.

Gimpel, Jean. *The Medieval Machine: The Industrial Revolution of the Middle Ages*. 2nd ed. Aldershot, Hants.: Wildwood House, 1988.

Glick, Thomas F. "Agriculture and Nutrition: The Mediterranean Region." In *Dictionary of the Middle Ages* 1, ed. Joseph R. Strayer. New York: Charles Scribner's Sons, 1982.

———. *Irrigation and Society in Medieval Valencia*. Cambridge, Mass: Harvard University Press, 1970.

Goffart, Walter. *Barbarians and Romans, A.D. 418–584: The Techniques of Accommodation*. Princeton, N.J.: Princeton University Press, 1980.

Goodall, Ian H. "The Medieval Blacksmith and His Products." In *Medieval Industry*, ed. D. W. Crossley. London: Council for British Archaeology, 1981.

Greisenegger, Wolfgang. *Die Realität im religiösen Theater des Mittelalters: Ein Beitrag zur Rezeptionsforschung.* Wiener Forschungen zur Theater- und Medienwissenschaft 1. Vienna: Braumüller, 1978.

Grigg, David. *The Dynamics of Agricultural Change: The Historical Experience.* London: Hutchinson, 1982.

Groenman-van Waateringe, W., and L. H. van Wijngaarden-Bakker, eds. *Farm Life in a Carolingian Village: A Model Based on·Botanical and Zoological Data from an Excavated Site.* Studies in Prae- en Protohistorie 1. Assen/Maastricht, Neth. and Wolfeboro, N.H.: Van Gorcum, 1987.

Guichard, Pierre. *L'Espagne et la Sicile musulmanes aux XIe et XIIe siècles.* Lyon: Presses Universitaires de Lyon, 1990.

Guillaumond, Catherine. "L'eau dans l'alimentation et la cuisine arabe du IXème au XIIIème siècles." In *L'Homme et l'eau en Méditerranée et au Proche Orient 2,* ed. P. Louis. Lyon: Maison de l'Orient, 1986.

Gurevich, Aron J. *Categories of Medieval Culture,* trans. G. L. Campbell. London: Routledge & Kegan Paul, 1985.

———. *Medieval Popular Culture: Problems of Belief and Perception.* Cambridge: Cambridge University Press, 1988.

Hall, David. "The Origins of Open-Field Agriculture: The Archaeological Fieldwork Evidence." In *The Origins of Open Field Agriculture,* ed. Trevor Rowley. London: Croom Helm, 1981.

Hallam, H. E. *Rural England, 1066–1348.* Fontana History of England. Glasgow: Fontana Paperbacks, 1981.

———. "The Worker's Diet." In *The Agrarian History of England and Wales 2: 1042–1350,* ed. H. E. Hallam. Cambridge: Cambridge University Press, 1988.

Hanawalt, Barbara A. "Economic Influences on the Pattern of Crime in England, 1300–1348." *American Journal of Legal History* 18 (1974): 281–97.

———. *The Ties That Bound: Peasant Families in Medieval England.* New York: Oxford University Press, 1986.

Hansen, Wilhelm. *Kalenderminiaturen der Stundenbücher: Mittelalterliches Leben im Jahreslauf.* Munich: Georg D. W. Callwey, 1984.

Hare, J. N. "Change and Continuity in Wiltshire Agriculture in the Later Middle Ages." In *Agricultural Improvement: Medieval and Modern,* ed. Walter Minchinton. Exeter Papers in Economic History 14. Exeter: University of Exeter, 1981.

Harvey, Barbara F. "Introduction: The 'Crisis' of the Early-Fourteenth Century." In *Before the Black Death: Studies in the "Crisis" of the Early Fourteenth Century,* ed. Bruce M. S. Campbell. Manchester: Manchester University Press, 1991.

———. "The Population Trend in England between 1300 and 1348." *Transactions of the Royal Historical Society* 5th ser 16 (1966): 23–42.

Harvey, P. D. A. *The Peasant Land Market in Medieval England.* Oxford: Clarendon, 1984.

Hatcher, John. "Farming Techniques: South-Western England." In *The Agrarian History of England and Wales 2: 1042–1350,* ed. H. E. Hallam. Cambridge: Cambridge University Press, 1988.

———. *Plague, Population and the English Economy: 1348–1530.* Studies in Economic and Social History, Prepared for the Economic History Society. London: Macmillan, 1977.

Heidinga, H. A. *Medieval Settlement and Economy North of the Lower Rhine: Archeology and History of Kootwijk and the Veluwe (the Netherlands)*. Cingula 9. Assen/ Maastricht, Neth. and Wolfeboro, N.H.: Van Gorcum, 1987.

Heine, Peter. *Kulinarische Studien: Untersuchungen zur Kochkunst in arabisch-islamischen Mittelatter*. Wiesbaden: O. Harrassowitz, 1988.

Hémardinquer, J.-J. "L'introduction du maïs et la culture des sorghos dans l'ancienne France." *Bulletin Philologique et Historique* (1963): 429–59.

Henisch, Bridget Ann. *Fast and Feast: Food in Medieval Society*. University Park: Penn State Press, 1976.

Herlihy, David. "The Agrarian Revolution in Southern France and Italy, 801–1150." *Speculum* 33 (1958): 23–41.

———. "The Carolingian Mansus." *Economic History Review* 13 (1960/61): 79–89.

———. *Medieval Households*. Cambridge, Mass.: Harvard University Press, 1985.

———. "Three Patterns of Social Mobility in Medieval Society." *Journal of Interdisciplinary History* 3 (1973): 623–47. Reprinted in Herlihy, *The Social History of Italy and Western Europe, 700–1500: Collected Studies*. London: Variorum, 1978.

Herlihy, David and Christiane Klapisch-Zuber. *Tuscans and Their Families: A Study of the Florentine Catasto of 1427*. New Haven, Conn.: Yale University Press, 1985.

Hilton, R. H. *The English Peasantry in the Later Middle Ages*. Oxford: Clarendon, 1975.

Hocquet, Jean-Claude. "Le pain, le vin et la juste mesure à la table des moines carolingiens." *Annales: ESC* 40 (1985): 661–86.

Hodges, Richard. *Dark Age Economics: The Origins of Towns and Trade, A.D. 600–1000*. London: Duckworth, 1982.

Hogan, M. Patricia. "The Labor of Their Days: Work in the Medieval Village." *Studies in Medieval and Renaissance History* n.s. 8 (1986): 77–186.

Hollyman, K. J. *Le Développement du vocabulaire féodal en France pendant le haut moyen âge: Étude sémantique*. Geneva: Droz, 1957.

Holt, Richard. *The Mills of Medieval England*. Oxford: Basil Blackwell, 1988.

Illich, Ivan. *Gender*. New York: Pantheon, 1982.

Imamuddin, S. M. *Some Aspects of the Socio-economic and Cultural History of Muslim Spain, 711–1492 A.D.* Leiden: E. J. Brill, 1965.

Janssen, Walter. "Gewerbliche Produktion des Mittelalters als Wirtschaftsfaktor im ländlichen Raum." In *Das Handwerk in vor- und frühgeschichtlicher Zeit* 2: *Archäologische und philologische Beiträge*, ed. Herbert Jankuhn et al. Abhandlungen der Akademie der Wissenschaften in Göttingen, phil.-hist. Kl. 3rd ser. 123. Göttingen: Vandenhoeck & Ruprecht, 1983.

Jaritz, Gerhard. "Der Einfluß der politischen Verhältnisse auf die Entwicklung der Alltagskultur im spätmittelalterlichen Österreich." In *Bericht über den sechzehnten österreichischen Historikertag*. Veröffentlichungen des Verbandes österreichischer Geschichtsvereine 24. Vienna: Verband österreichischer Geschichtsvereine, 1985.

———. "Zur materiellen Kultur der Steiermark im Zeitalter der Gotik." In *Gotik in der Steiermark, Ausstellungskatalog*. Graz: Kulturreferat der Steiermarkischen Landesregierung, 1978.

——— . *Zwischen Augenblick und Ewigkeit: Einführung in die Alltagsgeschichte des Mittelalters.* Vienna: Böhlau, 1989.

Kaske, R. E. "The Character Hunger in *Piers Plowman*." In *Medieval English Studies Presented to George Kane*, ed. Edward Donald Kennedy, Ronald Waldron, and Joseph S. Wittig. Wolfeboro, N.H.: Boydell & Brewer, 1988.

Kaye, Harvey J. "From Feudalism to Capitalism: The Debate Goes On." *Peasant Studies* 13(3) (1986): 171–80.

Keene, Derek. "Medieval London and Its Region." *London Journal* 14 (1989): 99–111.

Kershaw, Ian. "The Great Famine and the Agrarian Crisis in England, 1315–1322." *Past and Present* 59 (1973): 3–50. Reprinted in *Peasants, Knights and Heretics: Studies in Medieval English Social History*, ed. R. H. Hilton. Cambridge: Cambridge University Press, 1976.

Kitzinger, Ernst. *The Mosaics of Monreale*. Palermo: S. F. Flaccovio, 1960.

Klapisch-Zuber, Christiane. *Women, Family, and Ritual in Renaissance Italy*, trans. Lydia Cochrane. Chicago: University of Chicago Press, 1985.

Kosminsky, E. A. *Studies in the Agrarian History of England in the Thirteenth Century*, trans. Ruth Kisch and ed. R. H. Hilton. Oxford: Basil Blackwell, 1956.

Kowaleski, Maryanne. "Town and Country in Late Medieval England: The Hide and Leather Trade." In *Work in Towns, 850–1850*, ed. Penelope J. Corfield and Derek Keene. Leicester: Leicester University Press, 1990.

Lamb, H. H. "An Approach to the Study of the Development of Climate and Its Impact on Human Affairs." In *Climate and History: Studies in Past Climates and Their Impact on Man*, ed. T. M. L. Wigley, M. J. Ingram, and G. Farmer. Cambridge: Cambridge University Press, 1981.

Langdon, John. "Agricultural Equipment." In *The Countryside of Medieval England*, ed. Grenville Astill and Annie Grant. Oxford: Basil Blackwell, 1988.

——— . *Horses, Oxen and Technological Innovation: The Use of Draught Animals in English Farming from 1066 to 1500*. Cambridge: Cambridge University Press, 1986.

——— . "Inland Water Transport in Medieval England." *Journal of Historical Geography* 19 (1993): 1–11.

Laslett, Peter and Richard Wall, eds. *Household and Family in Past Time*. Cambridge: Cambridge University Press, 1972.

Last, Martin. "Villikationen geistlicher Grundherren in Nordwestdeutschland in der Zeit vom 12. bis zum 14. Jahrhundert (Diözesen Osnabrück, Bremen, Verden, Minden, Hildesheim)." In *Die Grundherrschaft im späten Mittelalter* 1, ed. Hans Patze. Vorträge und Forschungen 27. Sigmaringen: Thorbecke, 1983.

Lefèbvre, Joel. *Les Fols et la folie: Étude sur les genres du comique et la création littéraire en Allemagne pendant la Renaissance*. Paris: C. Klincksieck, 1968.

Le Goff, Jacques. *Time, Work, and Culture in the Middle Ages*, trans. Arthur Goldhammer. Chicago: University of Chicago Press, 1980.

Le Roy Ladurie, Emmanuel. *Montaillou: The Promised Land of Error*, trans. Barbara Bray. New York: George Braziller, 1978.

——— . *Times of Feast, Times of Famine: A History of Climate since the Year 1000*, trans. Barbara Bray. Garden City, N.Y.: Doubleday, 1971; reprint New York: Farrar, Straus and Giroux, 1988.

Lomas, Richard. "The Black Death in County Durham." *Journal of Medieval History* 15 (1989): 127–40.

Lopez. Robert S. *The Commercial Revolution of the Middle Ages, 950–1350.* Englewood Cliffs, N.J.: Prentice-Hall, 1971.

Maddicott, John Robert. *The English Peasantry and the Demands of the Crown, 1294– 1341.* Past and Present Supplement 1. 1975. Reprinted in *Landlords, Peasants and Politics in Medieval England,* ed. T. H. Aston. Cambridge: Cambridge University Press, 1987.

Mâle, Emile. *The Gothic Image: Religious Art in France of the Thirteenth Century,* trans. Dora Nussey. New York: Harper, 1972.

Mane, Perrine. *Calendriers et techniques agricoles (France-Italie, XIIe–XIIIe siècles).* Paris: Sycomore, 1983.

————. "Comparaison des thèmes iconographiques des calendriers monumentaux et enluminés en France, aux XIIe et XIIIe s." *Cahiers de Civilisation Médiévale* 29 (1986): 257–64.

————. "Émergence du vêtement de travail à travers l'iconographie médiévale." In *Le Vêtement: Histoire, archéologie et symbolique vestimentaires au moyen âge,* ed. Michel Pastoureau. Paris: Léopard d'Or, 1989.

Manselli, Raoul. *La Religion populaire au moyen âge: Problèmes de méthode et d'histoire.* Conference Albert-le-Grand, 1973. Montréal: Institut d'Etudes Médiévales Albert-le-Grand, 1975.

Martin, Jochen, and Renate Zoepffel, eds. *Aufgaben, Rollen und Räume von Frau und Mann.* 2 vols. Munich: K. A. Freiburg, 1989.

Mate, Mavis. "The Agrarian Economy of South-East England before the Black Death: Depressed or Buoyant?" In *Before the Black Death: Studies in the "Crisis" of the Early Fourteenth Century,* ed. Bruce M. S. Campbell. Manchester: Manchester University Press, 1991.

————. "The Estates of Canterbury Cathedral Priory Before the Black Death, 1315–1348." *Studies in Medieval and Renaissance History* n.s. 8 (1986): 3–31.

————. "The Impact of War on the Economy of Canterbury Cathedral Priory, 1294–1340." *Speculum* 57 (1982): 761–78.

————. "Medieval Agrarian Practices: The Determining Factors." *Agricultural History Review* 33 (1985): 22–31.

————. "Profit and Productivity on the Estates of Isabella de Forz (1260–92)." *Economic History Review* 2nd ser. 33 (1980): 326–34.

Matossian, Mary Kilbourne. *Poisons of the Past: Molds, Epidemics, and History.* New Haven, Conn.: Yale University Press, 1989.

McKintosh, Marjorie Keniston. *Autonomy and Community: The Royal Manor of Havering, 1200–1500.* Cambridge: Cambridge University Press, 1986.

McKitterick, Rosamond. *The Carolingians and the Written Word.* Cambridge: Cambridge University Press, 1989.

————. *The Frankish Church and the Carolingian Reforms, 789–895.* London: Royal Historical Society, 1977.

McNeill, John T. and Helena M. Gamer. *Medieval Handbooks of Penance.* New York: Columbia University Press, 1965.

Meiss, Millard. *French Painting in the Time of Jean de Berry: The Late Fourteenth*

Century and the Patronage of the Duke. 2nd ed., 2 vols. London: Phaidon Press, 1969.

Ménard, Philippe. *Les Fabliaux: Contes à rire du moyen âge.* Paris: Presses Universitaires de France, 1983.

——. *Le Rire et le sourire dans le roman courtois en France au moyen âge (1150–1250).* Geneva: Droz, 1969.

Metz, Wolfgang. "Die Agrarwirtschaft im karolingischen Reiche." In *Karl der Grosse: Lebenswerk und Nachleben* 1, ed. Helmut Beumann. Düsseldorf: L. Schwann, 1965.

Middleton, Christopher. "The Sexual Division of Labour in Feudal England." *New Left Review* 113–114 (1979): 147–68.

Miller, Edward. "Farming Techniques: Northern England." In *The Agrarian History of England and Wales* 2: *1042–1350*, ed. H. E. Hallam. Cambridge: Cambridge University Press, 1988.

——, ed. *The Agrarian History of England and Wales* 3: *1348–1500.* Cambridge: Cambridge University Press, 1991.

Miller, Edward and John Hatcher. *Medieval England: Rural Society and Economic Change 1086–1348.* London: Longman, 1978.

Mitterauer, Michael and Reinhard Sieder, eds. *Historische Familienforschung.* Frankfurt: Suhrkamp, 1982.

Mohl, Ruth. *The Three Estates in Medieval and Renaissance Literature.* New York: Columbia University Press, 1933.

Mollat, Michel. *The Poor in the Middle Ages: An Essay in Social History*, trans. Arthur Goldhammer. New Haven, Conn.: Yale University Press, 1986.

Montanari, Massimo. *Alimentazione e cultura nel Medioevo.* Rome: Laterza, 1988.

——. *The Culture of Food*, trans. Carl Ipsen. Oxford: Basil Blackwell, 1994.

——. "Rural Food in Late Medieval Italy." In *Bäuerliche Sachkultur des Spätmittelalters.* Veröffentlichungen des Instituts für mittelalterliche Realienkunde Österreichs 7, Sitzungsberichte der österreichischen Akademie der Wissenschaften, phil.-hist. Kl. 439. Vienna: Verlag der österreichischen Akademie der Wissenschaften, 1984.

Morris, C. A. "Early Medieval Separate-Bladed Shovels from Ireland." *Journal of the Royal Society of Antiquaries of Ireland* 111 (1981): 50–69.

——. "A Group of Early Medieval Spades." *Medieval Archaeology* 24 (1980): 205–10.

Moxey, Keith P. F. *Peasants, Warriors and Wives: Popular Imagery in the Reformation.* Chicago: University of Chicago Press, 1989.

Moyne, Ernest J. *Raising the Wind: The Legend of Lapland and Finland Wizards in Literature.* Newark: University of Delaware Press, 1981.

Nada Patrone, Anna Maria. *Il cibo del ricco ed il cibo del povero: Contributo alla storia qualitativa dell'alimentazione—l'area pedemontana negli ultimi secoli del Medioevo.* Biblioteca di Studi Piemontesi 10. Torino: Centro Studi Piemontesi, 1981.

Naidoff, B. D. "A Man to Work the Soil: A New Interpretation of Genesis 2–3." *Journal for the Study of the Old Testament* 2–3 (1978): 2–14.

Newman, Francis X., ed. *Social Unrest in the Late Middle Ages.* Papers of the Fifteenth Annual Conference of the Center for Medieval and Early Renaissance Studies. Binghamton, N.Y.: Medieval and Renaissance Texts and Studies, 1986.

North, Douglass C. *Structure and Change in Economic History*. New York: Norton, 1981.

Nykrog, Per. *Les Fabliaux: Étude d'histoire littéraire et de stylistique médiévales*. Copenhagen, 1957; nouv. ed. Geneva: Droz, 1973.

Oexle, Otto Gerhard. "Deutungsschemata der sozialen Wirklichkeit im frühen und hohen Mittelalter." In *Mentalitäten im Mittelalter: Methodische und inhaltliche Probleme*, ed. František Graus. Vorträge und Forschungen 35. Sigmaringen: Thorbecke, 1987.

Ormrod, W. M. "The Crown and the English Economy, 1290–1348." In *Before the Black Death: Studies in the "Crisis" of the Early Fourteenth Century*, ed. Bruce M. S. Campbell. Manchester: Manchester University Press, 1991.

Oschinsky, Dorothea, ed. *Walter of Henley and Other Treatises on Estate Management and Accounting*. Oxford: Clarendon, 1971.

Ostling, Christine. "The Ploughing Adam in Medieval Church Paintings." In *Man and Picture: Papers from the First International Symposium for Ethnological Picture Research in Lund, 1984*, ed. Nils-Avrid Bringéus. Stockholm: Almqvist & Wiksell, 1986.

Overton, Mark and Bruce M. S. Campbell. "Productivity Change in European Agricultural Development." In *Land, Labour and Livestock: Historical Studies in European Agricultural Productivity*, ed. Campbell and Overton. Manchester: Manchester University Press, 1991.

Ovitt, George. *The Restoration of Perfection: Labor and Technology in Medieval Culture*. New Brunswick, N.J.: Rutgers University Press, 1987.

Owst, G. R. *Literature and Pulpit in Medieval England*. 2nd rev. ed. Oxford: Basil Blackwell, 1961.

Pächt, Otto. "A Cycle of English Frescoes in Spain." *Burlington Magazine* 103 (1961): 166–75.

Papanicolaou, Linda Morey. "The Iconography of the Genesis Window of the Cathedral of Tours." *Gesta* 20 (1981): 179–89.

Parker, Vanessa. *The Making of Kings Lynn*. London: Phillimore, 1971.

Pearsall, Derek. "Poverty and Poor People in *Piers Plowman*." In *Medieval English Studies Presented to George Kane*, ed. Edward Donald Kennedy, Ronald Waldron, and Joseph S. Wittig. Wolfeboro, N.H.: Boydell & Brewer, 1988.

Percival, John. "Seigneurial Aspects of Late Roman Estate Management." *English Historical Review* 84 (1969): 449–73.

Peri, Illuminato. *Città e campagna in Sicilia*. Palermo: Presso l'Accademia, 1953–56.

Perroy, Edouard. "A l'origine d'une économie contractée: Les crises du XIVe siècle." *Annales: Économies, Sociétés, Civilisations* 3 (1949): 167–82. Trans. as "At the Origin of a Contracted Economy: The Crises of the 14th Century." In *Essays in French Economic History*, ed. Rondo Cameron. Homewood, Ill.: R. D. Irwin, 1970.

———. *La Terre et les paysans en France aux XIIe et XIIIe siècles*. Paris: Société d'Édition d'Enseignement Supérieur, 1973.

Persson, Karl Gunnar. "Labour Productivity in Medieval Agriculture: Tuscany and the 'Low Countries.'" In *Land, Labour and Livestock: Historical Studies in European Agricultural Productivity*, ed. Bruce M. S. Campbell and Mark Overton. Manchester: Manchester University Press, 1991.

———. *Pre-industrial Economic Growth: Social Organization and Technological Progress in Europe*. Oxford: Basil Blackwell, 1988.

Poole, Reginald Lane. *Illustrations of the History of Medieval Thought and Learning*. 2nd rev. ed. London, 1920; reprinted New York: Dover, 1960.

Postan, M. M. "Medieval Agrarian Society in Its Prime, Pt. 7: England." In *The Cambridge Economic History of Europe* 1: *The Agrarian Life of the Middle Ages*, 2nd ed., ed. Postan. Cambridge: Cambridge University Press, 1966.

———. *The Medieval Economy and Society: An Economic History of Britain 1100–1500*. London: Weidenfeld & Nicolson, 1972.

Postan, M. M. and J. Z. Titow. "Heriots and Prices on Winchester Manors." *Economic History Review* 2nd ser. 11 (1958–59): 392–417; reprinted in Postan, *Essays on Medieval Agriculture and General Problems of the Medieval Economy*. Cambridge: Cambridge University Press, 1973.

Poulin, Joseph-Claude. "Entre magie et religion: Recherches sur les utilisations marginales de l'écrit dans la culture populaire du haut moyen âge." In *La Culture populaire au moyen âge*, ed. Pierre Boglioni. Montréal: L'Aurore, 1979.

Power, John P. and Bruce M. S. Campbell, "Cluster Analysis and the Classification of Medieval Demesne-Farming Systems." *Transactions of the Institute of British Geographers* n.s. 17 (1992): 227–45.

Raftis, J. Ambrose. *The Estates of Ramsey Abbey*. Toronto: Pontifical Institute of Mediaeval Studies, 1957.

———. "Farming Techniques: The East Midlands." In *The Agrarian History of England and Wales* 2: *1042–1350*, ed. H. E. Hallam. Cambridge: Cambridge University Press, 1988.

———. *Tenure and Mobility: Studies in the Social History of the Medieval English Village*. Toronto: Pontifical Institute of Mediaeval Studies, 1964.

Raftis, J. Ambrose, ed. *Pathways to Medieval Peasants*. Toronto: Pontifical Institute of Mediaeval Studies, 1981.

Raupp, Hans-Joachim. *Bauernsatiren: Entstehung und Entwicklung des bäuerlichen Genres in der deutschen und niederländischen Kunst ca. 1470–1570*. Niederzier: Lukassen, 1986.

Ravensdale, J. R. *Liable to Floods: Village Landscape on the Edge of the Fens* A.D. *450–850*. London: Cambridge University Press, 1974.

Razi, Zvi. *Life, Marriage and Death in a Medieval Parish: Economy, Society and Demography in Halesowen, 1270–400*. Cambridge: Cambridge University Press, 1980.

Rey-Flaud, Henri. *Pour une dramaturgie du moyen âge*. Paris: Presses Universitaires de France, 1980.

Riché, Pierre. *Daily Life in the World of Charlemagne*, trans. Jo Ann McNamara. Philadelphia: University of Pennsylvania Press, 1978.

Richter, Michael. *The Formation of the Medieval West*. Dublin: Four Courts Press, 1994.

Rösener, Werner. *Peasants in the Middle Ages*, trans. Alexander Stützer. Urbana: University of Illinois Press, 1992.

Rouche, Michel. "The Early Middle Ages in the West." In *A History of Private Life* 1: *From Pagan Rome to Byzantium*, ed. Paul Veyne and trans. Arthur Goldhammer. Cambridge, Mass.: Harvard University Press, 1987.

————. "La faim à l'époque carolingienne: Essai sur quelques types de rations alimentaires." *Revue Historique* 250 (1973): 295–320.

————. "Les repas de fête à l'époque carolingienne." In *Manger et boire au moyen âge: Actes de Colloque de Nice (15–17 octobre 1982)*, ed. Denis Menjot. Paris: Belles Lettres, 1984.

Ruiz, Teofilo F. *Crisis and Continuity: Land and Town in Late Medieval Castile*. Philadelphia: University of Pennsylvania Press, 1994.

Rutz, Henry J. and Benjamine S. Orlove, eds. *The Social Economy of Consumption*. Lanham, Md.: University Press of America, 1989.

Ryder, M. L. "The History of Sheep Breeds in Britain." *Agricultural History Review* 12(1) (1964): 1–12; (2): 65–82.

————. "Livestock." In *The Agrarian History of England and Wales* 1: *Prehistory*, ed. Stuart Piggott. Cambridge: Cambridge University Press, 1981.

Schiller, Gertrud. *Iconography of Christian Art*, trans. Janet Seligman. Greenwich, Conn.: New York Graphic Society, 1971.

Schulze-Busacker, Elisabeth. *Proverbes et expressions proverbiales dans la littérature narrative du moyen âge français*. Paris: Champion, 1985.

Schwineköper, Berent. "Die mittelalterliche Dorfgemeinde in Elbostfalen und in den benachbarten Markengebieten." In *Die Anfänge der Landgemeinde und ihr Wesen* 2. Vorträge und Forschungen 8. Konstanz: Thorbecke, 1964.

Scott, Kathleen L. "The Illustrations of *Piers Plowman* in Bodleian Library MS. Douce 104." *The Yearbook of Langland Studies* 4 (1990): 1–86.

Searle, Eleanor. *Lordship and Community: Battle Abbey and Its Banlieu, 1066–1538*. Toronto: Pontifical Institute of Mediaeval Studies, 1974.

Seavoy, Ronald E. *Famine in Peasant Societies*. New York: Greenwood, 1986.

Segalen, Martine. *Love and Power in the Peasant Family: Rural France in the Nineteenth Century*, trans. Sarah Mathews. Oxford: Basil Blackwell, 1983.

Sen, Amartya. *Poverty and Famine: An Essay on Entitlement and Deprivation*. Oxford: Clarendon, 1981.

Shahar, Shulamith. *The Fourth Estate: A History of Women in the Middle Ages*, trans. Chaya Galai. London: Methuen, 1983.

Sheail, John. *Rabbits and Their History*. Newton Abbot: David & Charles, 1971.

Shrewsbury, J. F. D. *A History of Bubonic Plague in the British Isles*. London: Cambridge University Press, 1970.

Sivéry, Gérard. *Terroirs et communautés rurales dans l'Europe occidentale au moyen âge*. Lille: Presses Universitaires de Lille, 1990.

Smith, Richard M. "Demographic Developments in Rural England, 1300–48: A Survey." In *Before the Black Death: Studies in the "Crisis" of the Early Fourteenth Century*, ed. Bruce M. S. Campbell. Manchester: Manchester University Press, 1991.

————. "Families and Their Land in an Area of Partible Inheritance: Redgrave, Suffolk, 1260–1320." In *Land, Kinship, and Life-Cycle*, ed. Richard M. Smith. Cambridge: Cambridge University Press, 1984.

Sorokin, Pitirim A. *Hunger as a Factor in Human Affairs*, trans. Elena P. Sorokin and ed. T. Lynn Smith. Gainesville: University Presses of Florida, 1975.

Specht, Henrik. *Poetry and the Iconography of the Peasant: The Attitude to the Peasant in Late Medieval English Literature and in Contemporary Calendar Illustration.* Copenhagen: Department of English, University of Copenhagen, 1983.

Stephenson, M. J. "Wool Yields in the Medieval Economy." *Economic History Review* 2nd ser. 41 (1988): 368–91.

TeBrake, William H. *Medieval Frontier: Culture and Ecology in Rijnland.* College Station: Texas A&M University Press, 1985.

Thirsk, Joan. "Field Systems of the East Midlands." In *Studies of Field Systems in the British Isles*, ed. Alan R. H. Baker and Robin A. Butlin. Cambridge: Cambridge University Press, 1973.

Thornton, Christopher. "The Determinants of Land Productivity on the Bishop of Winchester's Demesne of Rimpton, 1208 to 1403." In *Land, Labour and Livestock: Historical Studies in European Agricultural Productivity*, ed. Bruce M. S. Campbell and Mark Overton. Manchester: Manchester University Press, 1991.

Titow, J. Z. "Le climat à travers les rôles de compatabilité de l'évêché de Winchester (1350–1450)." *Annales: Économies, Sociétés, Civilisations* 25 (1970): 312–50.

————. *English Rural Society, 1200–1350.* London: Allen & Unwin, 1969.

————. *Winchester Yields: A Study in Medieval Agricultural Productivity.* Cambridge: Cambridge University Press, 1972.

Twigg, Graham. *The Black Death: A Biological Reappraisal.* London: Batsford; New York: Schocken Books, 1984.

Veale, E. M. "The Rabbit in England." *Agricultural History Review* 5 (1957): 85–90.

Verdier, Philippe. "Woman in the Marginalia of Gothic Manuscripts and Related Works." In *The Role of Woman in the Middle Ages*, ed. Rosmarie Thee Morewedge. Papers of the Sixth Annual Conference of the Center for Medieval and Early Renaissance Studies, State University of New York at Binghamton, May 6–7, 1972. Albany: State University of New York Press, 1975.

Verhulst, Adriaan. "The 'Agricultural Revolution' of the Middle Ages Reconsidered." In *Law, Custom and the Social Fabric in Medieval Europe: Essays in Honor of Bryce Lyon*, ed. S. Bachrach and D. Nicholas. Kalamazoo, Mich.: Medieval Institute Publications, 1990. Reprinted in Verhulst, *Rural and Urban Aspects of Early Medieval Northwest Europe.* Aldershot, Hants.: Variorum, 1992.

————. "La genèse du régime domanial classique en France au haut moyen âge." In *Agricoltura e mondo rurale in Occidente nell'alto Medioevo.* Centro italiano di studi sull'alto Medioevo. Settimane di Studio 13. Spoleto: Presso la sede del Centro, 1966.

————. "Karolingische Agrarpolitik: Das Capitulare de Villis und die Hungersnöte von 792/93 und 805/06." *Zeitschrift für Agrargeschichte und Agrarsoziologie* 13 (1965): 179–85.

Vogel, Cyrille. "Pratiques superstitieuses au début au XIe siècle d'après le *Corrector sive Medicus* de Burchard, évêque de Worms (965–1025)." In *Études de civilization médiévale (IXe–XIIe siècles): Mélanges offerts à Edmond-René Labande.* Poitiers: C.E.S.C.M., 1974.

Wallace-Hadrill, J. M. *The Frankish Church.* Oxford: Clarendon, 1983.

Watson, Andrew M. *Agricultural Innovation in the Early Islamic World: The Diffusion of Crops and Farming Techniques, 700–1100.* Cambridge: Cambridge University Press, 1983.

———. "The Arab Agricultural Revolution and Its Diffusion." *Journal of Economic History* 34 (1974): 8–35.

———. "A Medieval Green Revolution." In *The Islamic Middle East, 700–1900,* ed. A. L. Udovitch. Princeton, N.J.: Darwin, 1981.

———. "The Rise and Spread of Old World Cotton." In *Studies in Textile History in Memory of Harold B. Burnham,* ed. Veronika Gervers. Toronto: Royal Ontario Museum, 1977.

Webster, James Carson. *The Labors of the Months in Antique and Mediaeval Art to the End of the Twelfth Century.* Princeton, N.J.: Princeton University Press, 1938.

Wenskus, Reinhard, Herbert Jankuhn, and Klaus Grinda, eds. *Wort und Begriff "Bauer."* Abhandlungen der Akademie der Wissenschaften, Göttingen, phil.-hist. Kl. 3.F. 89. Göttingen: Vandenhoeck & Ruprecht, 1975.

White, K. D. *Agricultural Implements of the Roman World.* Cambridge: Cambridge University Press, 1967.

White, Lynn, Jr. *Medieval Technology and Social Change.* New York: Oxford University Press, 1962.

———, ed. *The Transformation of the Roman World.* Center for Medieval and Renaissance Studies, University of California at Los Angeles, Contributions 3. Berkeley: University of California Press, 1966.

Williams, Jane Welch. *Bread, Wine and Money: The Windows of the Trades at Chartres Cathedral.* Chicago: University of Chicago Press, 1993.

Wolfram, Herwig. *Die Geburt Mitteleuropas: Geschichte Österreichs vor seiner Entstehung, 378–907.* Vienna: Kremayr & Scheriau, 1987.

Zink, Michel. *La Pastourelle: Poésie et folklore au moyen âge.* Paris: Bordas, 1972.

Contributors

Sándor Bökönyi was, until his death on December 25, 1994, Research Professor of the Archaeological Institute of the Hungarian Academy of Sciences in Budapest and was its Director from 1981 through 1993. He was the author of many books and articles on the history of domestic animals. His publications in English include *History of Domestic Mammals in Central and Eastern Europe* (1974), *Animal Husbandry and Hunting in Tac-Gorsium* (1984), and "Animals, Draft" and "Animals, Food" in the *Dictionary of the Middle Ages* (1982).

Herman Braet is Professor of French and Provençal Literature and Philology at the Universities of Antwerp and Louvain. He is the editor of the medieval text series *Ktemata*. His publications include *Le Songe dans la chanson de geste* (1975), *Deux Lais Féeriques* (1980), and a new annotated edition of Beroul's *Tristran et Iseut* (in collaboration with Guy Raynaud de Lage, 1989–90).

Karl Brunner is Professor of Medieval History and Auxiliary Sciences at the Institut für österreichische Geschichtsforschung in Vienna and the author of several books and articles on early medieval civilization, including *Oppostionelle Gruppen im Karolingerreich* (1979) and *Landherr, Bauer, Ackerknecht* (1985).

Michael Camille is Professor of Art History at the University of Chicago. He is the author of *The Gothic Idol: Ideology and Image-Making in Medieval Art* (1989) and *Image on the Edge: The Margins of Medieval Art* (1992).

Bruce M. S. Campbell is Reader in Economic and Social History at The Queen's University of Belfast. He has written extensively on the English medieval economy with particular reference to agriculture. In collaboration with Derek Keene he is currently researching the food supply of fourteenth-century London, and with John Power is preparing an agricultural and land-use atlas of England before the Black Death. He is the editor of *Before the Black Death: Studies in the "Crisis" of the Early Fourteenth Century* (1991), and co-editor, with Mark Overton, of *Land, Labour and Livestock: Historical Studies in European Agricultural Productivity* (1991).

Alfred W. Crosby is Professor of History, Geography, and American Studies at the University of Texas at Austin. He is the author of several books, two of which pertain to world agriculture: *The Columbian Exchange: Biological and Cultural Consequences of 1492* (1972) and *Ecological Imperialism: The Biological Expansion of Europe, 900–1900* (1986). The latter won the Ralph Waldo Emerson Prize of the Phi Beta Kappa Society.

Jane B. Dozer-Rabedeau received her Ph.D. in French medieval studies from UCLA. She has been a faculty member at the University of Washington, Seattle University, and Tufts University. She has presented numerous conference papers in the United States and Canada and has published "Mimesis and *li jeus de le fuellie*," *Tréteaux*, Bulletin de la Société Internationale pour l'Étude du Théâtre Médiéval 3 (1982). She is currently an independent medievalist in San Jose, California.

Paul Edward Dutton is Professor of History and Humanities at Simon Fraser University. He is the author of *The Politics of Dreaming in the Carolingian Empire* (1993) and the editor of *Carolingian Civilization: A Reader* (1993) and *The "Glosae super Platonem" of Bernard of Chartres* (1991).

Robert Worth Frank, Jr., is Professor Emeritus of English at The Pennsylvania State University and former head of the department. He is co-founder and editor of the *Chaucer Review* and author of *"Piers Plowman" and the Scheme of Salvation* and *Chaucer and "The Legend of Good Women."* He has held ACLS and Guggenheim fellowships and is a Visiting Fellow of Clare Hall, Cambridge. He is a past president of the New Chaucer Society. His recent work is in miracle literature, Chaucerian pathos, and *Piers Plowman*.

Bridget Ann Henisch is a former Lecturer in Medieval English Literature at the University of Reading, U.K., and is currently at work on a book about the "occupations of the months" in the medieval calendar tradition. She is the author of *Fast and Feast: Food in Medieval Society* (1976) and *Cakes and Characters: An English Christmas Tradition* (1984). Together with Heinz Henisch, she also has written *The Photographic Experience, 1839–1914: Images and Attitudes* (1994).

Gerhard Jaritz is Senior Research Fellow at the Institut für Realienkunde des Mittelalters und der frühen Neuzeit of the Austrian Academy of Arts and Sciences. He is the author of *Zwischen Augenblick und Ewigkeit* (1989) and is the editor of *Umweltbewältigung: Die historische Perspektive* (1994).

Ludolf Kuchenbuch is Professor of Ancient and Medieval History at Fern-

Universität- Gesamthochschule-in-Hagen, Germany. He is the author of *Feudalismus: Materialien zur Theorie und Geschichte* (1977), *Bäuerliche Gesellschaft und Kosterherrschaft im 9. Jahrhundert: Studien zur Sozialstruktur der Familia der Abtei Prüm* (1978), and *Grundherrschaft im früheren Mittelalter* (1991).

Del Sweeney received her Ph.D. in medieval history from Cornell University. She has been a faculty member at Queens College (CUNY), the University of Michigan at Flint, and Wayne State University and is now Assistant to the Dean, School of Nursing, at the University of Maryland at Baltimore.

Andrew M. Watson is Professor of Economics at the University of Toronto. He is the author of a number of studies on the economic history of medieval Christendom and the medieval Islamic world including *Agricultural Innovation in the Early Islamic World: The Diffusion of Crops and Farming Techniques* (1983).

Jane Welch Williams is Associate Professor of Medieval Art History and Director of the Art History Program at the University of Arizona, Tucson. She is the author of *Bread, Wine and Money: The Windows of the Trades at Chartres Cathedral* (1993). She is now working on a book about the images of women in Chartres Cathedral.

Index

University of Pennsylvania Press
MIDDLE AGES SERIES
Ruth Mazo Karras and Edward Peters, General Editors

F. R. P. Akehurst, trans. *The* Coutumes de Beauvaisis *of Philippe de Beaumanoir*. 1992

Peter L. Allen. *The Art of Love: Amatory Fiction from Ovid to the* Romance of the Rose. 1992

David Anderson. *Before the Knight's Tale: Imitation of Classical Epic in Boccaccio's* Teseida. 1988

Benjamin Arnold. *Count and Bishop in Medieval Germany: A Study of Regional Power, 1100–1350.* 1991

Mark C. Bartusis. *The Late Byzantine Army: Arms and Society, 1204–1453.* 1992

Thomas N. Bisson, ed. *Cultures of Power: Lordship, Status, and Process in Twelfth-Century Europe.* 1995

Uta-Renate Blumenthal. *The Investiture Controversy: Church and Monarchy from the Ninth to the Twelfth Century.* 1988

Gerald Bond. *The Loving Subject: Desire, Eloquence, and Power in Romanesque France.* 1995

Daniel Bornstein, trans. *Dino Compagni's* Chronicle *of Florence*. 1986

Maureen Boulton. *The Song in the Story: Lyric Insertions in French Narrative Fiction, 1200–1400.* 1993

Betsy Bowden. *Chaucer Aloud: The Varieties of Textual Interpretation.* 1987

Charles R. Bowlus. *Franks, Moravians, and Magyars: The Struggle for the Middle Danube, 788–907.* 1995

James William Brodman. *Ransoming Captives in Crusader Spain: The Order of Merced on the Christian-Islamic Frontier.* 1986

Kevin Brownlee and Sylvia Huot, eds. *Rethinking the* Romance of the Rose: *Text, Image, Reception.* 1992

Matilda Tomaryn Bruckner. *Shaping Romance: Interpretation, Truth, and Closure in Twelfth-Century French Fictions.* 1993

Otto Brunner (Howard Kaminsky and James Van Horn Melton, eds. and trans.). *Land and Lordship: Structures of Governance in Medieval Austria.* 1992

Robert I. Burns, S.J., ed. *Emperor of Culture: Alfonso X the Learned of Castile and His Thirteenth-Century Renaissance.* 1990

David Burr. *Olivi and Franciscan Poverty: The Origins of the* Usus Pauper *Controversy*. 1989

David Burr. *Olivi's Peaceable Kingdom: A Reading of the Apocalypse Commentary.* 1993

Thomas Cable. *The English Alliterative Tradition.* 1991

Anthony K. Cassell and Victoria Kirkham, eds. and trans. *Diana's Hunt/Caccia di Diana: Boccaccio's First Fiction.* 1991

John C. Cavadini. *The Last Christology of the West: Adoptionism in Spain and Gaul, 785–820.* 1993

Brigitte Cazelles. *The Lady as Saint: A Collection of French Hagiographic Romances of the Thirteenth Century.* 1991

Karen Cherewatuk and Ulrike Wiethaus, eds. *Dear Sister: Medieval Women and the Epistolary Genre.* 1993

Anne L. Clark. *Elisabeth of Schönau: A Twelfth-Century Visionary.* 1992

Willene B. Clark and Meradith T. McMunn, eds. *Beasts and Birds of the Middle Ages: The Bestiary and Its Legacy.* 1989

Richard C. Dales. *The Scientific Achievement of the Middle Ages.* 1973

Charles T. Davis. *Dante's Italy and Other Essays.* 1984

William J. Dohar. *The Black Death and Pastoral Leadership: The Diocese of Hereford in the Fourteenth Century.* 1994

Katherine Fischer Drew, trans. *The Burgundian Code.* 1972

Katherine Fischer Drew, trans. *The Laws of the Salian Franks.* 1991

Katherine Fischer Drew, trans. *The Lombard Laws.* 1973

Nancy Edwards. *The Archaeology of Early Medieval Ireland.* 1990

Richard K. Emmerson and Ronald B. Herzman. *The Apocalyptic Imagination in Medieval Literature.* 1992

Theodore Evergates. *Feudal Society in Medieval France: Documents from the County of Champagne.* 1993

Felipe Fernández-Armesto. *Before Columbus: Exploration and Colonization from the Mediterranean to the Atlantic, 1229–1492.* 1987

Jerold C. Frakes. *Brides and Doom: Gender, Property, and Power in Medieval German Women's Epic.* 1994

R. D. Fulk. *A History of Old English Meter.* 1992

Patrick J. Geary. *Aristocracy in Provence: The Rhône Basin at the ,awn of the Carolingian Age.* 1985

Peter Heath. *Allegory and Philosophy in Avicenna (Ibn Sînâ), with a Translation of the Book of the Prophet Muḥammad's Ascent to Heaven.* 1992

J. N. Hillgarth, ed. *Christianity and Paganism, 350–750: The Conversion of Western Europe.* 1986

Richard C. Hoffmann. *Land, Liberties, and Lordship in a Late Medieval Countryside: Agrarian Structures and Change in the Duchy of Wrocław.* 1990

Robert Hollander. *Boccaccio's Last Fiction: Il Corbaccio.* 1988

John Y. B. Hood. *Aquinas and the Jews.* 1995

Edward B. Irving, Jr. *Rereading Beowulf.* 1989

Richard A. Jackson, ed. Ordines Coronationis Franciae: *Texts and Ordines for the Coronation of Frankish and French Kings and Queens in the Middle Ages,* Vol. I. 1995

C. Stephen Jaeger. *The Envy of Angels: Cathedral Schools and Social Ideals in Medieval Europe, 950–1200.* 1994

C. Stephen Jaeger. *The Origins of Courtliness: Civilizing Trends and the Formation of Courtly Ideals, 939–1210.* 1985

Donald J. Kagay, trans. *The Usatges of Barcelona: The Fundamental Law of Catalonia.* 1994

Richard Kay. *Dante's Christian Astrology.* 1994

Ellen E. Kittell. *From* Ad Hoc *to Routine: A Case Study in Medieval Bureaucracy.* 1991

Alan C. Kors and Edward Peters, eds. *Witchcraft in Europe, 1100–1700: A Documentary History.* 1972

Barbara M. Kreutz. *Before the Normans: Southern Italy in the Ninth and Tenth Centuries.* 1992

Michael P. Kuczynski. *Prophetic Song: The Psalms as Moral Discourse in Late Medieval England.* 1995

E. Ann Matter. *The Voice of My Beloved: The Song of Songs in Western Medieval Christianity.* 1990

Shannon McSheffrey. *Gender and Heresy: Women and Men in Lollard Communities, 1420–1530.* 1995

A. J. Minnis. *Medieval Theory of Authorship.* 1988

Lawrence Nees. *A Tainted Mantle: Hercules and the Classical Tradition at the Carolingian Court.* 1991

Lynn H. Nelson, trans. *The Chronicle of San Juan de la Peña: A Fourteenth-Century Official History of the Crown of Aragon.* 1991

Barbara Newman. *From Virile Woman to WomanChrist: Studies in Medieval Religion and Literature.* 1995

Joseph F. O'Callaghan. *The Cortes of Castile-León, 1188–1350.* 1989

Joseph F. O'Callaghan. *The Learned King: The Reign of Alfonso X of Castile.* 1993

Odo of Tournai (Irven M. Resnick, trans.). *Two Theological Treatises:* On Original Sin *and* A Disputation with the Jew, Leo, Concerning the Advent of Christ, the Son of God. 1994

David M. Olster. *Roman Defeat, Christian Response, and the Literary Construction of the Jew.* 1994

William D. Paden, ed. *The Voice of the Trobairitz: Perspectives on the Women Troubadours.* 1989

Edward Peters. *The Magician, the Witch, and the Law.* 1982

Edward Peters, ed. *Christian Society and the Crusades, 1198–1229: Sources in Translation, including* The Capture of Damietta *by Oliver of Paderborn.* 1971

Edward Peters, ed. *The First Crusade: The* Chronicle of Fulcher of Chartres *and Other Source Materials.* 1971

Edward Peters, ed. *Heresy and Authority in Medieval Europe.* 1980

James M. Powell. *Albertanus of Brescia: The Pursuit of Happiness in the Early Thirteenth Century.* 1992

James M. Powell. *Anatomy of a Crusade, 1213–1221.* 1986

Susan A. Rabe. *Faith, Art, and Politics at Saint-Riquier: The Symbolic Vision of Angilbert.* 1995

Jean Renart (Patricia Terry and Nancy Vine Durling, trans.). *The Romance of the Rose or Guillaume de Dole.* 1993

Michael Resler, trans. Erec *by Hartmann von Aue.* 1987

Pierre Riché (Michael Idomir Allen, trans.). *The Carolingians: A Family Who Forged Europe.* 1993

Pierre Riché (Jo Ann McNamara, trans.). *Daily Life in the World of Charlemagne.* 1978

Jonathan Riley-Smith. *The First Crusade and the Idea of Crusading.* 1986

Joel T. Rosenthal. *Patriarchy and Families of Privilege in Fifteenth-Century England.* 1991

Teofilo F. Ruiz. *Crisis and Continuity: Land and Town in Late Medieval Castile.* 1994

James A. Rushing, Jr. *Images of Adventure: Ywain in the Visual Arts.* 1995

Steven D. Sargent, ed. and trans. *On the Threshold of Exact Science: Selected Writings of Anneliese Maier on Late Medieval Natural Philosophy.* 1982

James A. Schultz. *The Knowledge of Childhood in the German Middle Ages, 1100–1350.* 1995

Pamela Sheingorn, ed. and trans. *The Book of Sainte Foy.* 1995

Robin Chapman Stacey. *The Road to Judgment: From Custom to Court in Medieval Ireland and Wales.* 1994

Sarah Stanbury. *Seeing the* Gawain-Poet: *Description and the Act of Perception.* 1992

Robert D. Stevick. *The Earliest Irish and English Bookarts: Visual and Poetic Forms Before A.D. 1000.* 1994

Thomas C. Stillinger. *The Song of Troilus: Lyric Authority in the Medieval Book.* 1992

Susan Mosher Stuard. *A State of Deference: Ragusa/Dubrovnik in the Medieval Centuries.* 1992

Susan Mosher Stuard, ed. *Women in Medieval History and Historiography.* 1987

Susan Mosher Stuard, ed. *Women in Medieval Society.* 1976

Jonathan Sumption. *The Hundred Years War: Trial by Battle.* 1992

Ronald E. Surtz. *The Guitar of God: Gender, Power, and Authority in the Visionary World of Mother Juana de la Cruz (1481–1534).* 1990

Ronald E. Surtz. *Writing Women in Late Medieval and Early Modern Spain: The Mothers of Saint Teresa of Avila.* 1995

Del Sweeney, ed. *Agriculture in the Middle Ages.* 1995

William H. TeBrake. *A Plague of Insurrection: Popular Politics and Peasant Revolt in Flanders, 1323–1328.* 1993

Patricia Terry, trans. *Poems of the Elder Edda.* 1990

Hugh M. Thomas. *Vassals, Heiresses, Crusaders, and Thugs: The Gentry of Angevin Yorkshire, 1154–1216.* 1993

Ralph V. Turner. *Men Raised from the Dust: Administrative Service and Upward Mobility in Angevin England.* 1988

Mary F. Wack. *Lovesickness in the Middle Ages: The* Viaticum *and Its Commentaries.* 1990

Benedicta Ward. *Miracles and the Medieval Mind: Theory, Record, and Event, 1000–1215.* 1982

Suzanne Fonay Wemple. *Women in Frankish Society: Marriage and the Cloister, 500–900.* 1981

Kenneth Baxter Wolf. *Making History: The Normans and Their Historians in Eleventh-Century Italy.* 1995

Jan M. Ziolkowski. *Talking Animals: Medieval Latin Beast Poetry 750–1150.* 1993

This book has been set in Linotron Galliard. Galliard was designed for Mergenthaler in 1978 by Matthew Carter. Galliard retains many of the features of a sixteenth-century typeface cut by Robert Granjon but has some modifications that give it a more contemporary look.

Printed on acid-free paper.